T0173786

DEFOAMING

SURFACTANT SCIENCE SERIES

FOUNDING EDITOR

MARTIN J. SCHICK
1918–1998

SERIES EDITOR

ARTHUR T. HUBBARD
Santa Barbara Science Project
Santa Barbara, California

ADVISORY BOARD

DANIEL BLANKSCHTEIN
Department of Chemical
Engineering
Massachusetts Institute of
Technology
Cambridge, Massachusetts

S. KARABORNI
Shell International Petroleum
Company Limited
London, England

LISA B. QUENCER
The Dow Chemical Company
Midland, Michigan

JOHN F. SCAMEHORN
Institute for Applied Surfactant
Research
University of Oklahoma
Norman, Oklahoma

P. SOMASUNDARAN
Henry Krumb School of Mines
Columbia University
New York, New York

ERIC W. KALER
Department of Chemical
Engineering
University of Delaware
Newark, Delaware

CLARENCE MILLER
Department of Chemical
Engineering
Rice University
Houston, Texas

DON RUBINGH
The Procter & Gamble Company
Cincinnati, Ohio

BEREND SMIT
Shell International Oil Products
B.V.
Amsterdam, The Netherlands

JOHN TEXTER
Strider Research Corporation
Rochester, New York

1. Nonionic Surfactants, *edited by Martin J. Schick* (see also Volumes 19, 23, and 60)

2. Solvent Properties of Surfactant Solutions, *edited by Kozo Shinoda* (see Volume 55)

3. Surfactant Biodegradation, *R. D. Swisher* (see Volume 18)

4. Cationic Surfactants, *edited by Eric Jungermann* (see also Volumes 34, 37, and 53)

5. Detergency: Theory and Test Methods (in three parts), *edited by W. G. Cutler and R. C. Davis* (see also Volume 20)

6. Emulsions and Emulsion Technology (in three parts), *edited by Kenneth J. Lissant*

7. Anionic Surfactants (in two parts), *edited by Warner M. Linfield* (see Volume 56)

8. Anionic Surfactants: Chemical Analysis, *edited by John Cross*

9. Stabilization of Colloidal Dispersions by Polymer Adsorption, *Tatsuo Sato and Richard Ruch*

10. Anionic Surfactants: Biochemistry, Toxicology, Dermatology, *edited by Christian Gloxhuber* (see Volume 43)

11. Anionic Surfactants: Physical Chemistry of Surfactant Action, *edited by E. H. Lucassen-Reynders*

12. Amphoteric Surfactants, *edited by B. R. Bluestein and Clifford L. Hilton* (see Volume 59)

13. Demulsification: Industrial Applications, *Kenneth J. Lissant*

14. Surfactants in Textile Processing, *Arved Datyner*

15. Electrical Phenomena at Interfaces: Fundamentals, Measurements, and Applications, *edited by Ayao Kitahara and Akira Watanabe*

16. Surfactants in Cosmetics, edited by Martin M. Rieger (see Volume 68)

17. Interfacial Phenomena: Equilibrium and Dynamic Effects, *Clarence A. Miller and P. Neogi*

18. Surfactant Biodegradation: Second Edition, Revised and Expanded, *R. D. Swisher*

19. Nonionic Surfactants: Chemical Analysis, *edited by John Cross*

20. Detergency: Theory and Technology, *edited by W. Gale Cutler and Erik Kissa*

21. Interfacial Phenomena in Apolar Media, *edited by Hans-Friedrich Eicke and Geoffrey D. Parfitt*

22. Surfactant Solutions: New Methods of Investigation, *edited by Raoul Zana*

23. Nonionic Surfactants: Physical Chemistry, *edited by Martin J. Schick*

24. Microemulsion Systems, *edited by Henri L. Rosano and Marc Clausse*

25. Biosurfactants and Biotechnology, *edited by Naim Kosaric, W. L. Cairns, and Neil C. C. Gray*

26. Surfactants in Emerging Technologies, *edited by Milton J. Rosen*

27. Reagents in Mineral Technology, *edited by P. Somasundaran and Brij M. Moudgil*

28. Surfactants in Chemical/Process Engineering, *edited by Darsh T. Wasan, Martin E. Ginn, and Dinesh O. Shah*

29. Thin Liquid Films, *edited by I. B. Ivanov*

30. Microemulsions and Related Systems: Formulation, Solvency, and Physical Properties, *edited by Maurice Bourrel and Robert S. Schechter*

31. Crystallization and Polymorphism of Fats and Fatty Acids, *edited by Nissim Garti and Kiyotaka Sato*

32. Interfacial Phenomena in Coal Technology, *edited by Gregory D. Botsaris and Yuli M. Glazman*

33. Surfactant-Based Separation Processes, *edited by John F. Scamehorn and Jeffrey H. Harwell*

34. Cationic Surfactants: Organic Chemistry, *edited by James M. Richmond*

35. Alkylene Oxides and Their Polymers, *F. E. Bailey, Jr., and Joseph V. Koleske*

36. Interfacial Phenomena in Petroleum Recovery, *edited by Norman R. Morrow*

37. Cationic Surfactants: Physical Chemistry, *edited by Donn N. Rubingh and Paul M. Holland*

38. Kinetics and Catalysis in Microheterogeneous Systems, *edited by M. Grätzel and K. Kalyanasundaram*

39. Interfacial Phenomena in Biological Systems, *edited by Max Bender*

40. Analysis of Surfactants, *Thomas M. Schmitt* (see Volume 96)

41. Light Scattering by Liquid Surfaces and Complementary Techniques, *edited by Dominique Langevin*

42. Polymeric Surfactants, *Irja Piirma*

43. Anionic Surfactants: Biochemistry, Toxicology, Dermatology. Second Edition, Revised and Expanded, *edited by Christian Gloxhuber and Klaus Künstler*

44. Organized Solutions: Surfactants in Science and Technology, *edited by Stig E. Friberg and Björn Lindman*

45. Defoaming: Theory and Industrial Applications, *edited by P. R. Garrett*

46. Mixed Surfactant Systems, *edited by Keizo Ogino and Masahiko Abe*

47. Coagulation and Flocculation: Theory and Applications, *edited by Bohuslav Dobiás*

48. Biosurfactants: Production Properties Applications, edited by Naim Kosaric

49. Wettability, *edited by John C. Berg*

50. Fluorinated Surfactants: Synthesis Properties Applications, *Erik Kissa*

51. Surface and Colloid Chemistry in Advanced Ceramics Processing, *edited by Robert J. Pugh and Lennart Bergström*

52. Technological Applications of Dispersions, *edited by Robert B. McKay*

53. Cationic Surfactants: Analytical and Biological Evaluation, *edited by John Cross and Edward J. Singer*

54. Surfactants in Agrochemicals, *Tharwat F. Tadros*

55. Solubilization in Surfactant Aggregates, *edited by Sherril D. Christian and John F. Scamehorn*

56. Anionic Surfactants: Organic Chemistry, *edited by Helmut W. Stache*

57. Foams: Theory, Measurements, and Applications, *edited by Robert K. Prud'homme and Saad A. Khan*

58. The Preparation of Dispersions in Liquids, *H. N. Stein*

59. Amphoteric Surfactants: Second Edition, *edited by Eric G. Lomax*

60. Nonionic Surfactants: Polyoxyalkylene Block Copolymers, *edited by Vaughn M. Nace*

61. Emulsions and Emulsion Stability, *edited by Johan Sjöblom*

62. Vesicles, *edited by Morton Rosoff*

63. Applied Surface Thermodynamics, *edited by A. W. Neumann and Jan K. Spelt*

64. Surfactants in Solution, *edited by Arun K. Chattopadhyay and K. L. Mittal*

65. Detergents in the Environment, *edited by Milan Johann Schwuger*

66. Industrial Applications of Microemulsions, *edited by Conxita Solans and Hironobu Kunieda*

67. Liquid Detergents, *edited by Kuo-Yann Lai*

68. Surfactants in Cosmetics: Second Edition, Revised and Expanded, *edited by Martin M. Rieger and Linda D. Rhein*

69. Enzymes in Detergency, *edited by Jan H. van Ee, Onno Misset, and Erik J. Baas*

70. Structure-Performance Relationships in Surfactants, *edited by Kunio Esumi and Minoru Ueno*

71. Powdered Detergents, *edited by Michael S. Showell*

72. Nonionic Surfactants: Organic Chemistry, *edited by Nico M. van Os*

73. Anionic Surfactants: Analytical Chemistry, Second Edition, Revised and Expanded, *edited by John Cross*

74. Novel Surfactants: Preparation, Applications, and Biodegradability, *edited by Krister Holmberg*

75. Biopolymers at Interfaces, *edited by Martin Malmsten*

76. Electrical Phenomena at Interfaces: Fundamentals, Measurements, and Applications, Second Edition, Revised and Expanded, *edited by Hiroyuki Ohshima and Kunio Furusawa*

77. Polymer-Surfactant Systems, *edited by Jan C. T. Kwak*

78. Surfaces of Nanoparticles and Porous Materials, *edited by James A. Schwarz and Cristian I. Contescu*

79. Surface Chemistry and Electrochemistry of Membranes, *edited by Torben Smith Sørensen*

80. Interfacial Phenomena in Chromatography, *edited by Emile Pefferkorn*

81. Solid–Liquid Dispersions, *Bohuslav Dobiás, Xueping Qiu, and Wolfgang von Rybinski*

82. Handbook of Detergents, *editor in chief: Uri Zoller*
Part A: Properties, *edited by Guy Broze*

83. Modern Characterization Methods of Surfactant Systems, *edited by Bernard P. Binks*

84. Dispersions: Characterization, Testing, and Measurement, *Erik Kissa*

85. Interfacial Forces and Fields: Theory and Applications, *edited by Jyh-Ping Hsu*

86. Silicone Surfactants, *edited by Randal M. Hill*

87. Surface Characterization Methods: Principles, Techniques, and Applications, *edited by Andrew J. Milling*

88. Interfacial Dynamics, *edited by Nikola Kallay*

89. Computational Methods in Surface and Colloid Science, *edited by Malgorzata Borówko*

90. Adsorption on Silica Surfaces, *edited by Eugène Papier*

91. Nonionic Surfactants: Alkyl Polyglucosides, *edited by Dieter Balzer and Harald Lüders*

92. Fine Particles: Synthesis, Characterization, and Mechanisms of Growth, *edited by Tadao Sugimoto*

93. Thermal Behavior of Dispersed Systems, *edited by Nissim Garti*

94. Surface Characteristics of Fibers and Textiles, *edited by Christopher M. Pastore and Paul Kiekens*

95. Liquid Interfaces in Chemical, Biological, and Pharmaceutical Applications, *edited by Alexander G. Volkov*

96. Analysis of Surfactants: Second Edition, Revised and Expanded, *Thomas M. Schmitt*

97. Fluorinated Surfactants and Repellents: Second Edition, Revised and Expanded, *Erik Kissa*

98. Detergency of Specialty Surfactants, *edited by Floyd E. Friedli*

99. Physical Chemistry of Polyelectrolytes, *edited by Tsetska Radeva*

100. Reactions and Synthesis in Surfactant Systems, *edited by John Texter*

101. Protein-Based Surfactants: Synthesis, Physicochemical Properties, and Applications, *edited by Ifendu A. Nnanna and Jiding Xia*

102. Chemical Properties of Material Surfaces, *Marek Kosmulski*

103. Oxide Surfaces, *edited by James A. Wingrave*

104. Polymers in Particulate Systems: Properties and Applications, *edited by Vincent A. Hackley, P. Somasundaran, and Jennifer A. Lewis*

105. Colloid and Surface Properties of Clays and Related Minerals, *Rossman F. Giese and Carel J. van Oss*

106. Interfacial Electrokinetics and Electrophoresis, *edited by Ángel V. Delgado*

107. Adsorption: Theory, Modeling, and Analysis, *edited by József Tóth*

108. Interfacial Applications in Environmental Engineering, *edited by Mark A. Keane*

109. Adsorption and Aggregation of Surfactants in Solution, *edited by K. L. Mittal and Dinesh O. Shah*

110. Biopolymers at Interfaces: Second Edition, Revised and Expanded, *edited by Martin Malmsten*

111. Biomolecular Films: Design, Function, and Applications, *edited by James F. Rusling*

112. Structure–Performance Relationships in Surfactants: Second Edition, Revised and Expanded, *edited by Kunio Esumi and Minoru Ueno*

113. Liquid Interfacial Systems: Oscillations and Instability, *Rudolph V. Birikh, Vladimir A. Briskman, Manuel G. Velarde, and Jean-Claude Legros*

114. Novel Surfactants: Preparation, Applications, and Biodegradability: Second Edition, Revised and Expanded, *edited by Krister Holmberg*

115. Colloidal Polymers: Synthesis and Characterization, *edited by Abdelhamid Elaissari*

116. Colloidal Biomolecules, Biomaterials, and Biomedical Applications, *edited by Abdelhamid Elaissari*

117. Gemini Surfactants: Synthesis, Interfacial and Solution-Phase Behavior, and Applications, *edited by Raoul Zana and Jiding Xia*

118. Colloidal Science of Flotation, *Anh V. Nguyen and Hans Joachim Schulze*

119. Surface and Interfacial Tension: Measurement, Theory, and Applications, *edited by Stanley Hartland*

120. Microporous Media: Synthesis, Properties, and Modeling, *Freddy Romm*

121. Handbook of Detergents, *editor in chief: Uri Zoller*
Part B: Environmental Impact, *edited by Uri Zoller*

122. Luminous Chemical Vapor Deposition and Interface Engineering, *HirotsuguYasuda*

123. Handbook of Detergents, editor in chief: Uri Zoller
Part C: Analysis, *edited by Heinrich Waldhoff and Rüdiger Spilker*

124. Mixed Surfactant Systems: Second Edition, Revised and Expanded, *edited by Masahiko Abe and John F. Scamehorn*

125. Dynamics of Surfactant Self-Assemblies: Micelles, Microemulsions, Vesicles and Lyotropic Phases, *edited by Raoul Zana*

126. Coagulation and Flocculation: Second Edition, *edited by Hansjoachim Stechemesser and Bohulav Dobiáš*

DEFOAMING

Theory and Industrial Applications

Edited by
P. R. Garrett

CRC Press
Taylor & Francis Group
Boca Raton London New York

CRC Press is an imprint of the
Taylor & Francis Group, an **informa** business
A TAYLOR & FRANCIS BOOK

CRC Press
Taylor & Francis Group
6000 Broken Sound Parkway NW, Suite 300
Boca Raton, FL 33487-2742

First issued in paperback 2019

© 1992 by Taylor Francis Group, LLC
CRC Press is an imprint of Taylor & Francis Group, an Informa business

No claim to original U.S. Government works

ISBN-13: 978-0-8247-8770-7 (hbk)
ISBN-13: 978-0-367-40261-7 (pbk)

This book contains information obtained from authentic and highly regarded sources. Reasonable efforts have been made to publish reliable data and information, but the author and publisher cannot assume responsibility for the validity of all materials or the consequences of their use. The authors and publishers have attempted to trace the copyright holders of all material reproduced in this publication and apologize to copyright holders if permission to publish in this form has not been obtained. If any copyright material has not been acknowledged please write and let us know so we may rectify in any future reprint.

Except as permitted under U.S. Copyright Law, no part of this book may be reprinted, reproduced, transmitted, or utilized in any form by any electronic, mechanical, or other means, now known or hereafter invented, including photocopying, microfilming, and recording, or in any information storage or retrieval system, without written permission from the publishers.

For permission to photocopy or use material electronically from this work, please access www.copyright. com (http://www.copyright.com/) or contact the Copyright Clearance Center, Inc. (CCC), 222 Rosewood Drive, Danvers, MA 01923, 978-750-8400. CCC is a not-for-profit organization that provides licenses and registration for a variety of users. For organizations that have been granted a photocopy license by the CCC, a separate system of payment has been arranged.

Trademark Notice: Product or corporate names may be trademarks or registered trademarks, and are used only for identification and explanation without intent to infringe.

Library of Congress Cataloging-in-Publication Data

Catalog record is available from the Library of Congress

Visit the Taylor & Francis Web site at
http://www.taylorandfrancis.com

and the CRC Press Web site at
http://www.crcpress.com

Preface

Problems associated with excessive foaming occur in a surprisingly wide range of situations. Development of suitable antifoams has therefore been a necessity for the successful solution of foaming problems in industries as different as textile dyeing and detergent manufacture. Arguably, the problem of controlling excessive foam is generally of equal importance to that of generating stable foams. Yet this problem usually receives scant attention in the few textbooks that deal with the subject of foams. It seems timely, therefore, to redress the balance and devote a text exclusively to the subject of defoaming.

In many areas of surface and colloid science, the sheer complexity of phenomena often means that technology has actually led scientific understanding. The development of antifoams involving the interaction of one disperse phase (gas) with at least one other disperse phase (emulsion droplets or particles) is perhaps such an area. By including a chapter in this volume concerning the current theoretical understanding of the mode of action of antifoams, it is hoped that some of the association of the subject with the mysterious arts will be dispelled.

In many cases, the selection of an antifoam that will adequately control the foam is only the beginning of a solution to a foaming problem. Attendant considerations include the amount of antifoam required, the state of dispersion of the antifoam, and the possibility of the antifoam interacting adversely with some aspect of a process. These considerations may often be of overriding importance and may therefore affect the selection of the antifoam. They are also usually specific to particular applications. For this reason the volume is divided into chapters concerning different industrial applications of antifoams so that detailed descriptions of the problems and solutions specific to given industries may be dealt with. Not all such applications are covered in this one volume, but a sufficiently wide range is included to give some indication of the appropriate approaches for any foam control problem. Also, a measure of coherence is maintained throughout the book despite the varied nature of the industries considered so that nomenclature and definitions of terms are consistent.

This book would not have been possible without the timely contributions of the various authors. Their patience in enduring the many editorial changes, which a volume such as this entails, is gratefully acknowledged. Finally, gratitude is due to Unilever Research for the provision of secretarial assistance which greatly facilitated the preparation of this volume.

P. R. Garrett

Contents

Preface   iii

Contributors   vii

1 The Mode of Action of Antifoams   1
   P. R. Garrett

2 Antifoams for Nonaqueous Systems in the Oil Industry   119
   Ian C. Callaghan

3 Defoaming in the Pulp and Paper Industry   151
   S. Lee Allen, Lawrence H. Allen, and Ted H. Flaherty

4 Application of Antifoams in Pharmaceuticals   177
   Rolland Berger

5 High-Performance Antifoams for the Textile Dyeing Industry   193
   George C. Sawicki

6 Foam Control in Detergent Products   221
   Horst Ferch and Wolfgang Leonhardt

7 Antifoams for Paints   269
   Maurice R. Porter

8 Surfactant Antifoams   299
   Trevor G. Blease, J. G. Evans, L. Hughes, and Philippe Loll

Index   325

Contributors

Lawrence H. Allen   Pulp and Paper Research Institute of Canada, Point Claire, Quebec, Canada

S. Lee Allen   Pulp and Paper Research Centre, McGill University, Montreal, Quebec, Canada

Rolland Berger   Th. Goldschmidt AG, Chemische Fabriken, Essen, Germany

Trevor G. Blease   Research and Technology Department, ICI Surfactants, Wilton, Middlesborough, United Kingdom

Ian C. Callaghan   Colloid Science Branch, BP Research, Sunbury-on-Thames, Middlesex, United Kingdom

J. G. Evans   Research and Technology Department, ICI Surfactants, Wilton, Middlesborough, United Kingdom

Horst Ferch* Applied Technologies Silicas and Silicates, Degussa AG, Hanau, Germany

*Retired.

Ted H. Flaherty   Dorset Industrial Chemicals, Ltd., Chateauguay, Quebec, Canada

P. R. Garrett   Port Sunlight Laboratory, Uniliver Research, Bebington, Wirral, United Kingdom

L. Hughes   Research and Technology Department, ICI Surfactants, Wilton, Middlesborough, United Kingdom

Wolfgang Leonhardt   Applied Technologies Silicas and Silicates, Degussa AG, Hanau, Germany

Philippe Loll   ICI Surfactants, Everberg, Belgium

Maurice R. Porter   Maurice R. Porter and Associates, Sully, South Glamorgan, United Kingdom

George C. Sawicki   Department of Science and Technology, Dow Corning Europe, Brussels, Belgium

1

The Mode of Action of Antifoams

P. R. GARRETT Unilever Research, Bebington, Wirral, United Kingdom

I.	Introduction	2
II.	The Stability of Foams	3
	A. Surface tension gradients and foam film stability	4
	B. Disjoining forces and foam film stability	8
III.	Antifoams and Surface Activity	11
	A. Antifoam effects in homogeneous systems	11
	B. Duplex oil films, oil lenses, and antifoam mechanism	13
	C. Surface tension gradients and theories of antifoam mechanism in heterogeneous systems	19
IV.	Inert Hydrophobic Particles and Capillary Theories of Antifoam Mechanism for Aqueous Systems	30
	A. Early work	30
	B. Contact angles and particle bridging mechanism	33
	C. Particle geometry and contact angle conditions for foam film collapse	41
	D. Particle size and kinetics	53
	E. Capture of particles by bubbles	58
	F. Melting of hydrophobic particles and antifoam behavior	63
V.	Mixtures of Hydrophobic Particles and Oils as Antifoams for Aqueous Systems	66
	A. Antifoam synergy	66
	B. Spreading behavior of antifoam oils on water	82
	C. Role of the oil in synergistic oil/particle antifoams	87
	D. Hypotheses concerning the role of the particles in synergistic oil/particle antifoams	94

E. Capture of oil droplets by bubbles and the
    role of the particles   98
F. Antifoam dimensions and kinetics   111
Acknowledgments   113
References   113

I. INTRODUCTION

This chapter attempts a complete review of the various mechanisms proposed for the action of antifoams over the past half-century. It is a feature of this subject that these mechanisms, although plausible, are often speculative. Thus, unequivocal experimental evidence is often lacking. Indeed, the full theoretical implications of proposed mechanisms are also often not fully developed. In the main, all of this derives from the extreme complexity of the relevant phenomena. Foam is itself extremely complex, consisting of (usually) polydisperse gas bubbles separated by draining films. These films exhibit complicated hydrodynamics involving the distinct rheology of air-liquid surfaces and, for thin films, colloidal interaction forces. The nature of the foam film collapse processes which are intrinsic to a foam are still imperfectly understood.

Antifoams are usually hydrophobic, finely divided, insoluble materials. Their presence therefore further complicates the complexities associated with foam. Indeed commercial antifoams for aqueous solutions usually consist of hydrophobic particles dispersed in hydrophobic oils. The action of such antifoams concerns the effect of a dispersion (of antifoam in foaming liquid) of a dispersion (the antifoam) on yet a third dispersion (the foam).

Theories of antifoam mechanism appear to fall into two broad categories: those which require the antifoam to be surface active at the air-liquid surfaces of the foaming liquid, and those which do not. Theories which require the antifoam to be surface active associate antifoam behavior with an effect on surface tension in the foam films which leads to film rupture. Theories which do not require the antifoam to be surface active usually concern the hydrophobic nature of the antifoam. They suppose that dewetting of antifoam entities in foam films produces capillary instabilities which lead to film rupture.

The chapter is divided into four sections. The first concerns an outline of the main processes which are believed to contribute to the stability of foam films in the absence of antifoam. The second examines the relationship between antifoam effects and surface activity of the antifoam. The third section concerns inert hydrophobic particles and capillary theories of antifoam action in aqueous solutions. The last section concerns the mode of

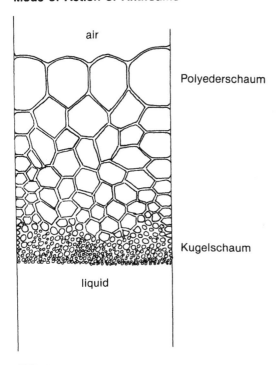

air

Polyederschaum

Kugelschaum

liquid

FIG. 1   Foam structure.

action of the mixtures of hydrophobic particles and oils which form the basis of many of the commercial antifoam concoctions proposed for aqueous foams.

II. THE STABILITY OF FOAMS

Before considering the mode of action of antifoams, we review the factors which contribute to the stability of a foam. A brief summary only is given here. For more complete accounts the reader is referred to the many reviews on the subject [1–5]. Our summary closely follows the excellent review by Lucassen [2].

The structure of a typical foam formed by, say, shaking a surfactant solution in a cylindrical vessel is shown schematically in Fig. 1. In the lower part of the foam, bubbles are spherical (so-called *kugelschaum*) and of small size with a relatively low gas volume fraction. As the liquid drains out of the foam, the bubbles distort to form polyhedra. This polyhedral foam *(polyederschaum)* consists of plane-parallel films joined by channels called plateau borders. The gas volume fraction is here relatively high and the density low so that polyhedra first form at the top of the foam column.

Throughout the foam there are differences in the sizes of adjacent bubbles. This will mean that differences in capillary pressure will exist between the adjacent bubbles so that gas will diffuse from small to large bubbles. The more soluble the gas in the continuous phase or the higher the partial vapor pressure of the continuous phase the faster this process of bubble disproportionation will proceed. There is no arrangement of bubbles in a foam which permits elimination of this process. Thus, for example, if there is free headspace above the foam the upper surface will consist of films of curved section. This will mean diffusion of gas out of the upper layer of bubbles into the headspace because of the capillary pressure implied by the curved surface.

As the films at the top of the foam thin, they become more susceptible to rupture by mechanical shock or vibration. Moreover, with some foams rupture of films at a certain thickness is spontaneous. Films at the top of the foam then tend to break first, and the foam collapses from the top downward in a catastrophic cascade.

A. Surface Tension Gradients and Foam Film Stability

Films formed by adjacent bubbles in a pure liquid are extremely unstable. Pure liquids therefore do not form foams. This arises in part because of the response of the films to any external force such as gravity. Consider, for example, a vertical plane-parallel film in a gravity field. There is no reason why any element of that film should move in response to the applied gravitational force with a velocity different from that of any adjacent element. No velocity gradients in a direction perpendicular to the plane of the film surface against the air will therefore exist. There will then be no viscous shear forces opposing the effect of gravity. The film will exhibit plug flow (resisted only by extensional viscous forces) with elements accelerated downward tearing it apart. The process is depicted in Fig. 2a.

This behavior can be drastically altered if we arrange for a tangential force to act in the plane of the liquid-air surface so that the surface is essentially rigid. In the case of a vertical plane-parallel film of a viscous liquid with such rigid surfaces, subject to gravity, a parabolic velocity profile will develop as shown in Fig. 2b. This means that velocity gradients will exist in a direction perpendicular to the film surfaces. A viscous stress will therefore be exerted at the air-liquid surface. This stress must be balanced by the tangential force acting in the plane of the surface. That force can only be a gradient of surface tension. This balance of viscous forces and surface tension gradients at the liquid-air surface can be written as

$$\frac{d\gamma_{AF}}{dy} = \eta_F \left(\frac{du_y}{dx}\right)_{x=0} \tag{1}$$

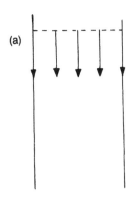

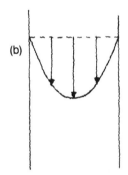

FIG. 2 Velocity profiles in draining foam films: (a) plug flow; (b) flow with parabolic velocity profile when film surfaces are immobile.

where γ_{AF} is the air-liquid surface tension of the foaming liquid, η_F is the viscosity, u_y is the velocity of flow in the y direction, y is the vertical distance, and x is the horizontal distance in the film.

Thus, we find that if the force of gravity (or indeed any other force such as that due to the capillary pressure caused by the curved plateau borders) is to be resisted by the film, then a surface tension gradient must exist at the air-liquid surface. In the case of a vertical film in the gravity field the gradient is [2]

$$\frac{d\gamma_{AF}}{dy} = \frac{\rho g h}{2} \tag{2}$$

where h is the film thickness, ρ is the liquid density, and g is the acceleration due to gravity. This gradient can only exist where differences of surface composition can occur. We therefore require the presence of more than one component in the film. Indeed, it is possible to speculate that in the case of, say, aqueous foams diffusion of water through the gas phase may rapidly remove any differences in concentration between different parts of a foam film if only one solute is present. In this case at least two solutes (or three components) would be required.

Surface tension gradients due to differences in the surface excess of soluble surface-active components may exist only when either the surface is not in equilibrium with the bulk composition or there are concomitant differences in bulk composition parallel to the surface. In the case of the former the magnitudes of the gradients are of course determined by the rate of transport of surfactant to the relevant surfaces. With concentrated surfactant solutions transport rates by diffusion will be rapid and surface tension gradients will tend to be eliminated. Thus, it has occasionally been reported that foamabilities decline at extremely high concentrations of surfactant in aqueous solution. Conversely, however, if foam films are denuded of surfactant because of extremely slow transport rates, then the maximum surface tension gradients which can be achieved will be small. Such films will therefore be susceptible to rupture when exposed to external stress. However, the complex problem of assessing both the effect of rate of transport on the surface tension gradients in foam films and the overall resultant impact upon foam film stability, when subject to an external stress, has not apparently been fully addressed.

Differences in bulk composition are possible in a thin foam film as a result of stretching the film. If the film is sufficiently thin, then any stretching causes a depletion of the bulk phase surfactant solution between the air-liquid surfaces of the foam film as more surfactant adsorbs on those surfaces. Distances perpendicular to the film are small so that, provided the stretching occurs reasonably slowly, equilibrium inside the film element may be always maintained. Depletion of bulk phase surfactant concentration will therefore necessarily mean an increase of the surface tension of the film as it is stretched. This will, however, only occur if reduction of surfactant concentration causes a concomitant increase in surface tension. In the case of a pure surfactant at concentrations above the critical micelle concentration (cmc), this may not always happen.

We find then that it is possible to generate a surface tension gradient in a foam film by stretching various elements of the film to different extents. The increase in surface tension due to stretching imparts an elasticity to the film. This property of foam films was first recognized by Gibbs [6] and is usually referred to as the Gibbs elasticity ϵ_G. It is defined as

$$\epsilon_G = \frac{2d\gamma_{AF}}{d \ln A} = -\frac{2d\gamma_{AF}}{d \ln h} \tag{3}$$

ϵ_G/ mNm^{-1}

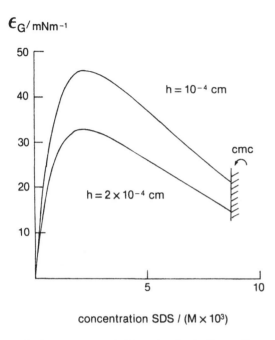

FIG. 3 Gibbs elasticities of submicellar sodium dodecylsulfate solutions. (From Ref. 2.)

where A is the film area and the factor 2 arises because of the two surfaces.

A plot of ϵ_G against concentration for a submicellar aqueous solution of sodium dodecylsulfate (SDS) is shown in Fig. 3 by way of example. Here we see that, except at very low concentrations, decreases in film thickness at constant concentration produce increases in Gibbs elasticity so that $(\partial \epsilon_G / \partial h)_c \leq 0$. Thus, as the film becomes thinner stretching will cause a relatively greater depletion of surfactant in the intralamellar liquid and the surface tension will rise to a greater extent.

The plot of Gibbs elasticity against concentration shown in Fig. 3 clearly reveals a maximum at concentration c_{max}. At extremely low concentrations of surfactant we find that upon stretching of the film there is essentially no contribution from the intralamellar liquid, and the surfactant behaves as an insoluble monolayer. Here with increase in surfactant concentration both the surface excess and the elasticity of the monolayer increase. However, further increases in the surfactant concentration will eventually mean that it significantly exceeds that required to compensate for stretching of the air-liquid surface, so $\epsilon_G \to 0$. These two opposing consequences of increasing concentration conspire to produce the maximum in a plot of Gibbs elasticity.

Lucassen [2] considers the effect of Gibbs elasticity on the development of stabilizing surface tension gradients in a foam film in the gravity field. He points out that if ϵ_G decreases as the film is stretched it will tend to be dynamically unstable. Under these circumstances any stretching force will tend to increase the area of the thinnest part of the film. Such situations will tend to prevail for films formed at concentrations on the low side of c_{max}. Thus, we can write

$$\frac{d\epsilon_G}{dh} = \left(\frac{\partial \epsilon_G}{\partial h}\right)_c + \left(\frac{\partial \epsilon_G}{\partial c}\right)_h \frac{dc}{dh} \tag{4}$$

where c is the concentration. We can therefore have $d\epsilon_G/dh > 0$ if $c < c_{max}$, because then $(\partial \epsilon_G/\partial c)_h > 0$, $dc/dh > 0$, and $(\partial \epsilon_G/\partial h)_c \to 0$ as $c \to 0$.

For vertical films prepared from concentrated solutions the requirement that the surface tension gradient satisfy Eq. 2 implies that rapid stretching will occur. In the case of micellar solutions of certain pure surfactants this may require achievement of submicellar concentrations in order that $\epsilon_G > 0$ and therefore $d\gamma_{AF}/dy > 0$. For vertical films prepared from extremely dilute surfactant solutions where $c < c_{max}$, Lucassen [2] shows that the magnitude of ϵ_G for thin elements at the top of the film may be less than that of thinner elements lower down. Any force acting on the film, such as an increase in weight as it grows, could mean catastrophic extension of the thinnest elements because of their lower Gibbs elasticities.

In summary, then, we find that surface tension gradients are necessary if freshly formed foam films are to survive. These gradients may occur if surface tensions depart from equilibrium values. This will happen when foam film air-liquid surfaces are expanded at rates which are fast so that equilibrium with the bulk surfactant concentration cannot be maintained. They may also occur when films are thin so that stretching may deplete intralamellar bulk phase to give rise to a Gibbs elasticity. Unfortunately there are few experimental observations which clearly reveal the importance of surface tension gradients in determining foam behavior. Perhaps the best examples are reported by Malysa et al. [7] and Prins [8].

B. Disjoining Forces and Foam Film Stability

Even though foam films may be stabilized by effective surface tension gradients they may still succumb to rupture due to the action of van der Waals forces. These have a significant effect on foam film stability at film thicknesses where the intralamellar molecules cease to have the properties of matter in bulk because of the perturbing influence of the two contiguous air-liquid surfaces. In effect, molecules in a thin film have a higher chemical potential (or vapor pressure) than matter in bulk at the same external pressure

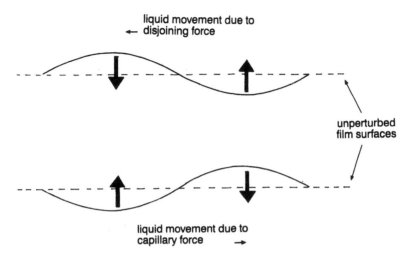

FIG. 4 Sinusoidal thickness perturbations in a thin liquid foam film.

because of the weaker attractive force field due to the limited number of molecules present. Such molecules will therefore tend to spontaneously transfer to any adjacent bulk phase. This process is said to be driven by a "disjoining pressure" Π which is negative when the film shows a tendency to become thinner. Clearly as the film becomes thinner the disjoining pressure becomes more negative so that

$$\frac{d\Pi}{dh} > 0 \tag{5}$$

A perturbation of the film thickness which produces thin and thick regions will tend to grow spontaneously because molecules in the thin region will transfer to the thick regions. However, any perturbation of a film to produce thick and thin regions must also inevitably increase the surface area of the film. This will increase the number of molecules in the relatively weak attractive force field close to the air-liquid surface. Such an increase in surface area will therefore be resisted by an opposing force—the surface tension.

Thickness perturbations, of a thermal or mechanical nature, may be considered to be of a wavelike nature. A symmetrical sinusoidal perturbation is shown in Fig. 4. Here the thin part of the film is subject to two opposing forces. Thus, a capillary pressure due to the surface tension tends to suck liquid back into the thin part of the film, and a disjoining force tends to push liquid away.

The magnitude of the capillary pressure is determined by the curvature of the film surface. For a given amplitude of the perturbation the curvature is determined by the wavelength. Thus, the shorter the wavelength the more marked the curvature and the stronger the capillary pressure. Vrij and Overbeek [9] deduce a critical wavelength λ_{crit} above which disjoining forces will dominate over the capillary pressure and the perturbation will spontaneously grow. The critical wavelength is

$$\lambda_{crit} = \left[\frac{2\pi^2\gamma_{AF}}{d\Pi/dh}\right]^{1/2} \tag{6}$$

The rate of growth of the perturbation will increase with increasing wavelength λ for $\lambda > \lambda_{crit}$ because the damping effect of the capillary pressure will decrease. However, for sufficiently long wavelengths the rate of growth will eventually begin to decline because of the increased distances over which the film liquid has to be moved against viscous resistance. An optimum wavelength of $\sqrt{2}\lambda_{crit}$ for the maximum rate of growth of a perturbation therefore exists [9].

Film collapse is supposed to occur when the amplitude of the fastest growing perturbation equals the thickness of the film. Vrij and Overbeek [9] have produced the simplest description of this process. They calculate the minimum total time for film drainage and subsequent growth of the fastest-growing perturbation. The average thickness of the film at the moment of rupture is the critical thickness h_{crit}. Experimental measurements of h_{crit} for microscopic foam films are in the region of a few tens of nanometers (see, for example, [10]).

Clearly this description ignores any contribution to disjoining forces from stabilizing interactions across a thin film due to adsorbed surfactant. These could arise from overlapping double layers or steric factors due to interaction of bulky surfactant head groups. Such forces could mean $\Pi > 0$ and $d\Pi/dh < 0$.

With many surfactants at sufficiently high concentrations (and therefore adsorption levels) the growth of perturbations at a particular critical thickness results not in foam film rupture but in the formation of spots of thin metastable film. These spots then grow to cover the whole film area. This film is thin so that destructive interference of reflected light occurs. It therefore appears black in reflected light.

The concentration at which this transition from rupture to black spot formation occurs is c_{black}. The transition probably occurs because of the effects of changes in the adsorption layer of surfactant on the thickness dependence of Π and $d\Pi/dh$. Foam film stability tends to increase markedly at concentrations above c_{black}.

Values for c_{black} in microscopic films have been measured for a number of surfactants [11]. In the case of aqueous solutions of surfactants c_{black} is generally significantly lower than the cmc. Thus, for example, for sodium dodecylsulfate, $c_{black} = 1.6 \times 10^{-6}$ M [11] and the cmc = 8.4×10^{-3} M [12], and for dodecyl hexaethyleneglycol $c_{black} = 4.9 \times 10^{-6}$ M [11] and the cmc = 8.7×10^{-5} M [13] at 25°C.

We have seen then that film rupture may occur because either surface tension gradients are not sufficiently high to enable the film to withstand stress or because $d\Pi/dh$ is always positive so that rupture is inevitable at a certain critical thickness. However, both phenomena are associated with low concentrations of surfactant (at least if we consider films formed slowly so that equilibrium between the air-liquid surface and the intralamellar liquid is maintained). Thus, we have $c_{max} \ll$ cmc for $d\epsilon_G/dh < 0$ and $c_{black} \ll$ cmc. Elimination of both causes of rupture should therefore be readily achieved at sufficiently high concentrations of surfactant. The poor discrimination in foam behavior often found with relatively concentrated aqueous micellar solutions of surfactants may well be attributable to that cause. Interesting differences in foam behavior are, however, often revealed when antifoam is added to the solution.

III. ANTIFOAMS AND SURFACE ACTIVITY

A. Antifoam Effects in Homogeneous Systems

One of the earliest generalizations concerning antifoams states that they must be present as undissolved particles (or droplets) in the liquid to be defoamed [14–16]. Indeed the presence of antifoam materials at concentrations lower than the solubility limit can even enhance foamability [17,18]. One well-known example concerns the foam-enhancing effect of dissolved polydimethylsiloxanes on hydrocarbon lube oils [17]. However, the generalization that antifoams must be present as undissolved entities has occasionally been repeated with some equivocation [18–20]. A number of authors in fact report experimental results which purport to show antifoam effects due to additives which are dissolved in the foaming solution [21–23]. Thus, Ross and Haak [21], for example, identify two types of antifoam behavior associated with the effect of materials like tributyl phosphate and methylisobutyl carbinol on the foam behavior of aqueous solutions of surfactants such as sodium dodecylsulfate and sodium oleate. Wherever the oil concentration exceeds the solubility limit, emulsified droplets of oil contribute to an effective antifoam action. However, if micelles of the surfactant are present, then the oil is solubilized to a certain limiting concentration. It is claimed [21,24] that a weak antifoam effect is associated with the presence of such solubilized oils in the micelles. The consequences of all this behavior are revealed if, for example, tributyl phosphate is added to micellar solutions of sodium oleate [21] at concentrations below the solubilization limit. A marked decrease in foamability is found immediately after dispersing the oil. As the oil becomes slowly solubilized, the foamability increases. However, even after the oil is completely solubilized the foamability is still apparently less than that intrinsic to the uncontaminated surfactant solution [21].

Ross and Haak [21] have also studied the effect of solubilized tributyl phosphate and methylisobutyl carbinol on the dynamic surface tensions of micellar solutions of sodium dodecyl sulfate and sodium oleate, using the oscillating jet technique. They claim that the rate of decrease of surface tension is significantly enhanced by the presence of these solubilisates, presumably because of their effect upon transport of micellar surfactant to the air-water surface. Ross and Haak [21] used these findings in seeking an explanation for the weak antifoam effect attributable to these solubilisates. They suggest that a major factor in determining foam film stability is restoration of foam film thickness after stretching. Thus, if a thick foam film is rapidly stretched, the surface tension will increase due to the finite time taken for surfactant to adsorb onto the expanded surface. The difference in surface tension between the stretched film element and the surrounding unperturbed film elements or plateau borders will lead to a Marangoni flow which will tend to restore the film thickness.

If, however, the rate of surface tension relaxation is too rapid, then the driving force for such restoration will disappear before it can occur to any significant extent. This will mean that the film can extend without effective resistance. According to Ross and Haak [21] the "resilience" is then lost. It is therefore implied by these authors that extension of the film would then proceed until rupture occurs. A similar argument had been advanced earlier by Burcik [25].

There are some obvious difficulties with this hypothesis. Thus, it concerns extremely thick films where only rate processes can give rise to surface tension increases on film stretching. However, as a film thins, such stretching will give rise to surface tension increases as a result of depletion of surfactant by adsorption (to give a finite Gibbs elasticity). This effect would appear even when adsorption is instantaneous. Therefore complete loss of resilience by increases in the rate of adsorption would be confined to extremely thick films—resilience would eventually be restored if the film was sufficiently stretched. Another difficulty concerns the time scale over which the supposed film stretching occurs. Thus, the dynamic surface tensions upon which the hypothesis is based were all measured for surface ages $< 3 \times 10^{-2}$ s (the range of the oscillating jet technique). However, suppose, for example, that the stretching occurs more slowly than Ross and Haak [21] imply so that the relevant time scale is, say, 10^{-1} s and solutions both with and without antifoam have reached equilibrium at that surface age. The adverse effect of solubilized tributyl phosphate and methylisobutyl carbinol on foamability would then clearly not concern their effect on dynamic surface tension over the irrelevant time scale probed by the oscillating jet method.

Absence of either independent evidence concerning the relevant time scale or of dynamic surface tensions over wider time scales means that the hypothesis of Ross and Haak [21] is inadequately tested. We should also note that, although rapid transport of surfactant to air-water surfaces is not proven to lead necessarily to low foamability or foam stability, it is easy to see that extremely slow transport would. Thus, for example, if bubbles of gas approach one another during foam generation, then film collapse and coalescence will inevitably occur if stabilizing surfactant is essentially absent from the relevant air-water surfaces due to slow transport (see Sec. II). Burcik [25] recognized this difficulty and compromised by suggesting that high foamability requires "moderate" rates of surface tension lowering.

We may conclude by noting that, in general, antifoam effects appear to require that the antifoam be undissolved in the foaming medium. Solubilization of the antifoam in micelles largely restores the foamability. However, there is some evidence to suggest that the solubilized antifoams may have a weak adverse effect on the foamability or foam stability of a micellar solution [21]. We should note, however, that dissolved antifoam can even enhance foamability [17,18].

B. Duplex Oil Films, Oil Lenses, and Antifoam Mechanism

Antifoams are often largely composed of oils which are present as undissolved droplets in foaming solutions. Presence in the form of undissolved droplets is not, however, a sufficient requirement. Thus, it has often been proposed that such oils should also be surface active and exhibit the property of spreading over the air-liquid surface of the foaming liquid [1,17,20,26,27]. The first proposal that spreading be associated with antifoam behavior was made by Leviton and Leighton [28] more than half a century ago. Subsequently there have been numerous attempts to correlate the antifoam behavior of indissolved oils with their spreading behavior [20,22,29–31].

Here we are concerned with examining the possibility that duplex film spreading by undissolved oils is a necessary aspect of their antifoam mechanism. Such spreading occurs when a thin film of oil with distinct upper and lower surfaces, having the surface tensions of the bulk oil, spreads over the surface of the substrate liquid. The driving force for this process arises from the relevant surface tensions.

If an oil which is dispersed as droplets in the foaming liquid is to spread at the air-liquid surface of that liquid it must first emerge into the surface. For this to happen spontaneously the entry coefficient E must be positive.

Here E is defined by

$$E = \gamma_{AF} + \gamma_{OF} - \gamma_{AO} \qquad (7)$$

where γ_{AF} is the air-liquid surface tension of the foaming solution, γ_{OF} is the antifoam oil-foaming solution interfacial tension, and γ_{AO} is the antifoam oil-air surface tension. Selection of the appropriate values of the three surface tensions should take account of the extent to which the antifoam and foaming solution have been preequilibrated. Thus, we could follow Ross [20] and define two other entry coefficients,

$$E' = \gamma_{AF} + \gamma'_{OF} - \gamma'_{AO} \qquad (8)$$

and

$$E'' = \gamma'_{AF} + \gamma'_{OF} - \gamma'_{AO} \qquad (9)$$

where the primes refer to surface tension values for the mutually saturated antifoam and foaming fluid: E is usually described as the initial entry coefficient, E' is the semi-initial entry coefficient where the antifoam droplets are saturated with respect to the foaming liquid and the foaming liquid-air surface is uncontaminated with antifoam, and E'' is the final entry coefficient where both fluids are mutually saturated.

If the oil is also to spread at the surface of the foaming liquid to form a duplex film, then the spreading coefficient S must also be positive. Here we may follow Harkins [32] and define an initial spreading coefficient

$$S = \gamma_{AF} - \gamma_{OF} - \gamma_{AO} \qquad (10)$$

a semi-initial spreading coefficient

$$S' = \gamma_{AF} - \gamma'_{OF} - \gamma'_{AO} \qquad (11)$$

and a final spreading coefficient

$$S'' = \gamma'_{AF} - \gamma'_{OF} - \gamma'_{AO} \qquad (12)$$

These three spreading coefficients are exactly analogous to the three entry coefficients in Eqs. 7–9.

If we have $E > 0$ and $S < 0$, then placing a drop of the oil on the surface of the foaming liquid will form an oil lens and not a duplex film. Harkins [32] claims that the final spreading coefficient S'' is always negative, so that an oil will not spread on the surface of the foaming liquid after it has become saturated with respect to that oil. This means that if $S > 0$ the resulting duplex film is unstable and will disproportionate to form oil lenses in equilibrium with antifoam-contaminated air-foaming liquid surfaces characterized by a surface tension γ'_{AF}. Although this type of behavior is often found in practice, Rowlinson and Widom [33] argue that it is not always found.

Thus, in disagreement with Harkins [32], they state that we should have S'' ≤ 0 and *not* $S'' < 0$, so that $S'' = 0$ is allowed. This means that stable duplex films can sometimes be found between mutually saturated fluids at equilibrium.

An additional complication associated with the relationships in Eqs. 7–12 concerns the effect of the finite time required for surfactant to transport to the relevant surfaces during foam generation. Thus, the surface tension of the foaming liquid may be higher than equilibrium even if the liquid is saturated with respect to the antifoam. We may then find, for example, that $S'' > 0$ under dynamic conditions because $(\gamma'_{AF})_{dynamic} > (\gamma'_{AF})_{equilibrium}$.

Ross [20] has advanced a hypothesis concerning the rupture of foam films by oil droplets for which $E > 0$ and $S > 0$. Such oil droplets will, as we have seen, emerge into the air-foaming liquid surface and spread as a duplex film. If this were to happen with an oil droplet simultaneously emerging into both surfaces of a foam film, a duplex film of oil would simultaneously spread over both surfaces of the foam film. This would squeeze out the original liquid in the film to produce a region composed entirely of antifoam oil where rupture would occur. Presumably this mechanism would cease to function if the foaming liquid and antifoam become mutually saturated so that we have $S'' < 0$ (if dynamic effects can be ignored). The antifoam effect would therefore decay with time. Further addition of antifoam oil would also be without effect.

Ross [20] interpreted some early results of Robinson and Woods [19], using this hypothesis. Those results concerned the effect of various undissolved oils on the foam behavior of both aqueous and nonaqueous solutions of surfactant. The oils included alkyl phosphates, alcohols (including diols), fatty acid esters, and polydimethylsiloxane. The solutions were of aerosol OT (sodium diethylhexyl sulfosuccinate) in either ethylene glycol or triethanolamine and sodium alkylbenzene sulfonate in water.

Robinson and Woods [19] observed that for these systems, wherever $E < 0$, no antifoam effect is found. This then represents clear evidence that a positive value of the initial entry coefficient is necessary for antifoam action. In addition, however, Ross [20] noted that of the 40 combinations of oils and solutions with $E > 0$ which exhibited antifoam behavior, 36 had $S > 0$ and 4 had $S < 0$. He claimed that this finding represented "definite" agreement with the proposed rupture mechanism which requires $E > 0$ and $S > 0$. That conclusion of course ignores the finding that antifoam effects *can* occur with $S < 0$. A rigorous conclusion would therefore be that either spreading (i.e., $S > 0$) is not a *necessary* aspect of the antifoam behavior of these oils or that two different mechanisms may operate corresponding to the two sets of conditions $E > 0$, $S > 0$ and $E > 0$, $S < 0$.

In a later paper Ross and Young [29] find other examples of systems for which $E > 0$ and $S < 0$ where antifoam effects are observed. Moreover,

Okasaki et al. [22] report that both petroleum ether and olive oil spread rapidly over the surface of 10^{-2} M sodium dodecylsulfate solution but do not yield antifoam effects with the same solution. By contrast these authors find that undissolved phenol does not spread on 10^{-2} M sodium dodecylsulfate solution but does function as an effective antifoam (albeit at the rather high antifoam concentration used in this work).

In a study of the antifoam effect of normal alcohols, Kruglyakov and Taube [31] failed to find a correlation between initial spreading coefficients S and antifoam effectiveness. Here for octanol (and higher alcohols) on aqueous solutions of a variety of surfactants the initial spreading coefficient was found to be negative despite optimal antifoam effect with that alcohol under the conditions studied. Indeed, in another paper Kruglyakov [34] states that among the "numerous systems investigated by Ross and other workers there are many cases where efficient foam breaking was observed at a negative spreading coefficient" (i.e., when $S < 0$).

Kruglyakov and Koretskaya [35] have extended this study of normal alcohols to include the effect of surfactant concentration. They find that the optimum chain length of alcohol for maximum antifoam effect increases with decreasing concentration of surfactant. This behavior is shown in Fig. 5 for solutions of a blend of ethoxylated nonyl phenols (in 0.1 M KCl). Here the limiting concentration of antifoam for the foam to exhibit instability is plotted against alcohol chain length for various concentrations of surfactant. Kruglyakov and Koretskaya [35] argue that the antifoam effect of these alcohols is due to the formation of "asymmetrical films" where an oil lens is present on one side of a foam film. The configuration is illustrated schematically in Fig. 6. Such lenses will form provided $E > 0$ and $S < 0$. Indeed, as we have seen, even if $S > 0$ the resulting duplex film may be unstable so that ultimately the conditions $E'' > 0$ and $S'' < 0$ may prevail and lenses may form.

Kruglyakov [34] argues that the aqueous asymmetrical film separating an oil droplet from the air-water surface is metastable in the same sense as a symmetrical air-water-air foam film provided $E'' > 0$. However, he also argues that the asymmetrical film will in general have a different stability from that of the symmetrical foam film with the same solution. This is supposed to arise in part because of differences in surface excess of surfactant at the oil-water and air-water surfaces. If the oil is polar, then the adsorption of the surfactant at the oil-water surface will be lower than at the air-water surface. In turn this will mean diminished stabilizing repulsive forces across the film due, for example, to overlapping electrostatic double layers if the surfactant is ionic. The stability of the film will therefore be diminished provided the entry coefficient is still positive. Foam film collapse is then supposed to occur when the relatively unstable aqueous film separating an

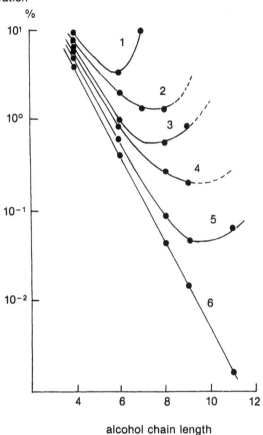

FIG. 5 Limiting alcohol concentrations for foam to exhibit instability at various concentrations of a blend of ethoxylated nonyl phenols (in 0.1 M KCl): (1) 1%, (2) 0.1%, (3) 0.5%, (4) 0.025%, (5) 0.005%, (6) 0.001%. (From Ref. 34.)

oil lens from the air-water surface in the foam film (see Fig. 2) breaks. Unfortunately, Kruglyakov [34,35] does not state precisely how the bridging oil lens formed by rupture of that asymmetric film can give rise to foam film collapse. It is, however, possible to deduce a detailed hypothesis as we describe below (see Sec. V.C.).

In interpreting the results shown in Fig. 5, Kruglyakov and Koretskaya [34,35] note that for all the alcohol-surfactant solution combinations the final entry coefficient E'' is everywhere positive. Ineffective antifoam behavior

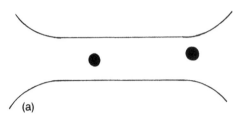

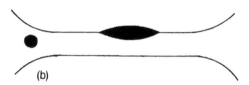

FIG. 6 Formation of "asymmetrical film" where oil lens is present in foam film. (From Refs. 34 and 35.)

for the lower-chain-length alcohols is ascribed to the higher solubility of these materials. Declining effectiveness of the higher-chain-length alcohols is ascribed to decline in polarity and increase in surfactant adsorption conspiring to enhance the stability of the oil-water-air films. That the optimum alcohol chain length for antifoam effectiveness declines with increasing surfactant concentration is then attributed to a concomitant increase in surfactant adsorption at the alcohol-water surface. This is supposed to enhance the stability of the oil-water-air film which diminishes the effectiveness of the longer-chain-length alcohols. Unfortunately Kruglyakov and Koretskaya [34,35] give no direct experimental evidence to confirm that the stability of the asymmetric films changes with alcohol chain length in a manner which correlates with changing antifoam effectiveness. Clearly, study of the kinetics of the process of emergence of alcohol droplets into the relevant air-water surfaces would be instructive.

It is argued by Kruglyakov [34] and others [36,37] that the well-known phenomenon of reduction in foamability of ethoxylated nonionic surfactants at temperatures above the cloud point (see [38], for example) may also concern the formation of oil lenses by nonionic-rich cloud phase droplets in a manner analogous to that proposed for the alcohols. Clear evidence that such droplets contribute to an antifoam effect is presented by Koretskaya [37], who showed that removal of the cloud phase droplets by filtration restores the foamability despite the reduction in overall surfactant concentration caused by such a procedure. The effectiveness of surfactant antifoams based upon such ethoxylated (or propoxylated) nonionics may therefore owe much to a low tendency of any other surfactant present to adsorb on the surface of cloud phase droplets. However, direct experimental evidence is lacking.

In summary, then, it would seem that spreading to form duplex films is not a necessary property of antifoam oils. Indeed, Kruglyakov and Koretskaya [34, 35] advance an alternative hypothesis which requires only that the entry coefficient be positive so that the oil may form lenses. This hypothesis is more general in that such lenses will often form even if the oil exhibits a positive initial spreading coefficient.

An interesting aspect of the hypothesis concerns the supposed effect of adsorbed surfactant on the stability of asymmetrical oil-foaming liquid-air films when that liquid is water. A consequence of this hypothesis is that if the oil is nonpolar with a high interfacial tension against water, then surfactant adsorption will be high and antifoam efficiency low even if $E > 0$ or $E'' > 0$. Hydrocarbons are examples of such nonpolar oils which are known to be relatively ineffective when used alone in aqueous ionic surfactant solutions [39,40]. The antifoam effectiveness of these oils in such solutions can, however, be enhanced significantly by addition of hydrophobic particles to the oil [39,40]. It would not be inconsistent with the hypothesis of Kruglyakov and Koretskaya [34,35] if we were to suppose that the role of such particles concerns the stability of the relevant oil-water-air films. We address the possibility in more detail below (see Sec. V.E.).

C. Surface Tension Gradients and Theories of Antifoam Mechanism in Heterogeneous Systems

We have disposed of the proposition that spreading from oil droplets to form duplex films is a necessary aspect of antifoam mechanism. However, certain undissolved materials which do spread to form duplex films (or even monolayers) may function because of that property. Here we examine two different approaches to antifoam mechanism which require such behavior. The first concerns elimination of stabilizing surface tension gradients as a result of displacement of the stabilizing surfactant monolayer by a spread film of antifoam. The second concerns the effect of the surface tension gradient, caused by spreading, on the thinning rate of foam films.

1. Elimination of Surface Tension Gradients

Kitchener [27] has argued that the role of "a completely effective antifoam agent must be to eliminate surface elasticity i.e. to produce a surface which has substantially constant tension when subjected to expansion." In effect, this is a corollary of the argument of Lucassen [2] (which we have outlined in Sec. II.A.) concerning the role of surface tension gradients in stabilizing foam films against external stress. It is also similar to the argument stated by Burcik [25]. As we have seen, Ross and Haak [21] have applied a variation of it to interpretation of antifoam behavior in homogeneous systems. Here we examine the application of this argument to heterogeneous systems, i.e., systems where undissolved antifoam materials are present.

Elimination of elasticity in heterogeneous systems is most probable when oil lenses in the surface are in equilibrium with a monolayer contaminated with the oil. Expansion of the surface will then mean more oil spreading out from the lenses, which will tend to maintain the surface tension. A monolayer contaminated with oil in equilibrium with oil lenses is of course the probable consequence of disproportionation of a nonequilibrium duplex film [32,33]. Obviously the effectiveness of the overall process will be determined by rate effects. Interpretation of measurements of the effect of antifoam materials on the relaxation of surface tension gradients therefore requires knowledge of the time scale of the relevant effects in foams.

Abe and Matsumura [41] have assessed the supposed role of the elimination of surface tension gradients under dynamic conditions in determining heterogeneous antifoam effectiveness. This work concerned the antifoam effect of alcohols on the foam behavior of an aqueous solution of sodium dodecylbenzenesulfonate. The alcohols included normal alcohols, branched alcohols, and diols.

These authors followed Ross and Haak [21] and measured the dynamic surface tension of dispersions of alcohols in solutions of sodium dodecylbenzenesulfonate, using the oscillating jet method. Here we remember that if the surface tension of the foaming liquid is lowered rapidly by the antifoam, then any rapid stretching of a foam film element will not result in an increase in surface tension of that element. There will therefore be no surface tension gradient between that element and the adjacent thicker elements. No reinforcing flow of liquid into the thin element will then occur. It is also obvious that the film will not be able to effectively withstand the effects of any external stress (due to gravity for example).

The limitations of this approach with respect to the time scale of the measurement of dynamic surface tension have already been outlined above (Sec. III.A.). Despite these limitations Abe and Matsumura [41] show a correlation between the dynamic surface tensions at surface ages of 2×10^{-2} s and foamability in the case of the normal alcohols. Thus, under these conditions the optimum antifoam is octanol, which produces the most rapid rate of surface tension reduction, as indicated by dynamic surface tension measurements. We should note here though that this correlation concerns measurement of dynamic surface tensions in the presence of $7.6 \times 10^{-3} M$ alcohol concentration for comparison with foamability in the presence of $3.84 \times 10^{-2} M$ alcohol (i.e., 5 g dm^{-3} in the case of octanol). However, at $7.68 \times 10^{-3} M$ the octanol is essentially ineffective as an antifoam (see Fig. 23 in Ref. 41) presumably because it is below the solubility limit. Moreover, increase in the octanol concentration above $7.68 \times 10^{-3} M$ has little effect on the dynamic surface tension at a surface age of 2×10^{-2} s. All of this would tend to suggest that whatever the significance of the correlation it has limited relevance for the basic proposition that rapid reduction of dynamic surface tension over the time scales measured by the oscillating jet will reduce foamability. Abe and Matsumura [41] in fact demonstrate the absence of any correlation between the rate of surface tension lowering (albeit again with different concentrations between surface tension and foamability measurements) for alkanediols. Here the optimum antifoam effectiveness is found with those compounds which yield the *slowest* rate of surface tension reduction.

Curiously, both Abe and Matsumura [41] and Kruglyakov and Koretskaya [35] have studied the effects of normal alcohols on the foam of aqueous sodium dodecylbenzenesulfonate solutions. Both find octanol to have the optimum chain length for antifoam effect at surfactant concentrations in the region of 1 g dm^{-3} (albeit at different temperatures and ionic strengths). They also find that the antifoam concentration for significant effect under these conditions is extremely high relative to that of the surfactant—in the region of 5–10 g dm^{-3} for octanol. Despite consensus concerning these experimental findings, the two groups of authors have adopted completely different hypotheses describing the mode of antifoam action of these alcohols. Surprisingly, Abe and Matsumura [41] do not cite the earlier work of Kruglyakov and Koretskaya [35].

The argument concerning elimination of surface elasticity (and therefore surface tension gradients) by antifoams has been extended to nonaqueous foaming liquids by Callaghan et al. [42]. These workers have studied the effect of polydimethylsiloxanes on the surface elasticity of crude oil. Polydimethylsiloxanes are used as antifoams to assist gas-oil separation during crude oil production and are apparently effective at the remarkably low concentration of 1 part in 10^6 (which presumably still exceeds the solubility limit). Callaghan et al. [42] find that polydimethylsiloxane diminishes the frequency-dependent dynamic dilational (elastic) modulus $|\epsilon| = d\gamma_{AO}(t)/d \ln A(t)$ relative to that found for the uncontaminated oil. Here $\gamma_{AO}(t)$ is the time-dependent air-crude oil surface tension, and $A(t)$ is the area of a constrained element of air-crude oil surface subject to time-dependent dilation. The effect is more marked the higher the molecular weight (or viscosity) of the polydimethylsiloxane. This correlates with an enhanced antifoam effectiveness found with increase in molecular weight. Callaghan et al. [42] argue that all of this represents evidence that the polydimethylsiloxane functions by elimination of surface elasticity and tension gradients. However, McKendrick et al. [43] find a frequency-independent *increase* in dynamic dilational modulus of a hydrocarbon oil upon addition of a fluorosilicone antifoam. Moreover, the correlations of Callaghan et al. [42] concern extremely long time scales in the range $10-10^3$ s. Whether these correlations are fortuitous or represent clear evidence for the proposition under test requires additional information concerning the relevance of those time scales. Absence of hard evidence concerning the magnitude of the relevant time scales is thrown into high relief if comparison is made between the approaches of Abe and Matsumura [41] and Callaghan et al. [42]. These workers have looked at time scales which are separated by more than three orders of magnitude. This is surprising, since they are both seeking evidence for the same basic proposition concerning the elimination of surface elasticity in foam films by antifoam oils.

2. Surface Tension Gradients Induced by Spreading Antifoam

We have seen that some authors form the view that antifoams *eliminate stabilizing* surface tension gradients in foam films. Here we consider the proposition that antifoams *cause destabilizing* surface tension gradients in foam films.

The proposed mechanism concerns spreading from an antifoam source present on the surface of a foam film. Such spreading is driven by a surface tension gradient between the spreading source and the leading edge of the spreading front. This surface tension gradient will act as a shear force dragging the underlying liquid away from the source to supposedly cause catastrophic thinning and foam film rupture. The process is shown schematically in Fig. 7. It is clearly a consequence of the well-known Marangoni effect, and we refer to it below as Marangoni spreading.

This mechanism was first proposed by Ewers and Sutherland [26]. These authors identified three categories of antifoam substance which could, in principle, function by this mechanism. These are:

"solids or liquids containing surface active material other than the substance stabilizing the film"

"liquids which contain the foam stabilizer in higher concentration than it is present in the foam"

"vapours of surface active liquids" (including the well-known example of the adverse effect of ether vapor on foam stability)

In the first category Ewers and Sutherland [26] included all spreading oils for which $S > 0$. In addition, they included monolayer spreading where a

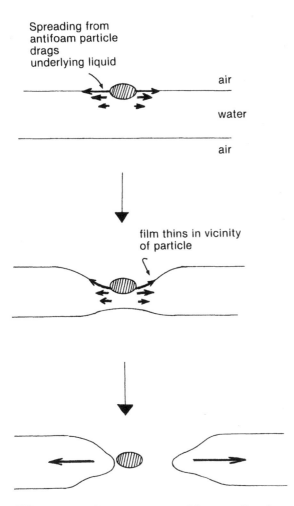

Spreading from
antifoam particle
drags
underlying liquid

air

water

air

film thins in vicinity
of particle

FIG. 7 Foam film rupture caused by spreading from antifoam particle.

solid or liquid forms a monolayer for which the initial spreading pressure S_m is

$$S_m = \gamma_{AF} - \gamma_m \tag{13}$$

where γ_m is the equilibrium surface tension of the spreading monolayer on a surface of the foaming liquid from which surfactant has been removed.

The underlying driving force for all of these categories of antifoam substance is the surface tension gradient $dS/d\chi$ or $dS_m/d\chi$, where χ is the spreading distance. Therefore, according to Ewers and Sutherland [26] the greater the magnitude of S (or S_m) the greater the force and the more effective the antifoam. Clearly here S could be measured on a nonequilibrium foam film where $(\gamma_{AF})_{dynamic} > (\gamma_{AF})_{equilibrium}$.

This approach of Ewers and Sutherland [26] was advanced with some claim to wide generality. However, the finding that many substances yield antifoam effects without causing destabilizing surface tension gradients [19,22,29,31,34,44] confounds exclusive generality. Moreover, a difficulty with the mechanism concerns the implicit assumption that the antifoam is always to be found in the thinnest and most vulnerable part of the foam. Thus, if the antifoam entity spreads from the plateau border, this will drag liquid into the adjacent foam films, which will stabilize those films by increasing their thicknesses.

Another limitation of the approach of Ewers and Sutherland [26] is that it is essentially qualitative. However, there have been some attempts to put it on a quantitative basis [17,45]. The first such attempt was due to Shearer and Akers [17], who were concerned with the antifoam effect of polydimethylsiloxanes on lube oils. Here the effect of antifoam on a monolayer raft of bubbles is modeled theoretically and the results compared with experiment.

In this approach, spreading from droplets of antifoam on the foaming liquid surface is supposed to augment gravity drainage from foam films. This gravity drainage is modeled by assuming that the boundaries of the foam films are immobile, in that Poiseuille flow prevails (with a parabolic velocity profile). A schematic illustration of the model where bubbles are supposed to form a raft at the surface of the foaming liquid is shown in Fig. 8a. Shearer and Akers [17] calculate that the time t_g taken for a thick foam film surrounding a bubble to drain to the thickness h_{crit}, at which rupture occurs, is

$$t_g \sim \frac{18\eta_F R_b}{\rho g h_{crit}^2} \tag{14}$$

Here ρ and η_F are respectively the density and viscosity of the foaming liquid, and R_b is the radius of the bubble. Equation 14 implies that the time t_g is independent of the height of the film above the planar liquid surface at which the bubble is resting. In actuality the film would become relatively thin at the top of the bubble. Failure to account for this effect is a consequence of incomplete consideration of continuity by Shearer and Akers [17] in the derivation of Eq. 14.

The treatment of film draining due to spreading given by Shearer and Akers [17] is summarized here to illustrate the essentials of the argument. Thus, consider one polydimethylsiloxone oil droplet of diameter a at the center of the foam film as indicated in Fig. 8b. The total spreading force is

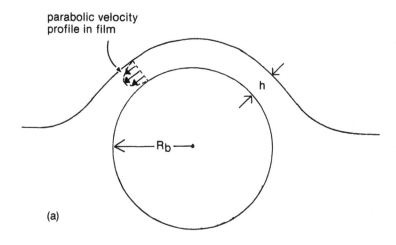

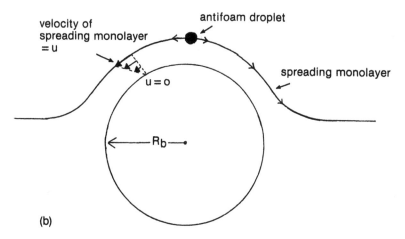

FIG. 8 Model of Shearer and Akers [17] describing effect of gravity drainage and spreading from an antifoam droplet on film thinning. (a) Foam film drainage under gravity from a bubble at a planar liquid surface (where $h \ll R_b$). (b) Foam film drainage induced by spreading from antifoam droplet.

assumed to be $\pi a S$ at the perimeter of the oil droplet, where a is the droplet diameter. It is also assumed that S becomes zero at the edge of the bubble in the plane of the liquid surface so that a linear gradient in S exists with respect to radial distance from the antifoam droplet. The shear stress τ is therefore crudely approximated as

$$\tau = \frac{aS}{k_a R_b^2} \tag{15}$$

where k_a is a geometrical constant which describes the area of the bubble surface above the plane of the liquid surface. This model of the shear stress obviously ignores one of the essential features of the spreading process, namely that the spreading front will expand over the substrate and the shear stress will decline with time. It will therefore tend to exaggerate the effects of spreading.

Shearer and Akers [17] argue that a linear velocity gradient u/h will exist in the film due to the imposed shear stress. Here h is the film thickness, and u is the velocity of the spreading surface (which is assumed to be constant throughout drainage). The opposite surface of the film is assumed to be stationary where $u=0$ (see Fig. 8b). Therefore it is possible to deduce from Eq. 15 that

$$u = \frac{aSh}{k_a \eta_F R_b^2} \tag{16}$$

Shearer and Akers [17] use a crude averaging procedure to allow for continuity in the derivation of Eq. 14. They also use a similar procedure to deduce from Eq. 16 that the time t_a taken for spreading to thin a thick film down to the critical rupture thickness h_{crit} is given by

$$t_a = \frac{k_b \eta_F R_b^3}{aSh_{crit}} \tag{17}$$

where k_b is a constant. Using parameters relevant for typical polydimethylsiloxane/lube oil combinations, Shearer and Akers [17] show that $t_a \ll t_g$, so that any contribution of gravity drainage to t_a may be neglected. Indeed their calculations suggest that t_a/t_g could be of order 10^{-3}. If the crude assumptions upon which they are based are not too misleading, then this would imply that spreading from polydimethylsiloxane droplets could significantly increase the rate of foam film rupture in lube oils.

Shearer and Akers [17] allow for the possible presence of several droplets of antifoam in a foam film by simply increasing the supposed spreading perimeter to $n\pi a$, where n is the number of droplets per film. If, however, the droplets are evenly dispersed over the surface of foam films, then spreading will occur in opposing directions. Flow of liquid from the film will be less than that predicted if the spreading perimeter is simply increased to $n\pi a$.

Equations 14 and 17 may be combined to give

$$\frac{t_a S}{t_g \rho} = \frac{k_c R_b^2 h_{crit}}{na} \tag{18}$$

where k_c is a constant and where allowance is made for the presence of n droplets in the film. The group $t_a S/t_g \rho$ is seen to be independent of the properties of the foaming liquid provided the bubble size can be controlled and h_{crit} is invariant. Indeed, log/log plots of this group against antifoam concentration for three different lube oils with a given polydimethylsiloxane were shown by Shearer and Akers [17] to lie on the same curve. Here t_g and t_a were obtained experimentally by assessing the half-life of bubble rafts. All of this is offered as experimental evidence for the validity of Eq. 18 and the model from which it is derived. In view of all the assumptions implicit in the approach of Shearer and Akers [17], this experimental accord with theory is remarkable.

Prins [45] has adopted a different approach to the theoretical description of the effect of spreading from heterogeneities on foam film stability. In contrast to Shearer and Akers [17] he recognizes that the spreading process involves a spreading front which advances over the substrate surface. This process has been the subject of a number of independent studies. Thus, Fay [46] and Hoult [47] show from dimensional analysis that an oil (for which $S > 0$) will spread as a duplex film over a clean water surface in a linear manner at a rate given by

$$\chi = k_s \left(\frac{S^2}{\eta_w \rho_w} \right)^{1/4} t^{3/4} \tag{19}$$

where χ is the spreading distance, k_s is a constant which has a value of about 1.33 for liquids such as polydimethylsiloxanes of viscosity ≤ 1000 mPa s [48], and η_w and ρ_w are, respectively, the viscosity and density of water. The rate of spreading is seen to be dependent upon S but is independent of the viscosity of the oil. Joos and Pintens [49] treat the process as a longitudinal disturbance and deduce theoretically that $k_s = 1.15$.

As we have seen, the spreading antifoam oil will cause the underlying liquid to move in the same direction. If the depth to which this movement penetrates is δ, we can write approximately

$$\frac{S}{\chi} = \eta_w \frac{d\chi/dt}{\delta} \tag{20}$$

Substituting Eq. 19 in Eq. 20, using the value $k_s = 1.15$ given by Joos and Pintens [49], yields for the penetration depth

$$\delta = \left(\frac{\eta_w t}{\rho_w} \right)^{1/2} \tag{21}$$

Prins [45] supposes that foam film collapse will occur when the penetration depth of a spreading duplex film equals the thickness of the foam film over which it is spreading. Clearly the depth will increase with time (eq. 21), and so some estimate of maximum time available is necessary in order to assess the maximum penetration depth δ_m. Prins estimates the maximum extent χ_m to which the spreading may occur in a radial sense from

$$\frac{\pi a^3}{6} = \pi \chi_m^2 d_m \tag{22}$$

where d_m is the thickness of the spreading film at the maximum extent and a is the diameter of the oil droplet. The time of spreading is then obtained somewhat approximately by substituting χ_m, derived for radial spreading, into Eq. 19, which is derived for linear spreading. Thus, Prins [45] combines Eqs. 19, 21, and 22 to obtain the maximum penetration depth:

$$\delta_m = 0.5a \left(\frac{\eta_w^2}{S\rho_w d_m} \right)^{1/3} \tag{23}$$

If foam film collapse occurs when the penetration depth equals the thickness of the film, then the larger the value of δ_m the thicker the films which may be ruptured. Prins [45] therefore equates high values of δ_m with high probability of foam film rupture. Equation 23 therefore implies that antifoam effectiveness will increase with increasing size of the antifoam droplet. At least in this respect there is agreement with Shearer and Akers [17], although for different reasons (see Eq. 18). Prins [45] has shown that increasing droplet size can increase antifoam effectiveness for small droplet sizes. However, this does not continue indefinitely because as the droplet size becomes large the probability of an antifoam entity being present in a foam film declines. An optimum particle size for high antifoam effectiveness is therefore obtained. Experimental results for soya bean oil antifoam in sodium caseinate solution illustrate this behavior and are reproduced in Fig. 9.

The crude nature of the model presented by Prins [45] gives rise to an awkward feature in Eq. 23. Thus, if $S = 0$ then we appear to have $\delta_m = \infty$. Equation 23 in fact means that antifoam effectiveness is predicted to increase as S decreases. Thus, Prins [45] presents a model of antifoam action, based upon spreading driven by surface tension forces, which predicts increasing antifoam effectiveness as those forces become weaker. Intuition would suggest an opposite conclusion (which is stated by Ewers and Sutherland [26] and deduced by Shearer and Akers [17]).

We are therefore left without an entirely satisfactory mathematical treatment of the antifoam mechanism proposed by Ewers and Sutherland [26]. Clearly the problem is difficult. However, a solution would permit more serious assessment of the significance of the mechanism particularly with

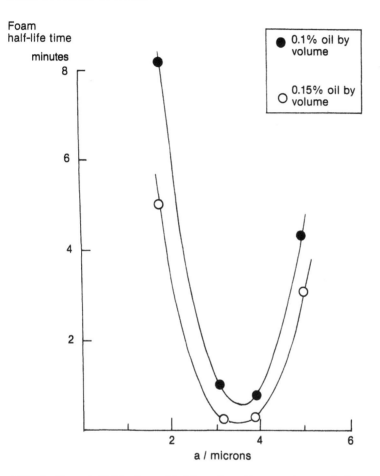

FIG. 9 Foam half-life time of 0.05 wt.% sodium caseinate solutions as a function of mean drop diameter a of soya bean oil. (From Ref. 45, reproduced by permission of the Royal Society of Chemistry).

regard to the effect of the magnitude of S (or S_m). Such a solution should take account of the effect of compression of the surfactant monolayer at the foaming liquid-air surface by the spreading front. This effect will clearly reduce the rate of spreading and will diminish the threat to foam film stability.

There are other problems with the mechanism. Spreading from antifoam entities present in the plateau borders could in principle reinforce foam film stability. Spreading from many different sources in the surface of a foam film would presumably rapidly eliminate surface tension gradients so that rupture induced by surface-tension-driven flow would become unlikely.

However, elimination of surface tension gradients can also in principle contribute to diminished foam film stability.

Another problem concerns the nonequilibrium nature of the process of spreading. Thus, if the antifoam and foaming liquid are mutually saturated, we can have $S'' < 0$ (and $S_m < 0$) [32,33]. Spreading will then only occur under nonequilibrium conditions where, for example, the surface tensions of foam films are markedly higher than equilibrium as a result of slow surfactant transport to the relevant surfaces.

We therefore have a mechanism which relies on a necessarily transient process. However, materials which have the relevant properties, such as spreading oils, will also be capable of participating in other mechanisms. The latter include elimination of surface tension gradients and formation of unstable bridging oil lenses (see Sec. V. C.). Unequivocal experimental verification of a role for this mechanism will therefore represent a formidable challenge.

IV. INERT HYDROPHOBIC PARTICLES AND CAPILLARY THEORIES OF ANTIFOAM MECHANISM FOR AQUEOUS SYSTEMS

A. Early Work

Solid particles which adhere to fluid-fluid surfaces so that part of their surface is exposed to one fluid and part to the other fluid exhibit a finite contact angle at the relevant surface. For almost three-quarters of a century it has been known that such particles may have a drastic effect on the behavior of dispersions formed by mixing fluids. One of the earliest publications is that of Pickering concerning the effect of particles on oil-water emulsion behavior [50]. A few years later Bartsch [51] described the stabilizing effect of hydrophobed mineral particles on froths formed by aqueous solutions of 3-methylbutanol.

Since then there have been many publications concerning the effect of particles on the formation and stability of foams, froths, and emulsions. Here we are mainly concerned with the destabilizing effect of particles on foams and froths. Practical examples include the effect of hydrophobed mineral particles and collector precipitates on the stability of mineral flotation froths [52,53]. Another is the use of wax [54] to control the foam of detergent formulations for automatic washing machines.

Arguably the earliest suggestion that an adverse effect of particles on aqueous froths may be attributable to low wettability (or high contact angle at the air-water surface, measured through water) was made by Mokrushin 40 years ago [55]. This concerned the effect of metal sulfides on froths formed by gelatin solutions. A few years later Dombrowski and Fraser [56] reported direct observation of the effect of wettability of particles on the disintegration of thin films of water or alcohol formed from spray nozzles. Here particles which were not wetted by the film liquid caused perforation of the film when the particle size was of the same order as the film thickness. Particles which were wetted by the film liquid had no effect on the manner of disintegration.

A later reference to the role of the wettability (or the contact angle) of dispersed solid material in determining froth stability is due to Livshitz and Dudenkov [57]. Here the destabilizing effect of hydrocarbon particles on froths is attributed to air-water contact angles >90° by analogy with the effect of particles on oil-water emulsion behavior. Thus the presence of particles with a contact angle >90° at the oil-water surface (measured through the water) in an oil-water mixture will produce a water-in-oil emulsion [58]. Livshitz and Dudenkov [57] argue that analogous behavior in the case of an air-water system is tantamount to destruction of the foam. This of course gives no mechanism for either supposedly analogous phenomenon. However, in a later paper, concerned with the effect of hydrophobed metal xanthates and oleate precipitates on flotation froths, Livshitz and Dudenkov [59] suggest that hydrophobic particles may bridge aqueous films. The more hydrophobic the particle, the closer are the wetting perimeters of each air-water surface at the particle. This will supposedly reduce the thickness of the film, which will result in acceleration of film rupture. In the case of a hydrophilic particle, on the other hand, contact angle formation does not take place and there is no film rupture. This mechanism appears to suggest that there should be a continuous improvement in defoaming ability with increasing contact angle.

A difficulty with establishing the role of wettability in determining the effectiveness of particles in breaking foams and froths is measurement of the relevant contact angle. This is particularly so for precipitated particles of sparingly soluble materials. Thus, representative smooth surfaces upon which contact angles can be readily measured may not be easily prepared. In recognition of this Dudenkov suggested that the solubility products of polyvalent metal salt precipitates of butyl xanthogenates and oleates should correlate with the contact angles [60]. Thus, the lower the solubility product the more hydrophobic is the precipitate for a given anionic hydrophobe. Whatever the merit of this supposed correlation, Dudenkov [60] produces results which indicate a sharp increase in the volume of air present in a froth for precipitates of oleate or xanthogenate with solubility products greater than a critical value. The latter decreases with increase in frother concentration. Examples of their results are presented in Fig. 10. These are interpreted by Dudenkov [60] as indicating the importance of wettability in de-

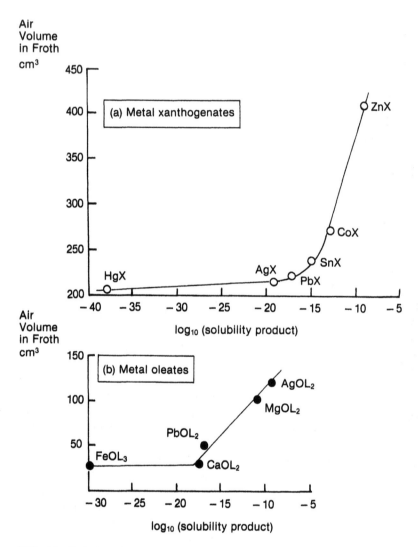

FIG. 10 Relationship between air volume in froth and solubility products of hydrophobic precipitates. (From Ref. 60.) (a) Metal butyl xanthogenates at 5×10^{-5} M in 30 mg dm^{-3} commercial frother solution. (b) Metal oleates at 2×10^{-5} M in 10 mg dm^{-3} commercial frother solution. (Reproduced by permission of Primary Sources.)

termining antifoam effectiveness. But the solubility products given by Dudenkov [60] may be suspect. Thus, the value given for calcium oleate is almost two orders of magnitude lower than that given by Irani and Callis [61].

Similar observations have been reported by Peper [62], who considered the effect of sodium soaps and fatty acids on the foamability and foam stability of dilute solutions of anionic surfactant solutions. Peper [62] found that in the presence of a stoichiometric excess of calcium chloride (for calcium soap formation) the adverse effect of sodium soaps on foam was in the order sodium stearate $\approx$ sodium palmitate > sodium oleate > sodium laurate, which follows the order of solubility products K_{sp} of the corresponding calcium soaps (calcium stearate, $K_{sp} \sim 2 \times 10^{-20}$; calcium palmitate, $K_{sp} \sim 6 \times 10^{-18}$; calcium oleate, $K_{sp} \sim 10^{-15}$; calcium laurate, $K_{sp} \sim 6 \times 10^{-13}$ at 25°C [61]). Curiously Peper [62] interprets his results in terms of formation of rigid islands of calcium soap monolayer interspersed with gaseous film of adsorbed surfactant. Peper [62] asserts that these islands will make the film unstable because of their "inflexible brittle nature." No theoretical arguments are given for why these should be unstable. Any role for the calcium soap precipitates, which will undoubtedly be found under these conditions, is ignored.

Peper [62] found that fatty acids reduce the surface tension of solutions of sodium dodecylbenzenesulfonate containing calcium chloride. The effect was found to be least pronounced for the fatty acid which forms the least soluble calcium soaps and the best antifoam. It is difficult to reconcile this finding with a spreading mechanism for antifoam action (see Sec. III.C.2.).

B. Contact Angles and Particle Bridging Mechanism

Attempts to relate the foam behavior of hydrophobic particles directly to contact angles followed more than a decade after the work of Dudenkov [60]. Of these a paper by Garrett [44] concerned finely divided polytetrafluoroethylene (PTFE) as antifoam material. PTFE has the merit of inertness so that any possible contribution from Marangoni spreading to the antifoam effect can be unambiguously eliminated. Moreover, reliable representative contact angles can be measured on polished PTFE plates.

The effect of finely divided ($\sim$5 μm) PTFE particles on the volume of foam generated by cylinder shaking for several aqueous surfactant solutions is reproduced in Fig. 11. An apparent correlation between volume of foam destroyed and receding contact angles is revealed. That the effects do not concern adsorption loss of surfactant onto PTFE surfaces can easily be demonstrated by filtering off the PTFE. Foamability of the solution is restored.

In seeking an explanation for the antifoam behavior of PTFE which is related to wettability, Garrett [44] followed Livshitz and Dudenkov [59] and

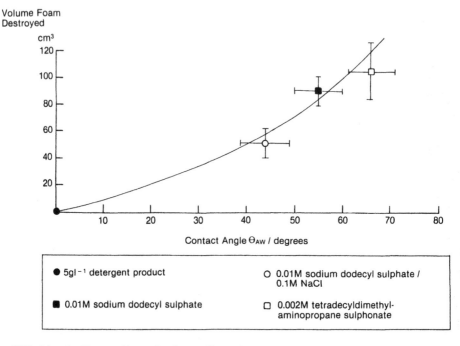

FIG. 11 Antifoam effect of polytetrafluorethylene particles as a function of receding contact angle. (From Ref. 44, reproduced by permission of Academic Press Inc.)

considered a bridging hydrophobic particle. Depending on the contact angle, two consequences may be distinguished for a hydrophobic spherical particle emerging into both air-water surfaces of a foam film. Thus, if the contact angle is $>90°$, there is no condition of mechanical equilibrium available to the particle. There will be an unbalanced capillary pressure ΔP in the vicinity of the particle given by the Laplace equation

$$\Delta P = \gamma_{AW}\left(\frac{1}{R_1} + \frac{1}{R_2}\right) \tag{24}$$

where R_1 and R_2 are the two radii of curvature describing the curved air-water surface in the vicinity of the particle and γ_{AW} is the air-water surface tension. The configuration is shown in Fig. 12 where the particle is assumed to be small so that the effect of gravity upon the shape of the air-water surface may be neglected. The two radii of curvative are of opposite sign so that $R_1 < 0$, and therefore the direction of the unbalanced capillary pressure depends upon the relative magnitudes of R_1 and R_2. The ratio R_2/R_1 is not obvious by inspection, although it has been simply assumed in later work

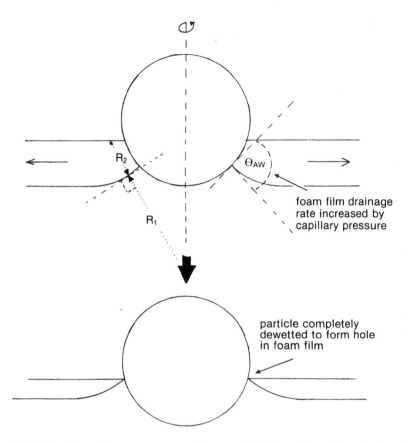

FIG. 12   Foam film rupture caused by a spherical particle with contact angle θ_{AW} > 90°.

by Frye and Berg [63] that

$$\frac{R_2}{R_1} \ll 1 \tag{25}$$

so that Eq. 24 becomes

$$\Delta P = \frac{\gamma_{AW}}{R_2} \tag{26}$$

The capillary force will therefore act in a direction to cause enhanced drainage out of the foam film. The process of drainage will continue until the two three-phase contact lines at the particle surface are coincident, whereupon a hole will form in the foam film. It has been shown by de Vries [64] that if the hole is sufficiently large it will spontaneously increase in size and lead to film collapse.

The second consequence of a particle emerging into both air-water surfaces of a foam film follows if the contact angle <90°. The resulting configurations are depicted in Fig. 13. Here the initial configuration is similar to that found for particles with contact angles >90° where, if inequality 25 is valid, an unbalanced capillary pressure will enhance drainage out of the foam film. However, as the film drains it will always attain some thickness where the film is planar and where the capillary pressure becomes zero. Further drainage will then result in a change in sign of both radii of curvature. If condition 25 is assumed, this will mean an unbalanced capillary pressure inhibiting further drainage of the film. It is therefore possible that a particle with this configuration will actually stabilize foam films. That hydrophobic particles may stabilize foams has in fact been occasionally reported in the literature [52,57,65,66].

The essentials of the mechanism of film collapse by spherical particles have been confirmed in a cinematographic study by Dippenaar [52], using hydrophobed glass spheres. Representations of high-speed movie frames are reproduced in Fig. 14 for a glass bead of 250 μm diameter and a contact angle of 102° in a distilled water film. Here the contact angle formed initially at the lower air-water surface is seen to be apparently less than the equilibrium value. The usual interpretation of such dynamic contact angles is that the equilibrium contact angle always prevails near the three-phase contact line and that the air-water surface exhibits extreme curvature in the same region to give an apparent macroscopic contact angle much lower than the equilibrium value. It is the capillary pressure implied by that curvature which produces the force driving film thinning. Collapse occurs about an order of magnitude faster than would have occurred if the particle had been absent [52]. This is seen to occur as the two three-phase contact lines become coincident. By contrast, a sphere of contact angle 74° had no effect on film rupture times and adopted a configuration similar to that of Fig. 13c.

A basic difficulty with application of this mechanism of collapse by hydrophobic particles to the results shown in Fig. 11 is that the contact angle for the PTFE particles is always <90°. However, the contact angles cited in Fig. 11 are equilibrium values. Foam generation is a nonequilibrium phenomenon where air-water surfaces will suffer rapid overall expansion. This will mean that the surface tension of a freshly generated foam film will in general be greater than the equilibrium surface tension of the solution from which it is formed because adsorption of surfactant is not instantaneous. This in turn should mean that the actual contact angle prevailing during foam generation will be greater than the equilibrium contact angle. Since the re-

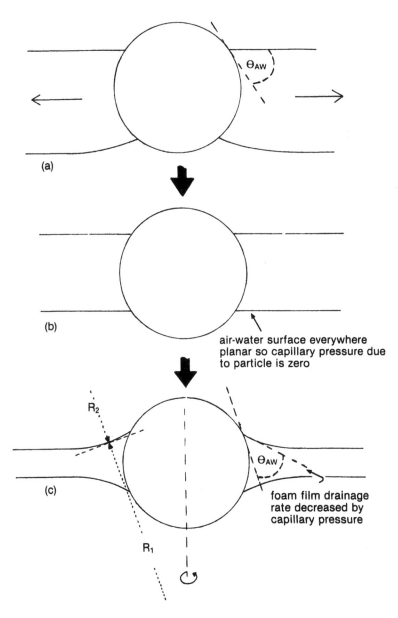

FIG. 13 Effect of spherical particle with contact angle $\theta_{AW} < 90°$ on foam film.

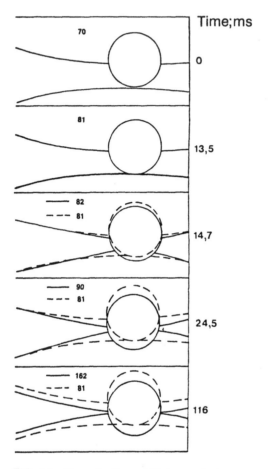

FIG. 14 Reproduction of high-speed cinematographic film frames showing interaction of hydrophobic glass bead ($\theta_{AW} = 102°$) with an aqueous film. (From Ref. 52, reproduced by permission of Elsevier Science Publishers).

ceding contact angle for PTFE against distilled water is in the region of 95° [44], dynamic contact angles >90° are in principle possible. However, such values would only be attained if all surfaces were almost denuded of surfactant. Survival of films with such surfaces for sufficient time to form a foam would seem unlikely anyway regardless of the presence of particles. These considerations suggest that contact angles >90° are unlikely to have prevailed in the foam experiments from which the results shown in Fig. 11 are taken.

In an attempt to explain the discrepancy, Garrett [44] suggested that the contact angle required for film collapse by a particle may be determined by

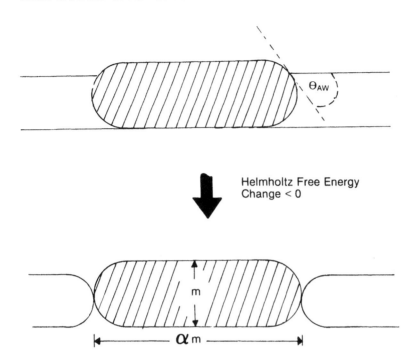

FIG. 15 Effect of particle geometry on foam collapse; disk-shaped geometry with various aspect ratios α. (From Ref. 44, reproduced by permission of Academic Press Inc.)

particle geometry. Thus, PTFE particles are modeled as disk shapes (with round edges) of various aspect ratios. Hole formation in a foam film by dewetting of particles is assumed to occur by the model shown in Fig. 15. The condition for hole formation is simply that the Helmholtz free-energy change accompanying the dewetting process shown in Fig. 15 should be negative. It follows from this model that the contact angle for film collapse decreases with increasing aspect ratios of the particles [44]. Thus, if the aspect ratios are sufficiently large, particles with contact angles <90° are predicted to cause film rupture by this mechanism. A major flaw with this approach is, however, neglect of the effects of viscous dissipation. Thus, for contact angles <90° an apparently stable configuration analogous to that shown in Fig. 13b is available for the particle. That hole formation is predicted despite this is due to the assumption that the free-energy gain in arriving at that configuration is all then available for doing work to produce complete dewetting of the particle.

A superior explanation for the role of particle geometry in producing foam collapse with contact angles <90°C emerges from the cinematographic study

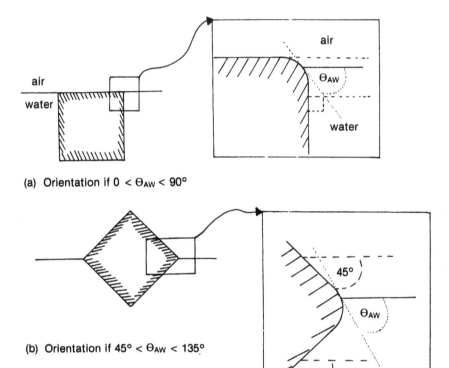

(a) Orientation if 0 < Θ_{AW} < 90°

(b) Orientation if 45° < Θ_{AW} < 135°

FIG. 16 Orientations of cubic particle at air-water surface.

by Dippenaar [52]. Here Dippenaar [52] examined the behavior of cubic particles of xanthanated galena with a contact angle of 80 ± 8° against water. Such particles were observed to adopt two different orientations, with about equal probability, at the air-water surface. These orientations are depicted in Fig. 16, where it is seen that the air-water surface intersects the particle edges for both orientations.

The behavior of air-water surfaces at edges is usually interpreted by supposing that the equilibrium contact angle is always satisfied at the submicroscopic level [67]. The edge is represented as having a submicroscopic circular cross section [67]. The air-water surface may then adopt an apparent single macroscopic angle at the edge for a range of contact angles, any of which may be satisfied at the submicroscopic level. All of this is illustrated in Fig. 16, where it is clear that the two orientations for a cubic particle require contact angles 0 < θ_{AW} < 90° and 45 < θ_{AW} < 135°, so that the xanthanated galena used by Dippenaar [52] with a contact angle of 80° may adopt either.

Dippenaar [52] observed that a hydrophobed galena particle with a horizontal orientation (Fig. 16a) had little effect on foam film stability. The behavior resembled that of spherical particles of contact angle < 90°. However, in the case of a particle with a diagonal orientation (Fig. 16b) two consequences arose when it contacted the second air-water surface of the thinning film. The particle either twisted to a horizontal orientation with two faces in the planes of the upper and lower film surfaces or retained its diagonal orientation. Particles which retained the diagonal orientation caused rapid film collapse. The sequence of events is illustrated in Fig. 17, where representations of the original high-speed cinematographic frames are reproduced.

Here the air-water surface is seen to move up the inclined left-hand face until the two three-phase contact lines become coincident on the edge, whereupon film collapse occurs. We therefore find that a cubic particle can give rise to film collapse, even though $\theta_{AW} < 90°$, by a mechanism exactly analogous to that shown for spheres in Fig. 12 and 14.

C. Particle Geometry and Contact Angle Conditions for Foam Film Collapse

1. Smooth Particles with Edges

It is possible to use the results of Dippenaar [52] to devise general rules concerning the role of particle geometry in determining film collapse. Thus, for particles with smooth curved surfaces, such as spheres, cylinders, and ellipsoids, the condition for film collapse by the process shown in Fig. 12 and 14 is

$$\theta_{AW} > 90° \tag{27}$$

where θ_{AW} is the receding contact angle at the air-water surface measured through the aqueous phase.

For smooth particles with edges the contact angle for rupture depends upon the geometry. The basic principles, made obvious by Dippenaar's work [52], are most clearly revealed if an axially symmetrical particle with three edges is considered. The geometry of this particle is shown in Fig. 18, whence it is seen that only one angle θ_p need be specified to define all edge angles (where $\theta_p < 90°$). If the diameter of the particle is large enough, it will float at an air-water surface with one of three possible configurations, depending upon the magnitude of θ_{AW}. These are illustrated in Fig. 18. In each case the rotational symmetry axis of the particle will be perpendicular to the air-water surface.

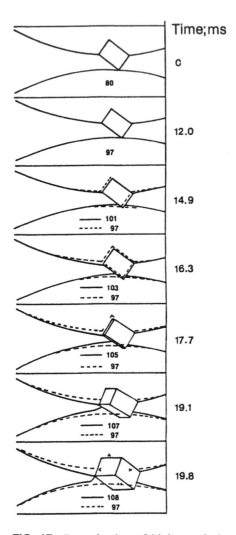

FIG. 17 Reproduction of high-speed cinematographic film frame showing interaction of xanthenated galena particle ($\theta_{AW} = 80 \pm 8°$) with aqueous film. (From Ref. 52, reproduced by permission of Elsevier Science Publishers.)

Configuration A corresponds to the condition

$$0 < \theta_{AW} < \theta_p \qquad (28)$$

configuration B to

$$\theta_p < \theta_{AW} < 180° - \theta_p \qquad (29)$$

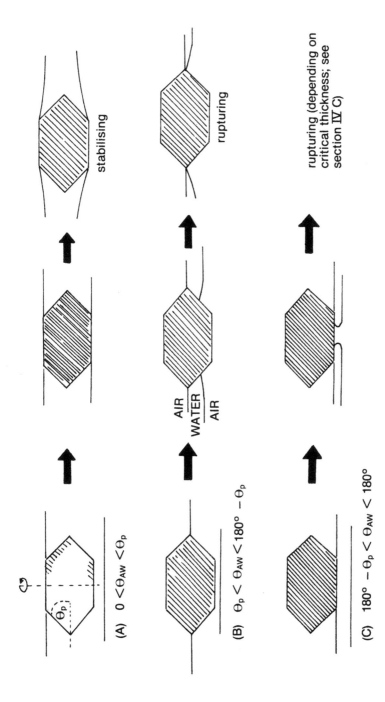

FIG. 18 Effect of axially symmetrical particle with edges on foam film stability.

and configuration C to

$$180° - \theta_p < \theta_{AW} < 180° \tag{30}$$

Since $\theta_p < 90°$, film rupture will not occur for a particle with the configuration A when bridging across a foam film occurs. This particle will interact with the film to produce a capillary pressure to oppose drainage. On the other hand, particles with configuration B will rupture films according to the process depicted in Fig. 18. Thus, condition 29 implies that particles with the selected geometry may rupture films when $\theta_{AW} < 90°$. Particles with configuration C will only affect film behavior if the thickness for spontaneous rupture of the aqueous film on the solid is greater than that for the aqueous film alone. Such a configuration is only possible if θ_p is in the region of about 70°–90° because values of θ_{AW} never usually exceed about 110°.

Particles which exhibit both smooth surfaces and edges will often be crystalline with straight edges. These are exemplified by the cubic crystals of galena used by Dippenaar [52]. This type of particle may be modeled by projecting the cross section shown in Fig. 18 in a direction perpendicular to the rotational axis. It is shown in Fig. 19. If the projection distance is long enough, then configurations in which the particle lies with that projection parallel to the air-water surface are energetically favored. However, by contrast with axially symmetrical particles, coincidence of both three-phase contact lines on an edge when such particles bridge films cannot cause complete dewetting of the particle. Growth of a hole in the foam film will occur from the edge and will not be axially symmetrical. That this will occur is of course confirmed by Dippenaar's observations with galena.

The number of possible configurations at the air-water surface of a straight-edged particle of given cross section can be much larger than that of an axially symmetric particle of the same cross section. For example, in the case of straight-edged particles of cross section defined by one angle θ_p shown in Fig. 19 the number of distinct configurations is 6 when $\theta_p = 30°$ and the contact angle $\theta_{AW} < 90°$. These are shown in Table 1, whence it is clear that not all of these configurations are mutually exclusive for a given contact angle. Most of them are "degenerate" in that the same configuration can be realized by using different edges and surfaces. Only one of the configurations may give rise to film collapse.

We may estimate the works of emergence W_i of the particle from the aqueous phase to form each of the possible configurations i shown in Table 1. Here W_i is given by

$$W_i = \gamma_{SA} \Delta A_{SA} - \gamma_{SW} \Delta A_{SW} - \gamma_{AW} \Delta A_{AW} \tag{31}$$

where γ_{SW}, γ_{SA}, and γ_{AW} are respectively the solid-water, solid-air, and air-

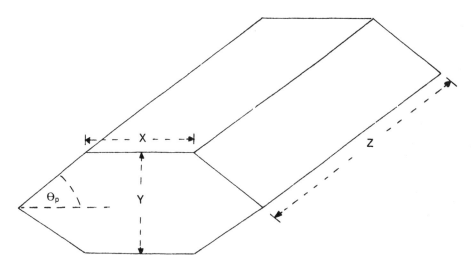

FIG. 19 Hypothetical particle with straight edges.

water surface tensions, and ΔA_{SW}, ΔA_{SA}, and ΔA_{AW} are the changes in solid-water, solid-air, and air-water surface areas accompanying emergence of the particle into the air-water surface. Since $\Delta A_{SA} = \Delta A_{SW}$ and we have the Young equation

$$\gamma_{SA} = \gamma_{SW} + \gamma_{AW} \cos \theta_{AW} \qquad (32)$$

then from Eq. 31 we can deduce

$$W_i = \gamma_{AW}(\Delta A_{SA} \cos \theta_{AW} - \Delta A_{AW}) \qquad (33)$$

It is a matter of elementary geometry to estimate ΔA_{SW} and ΔA_{AW} for the configurations in Table 1. By way of example, values of $Wi/\gamma_{AW}X^2$, where X is the cross-sectional dimension defined in Fig. 19, for each of these configurations are given in Table 1. These are calculated for $\theta_{AW} = 40°$ and particles of length $5X$. Here we see that the only configuration which can give rise to film collapse has the largest negative work of emergence. However, this is not the only configuration with a negative work of emergence, and it has the lowest degeneracy. Clearly, the effectiveness of the particle in causing film collapse requires consideration of the relative probabilities of achieving all the possible configurations.

There is no obvious rigorous approach which we may employ to calculate these probabilities. The problem does, however, resemble that addressed by classical statistical mechanics where the distribution over energy states of a system subject to thermal motion is considered. Thus, in general we have various states available to the particles ranging from free dispersion to adhe-

TABLE 1 Air-Water-Air Foam Film Collapse and the Configurations at the Air-

Possible configurations at air-water surface if $\theta_{AW} < 90°$	Conditions for realization of configurations	"Degeneracy"
$i = 1$ 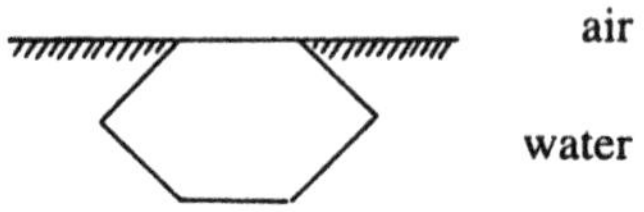 air water	$0 < \theta_{AW} < 30°$	2
$i = 2$ 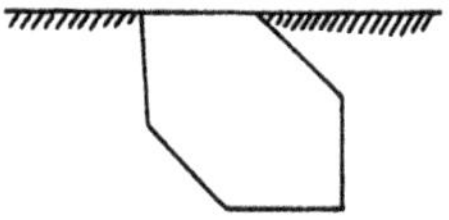	$0 < \theta_{AW} < 30°$	4
$i = 3$ 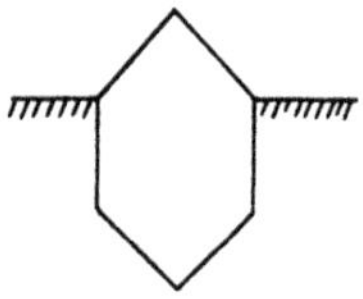	$60° < \theta_{AW} < 90°$	2
$i = 4$ 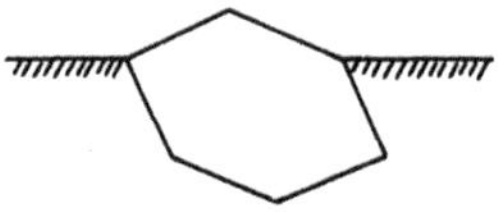	$15° < \theta_{AW} < 45°$	4
$i = 5$ 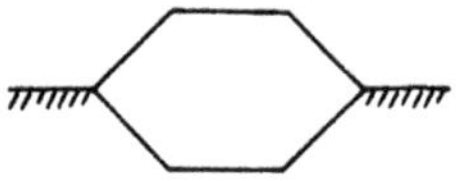	$30° < \theta_{AW} < 150°$	1
$i = 6$ 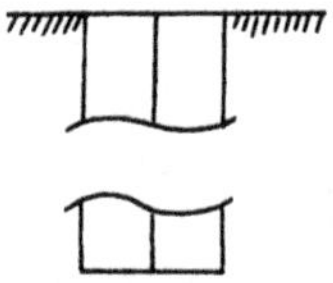	$0 < \theta_{AW} < 90°$	2

Particle geometry as shown in Fig. 19 with $\theta_p = 30°$ and $Y = X$. Z is the length of the particle defined in the figure.

Water Surface of a Particle with Several Edges

Foam film collapse if $\theta_{AW} < 90°$	Work of emergence W_i	$W_i/\gamma_{AW}X^2$ for $Z = 5X$ and $\theta_{AW} = 40°$
no	$W_i/\gamma_{AW} ZX = (\cos \theta_{AW} - 1)$	—
no	$W_i/\gamma_{AW} ZX = (\cos \theta_{AW} - 1)/2 \sin \theta_p$	—
no	$W_i/\gamma_{AW} ZX = (\cos \theta_{AW}/\sin\theta_p - 1)$	—
no	$W_i/\gamma_{AW} ZX = (1/(2 \sin \theta_p) + 1)\cos \theta_{AW} - [(1/(4 \sin^2 \theta_p)) + 1 - \cos (180 - \theta_p)/\sin \theta_p]^{1/2}$	-2.00
yes	$W_i/\gamma_{AW} ZX = (1 + 1/\sin\theta_p)\cos \theta_{AW} - (1 + 1/\tan \theta_p)$	-2.17
no	$W_i/\gamma_{AW}X^2 = (1 + 1/(2 \tan \theta_p)) (\cos \theta_{AW} - 1)$	-0.44

sion at the air-water surface with various configurations. However, during foam generation agitation will mean that the kinetic energies of the particles will in general far exceed those due to thermal motion alone. We can attempt to allow for this by according to the particles a "granular temperature" $\tilde{T}$ in a manner analogous to that described by, for example, Hopkins and Woodcock [68]. Thus, we may write

$$\tilde{T} = \frac{2E_T}{3(n_T - 1)k} \tag{34}$$

where n_T is the total number of particles, k is the Boltzmann constant, and $3(n_T - 1)$ is the number of degrees of freedom. Here E_T is the total kinetic energy of the particles so that

$$E_T = \frac{1}{2} \sum_{r=1}^{n_T} m_r v_r^2 \tag{35}$$

where m_r and v_r are respectively the velocity and mass of particle r. Therefore, $\tilde{T}$ is characteristic of the agitation conditions prevailing during foam generation. If $\tilde{T}$ can be accorded the same significance as temperature in the relevant context, then we can write a Boltzmann-type expression for the number of particles n_i with configuration i:

$$n_i \propto \exp(-W_i \beta) \tag{36}$$

where $\beta = 1/k\tilde{T}$. In the absence of agitation $\tilde{T} =, T$, so $\beta = 1/kT$.

The relative probability q_i of any configuration i occurring where the particle adheres to the air-water surface is therefore given by the familiar Boltzmann distribution

$$q_i = \frac{g_i \exp(-W_i \beta)}{\sum_{i=1}^{s} g_i \exp(-W_i \beta)} \tag{37}$$

where g_i is the "degeneracy" of configuration i and s is the number of possible configurations at the air-water surface. Low negative or even positive values for W_i, found for the configurations which do not lead to film collapse shown in Table 1, therefore imply low relative probabilities of occurrence.

The significance of β is perhaps easier to see if the relative proportion of configuration $i = 5$ (which will cause film collapse) and configuration $i = 4$, shown in Table 1, are compared. Thus, we have

$$n_4 = g_4 n_5 \exp[(W_5 - W_4)\beta] \tag{38}$$

The dimensions of typical antifoam particles are such that X (see Fig. 19) is in the region of 1–10 μm, so that $W_5 - W_4$ is in the region of $-(5 \times 10^{-13}$–$5 \times 10^{-15})$ J (for $\gamma_{AW} \sim 30$ mN m^{-1}). In the case of thermal motion alone $\beta = 1/kT \sim 2.5 \times 10^{20}$ J^{-1} at ambient temperatures, and we obtain $n_4/n_5 \sim 0$, so that the probability of finding configurations of the particles at the air-water surface which do not cause foam film collapse is essentially zero. However, where the agitation is intense, it is reasonable to suppose that the value of β could be considerably in excess of $1/kT$, so that configurations which do not cause foam collapse may become more probable. Increasing n_4/n_5 to, for example, 0.01 would require values of $1/\beta$ from $(2 \times 10^5$–$2 \times 10^7)$ kT, depending on the size of the particle.

In principle it is possible to deduce similar conditions to those depicted in Table 1 for any crystal habit. The resulting conditions may then be combined with contact angle measurements to deduce the antifoam potential of derived crystalline particles. Apart from the work of Dippenaar [52], concerning galena, no studies of the antifoam behavior of such well-defined hydrophobic crystalline particles have been made. However, Frye and Berg [63] have demonstrated the effect of sharp edges by comparison of the antifoam effectiveness of polydimethylsiloxane hydrophobed smooth spherical and ground-glass particles. Here ground glass formed shards with relatively few edges. Contact angles were adjusted by using different concentrations of several surfactants. The relative effects of the different types of particle are shown in Fig. 20. The antifoam effectiveness of ground-glass particles for a given contact angle is clearly much greater than that of smooth spheres.

That there is, however, an apparent indication of antifoam effects for smooth spheres with contact angles less than 90° is attributed by Frye and Berg [63] to the presence of a minor proportion of nonspherical particles with sharp edges among the spheres. These authors also found these hydrophobic ground-glass particles to be more effective than PTFE particles presumably because of the greater incidence of sharp edges with the former.

2.   Rough Particles with Many Edges

Sparingly soluble materials are often found precipitated from aqueous solution in the form of amorphous particles of ill-defined structure. Precipitated silica is an obvious extreme example where fractal structures are formed (of dimension about 2.4) [69]. Such materials may be intrinsically hydrophobic or be rendered hydrophobic by surface treatment.

In contrast to the ground-glass particles used by Frye and Berg [63], random rough amorphous and hydrophobic particles may in general adopt many different possible configurations at the air-water surface for a given contact angle. Some of these may be associated with film collapse if the requisite contact angle conditions are satisfied. However, the existence of many edges and asperities will increase the chances of a planar interface hinging at more

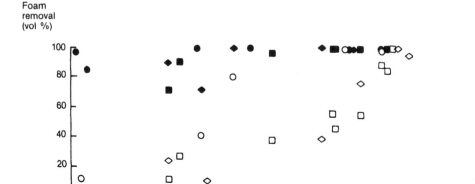

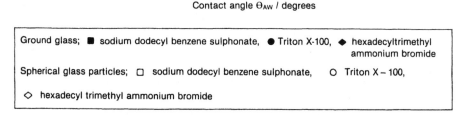

FIG. 20 Antifoam effect of hydrophobed glass particles. (From Ref. 63, reproduced by permission of Academic Press Inc.)

than one site on a particle. Stable bridging configurations in aqueous foam films may then occur with two planar air-water surfaces. The case of an axially symmetrical rough particle with regular rugosities is shown in Fig. 21 by way of example. Here only one of the allowed configurations which satisfy the contact angle θ_{AW} condition

$$\theta_r < \theta_{AW} < 180° - \theta_r \tag{39}$$

(where $\theta_r < 90°$ and is defined in Fig. 21) can give rise to film collapse. For that configuration the air-water surface is depicted in Fig. 21 as hinging on rugosity $ii = 8$ so that most of the particle is outside the aqueous phase. All other configurations would actually stabilize a foam film against drainage. Moreover, the probability of occurrence of the destabilizing configuration is relatively low if $\theta_{AW} < 90°$. This is revealed by consideration of the relative values of the work W_{ii}^+ of formation of each configuration. Here W_{ii}^+ may be calculated in a similar manner to W_i by using Eq. 33.

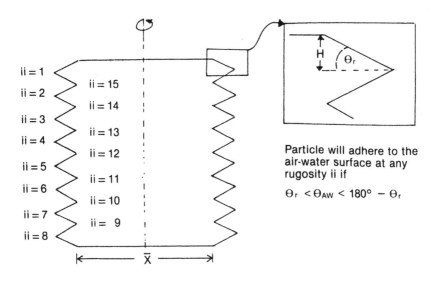

Particle will adhere to the air-water surface at any rugosity ii if

$$\Theta_r < \Theta_{AW} < 180° - \Theta_r$$

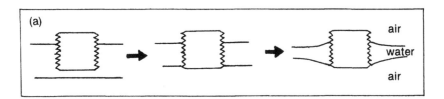

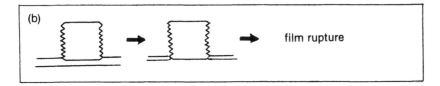

FIG. 21 Axially symmetrical particle with many edges: (a) example of configuration which stabilizes foam film, (b) configuration with air-water surface hinging on rugosity $ii = 8$ will rupture foam film.

TABLE 2 Work of Emergence into Air-Water Surface, W_{ii}^+ of Axially Symmetrical Particle with Regular Rugosities

Configuration with air-water surface hinging on rugosity ii	$W_{ii}^+ / \gamma_{AW} \, \bar{X}^2$		
	$\theta_{AW} = 45°$	$\theta_{AW} = 85°$	$\theta_{AW} = 95°$
$ii = 1$	-0.347	-1.291	-1.557
2	0.695	-1.163	-1.685
3	1.738	-1.034	-1.814
4	2.781	-0.905	-1.943
5	3.823	-0.777	-2.071
6	4.866	-0.648	-2.172
7	5.908	-0.520	-2.328
8	6.951	-0.391	-2.457

With $\theta_r = 30°$ and $H/\bar{X} = 0.1$; see Fig. 21 for definition of θ_r, H, and $\bar{X}$.

The geometry of the particle shown in Fig. 21 means that we can write, for W_{ii}^+,

$$\frac{W_{ii}^+}{\gamma_{AW}\bar{X}^2} = \pi\left[p_{ii}\left(\frac{H/\bar{X}}{\sin\theta_r}\right)\left(1 + \frac{H/\bar{X}}{\tan\theta_r}\right) + 0.25\right]\cos\theta_{AW}$$
$$- \pi\left[0.5 + \frac{H/\bar{X}}{\tan\theta_r}\right]^2 \tag{40}$$

where $\bar{X}$ is defined in Fig. 21 and where p_{ii} is the number of dewetted conical segments each of height H associated with the air-water surface hinging on rugosity ii. Values of $W_{ii}^+/\gamma_{AW}\bar{X}^2$ for different configurations of the particle are presented in Table 2 for three different contact angles by way of example.

Low negative values or even positive values for W_{ii}^+ found for the destabilizing configuration $ii = 8$ shown in Table 2 for $\theta_{AW} < 90°$ imply low relative probabilities of occurrence. Conversely, for $\theta_{AW} > 90°$ the destabilizing configuration has the highest negative value for W_{ii}^+, which implies the highest probability of occurrence.

In conclusion, it is tempting to generalize from this example that for rough particles with many edges and asperities the contact angle for destabilization cannot be specified with certainty so that antifoam effects will even occur with $\theta_{AW} < 90°$ (provided condition 39 is satisfied). However, destabilizing configurations for $\theta_{AW} < 90°$ will occur only with low probability, so that the overall antifoam behavior will be relatively weak or even negligible in those circumstances. Thus, for film collapse the particle must adopt a configuration at the air-water surface so that no rugosities exist below that surface at which the second air-water surface of a foam film could hinge. Such configurations of necessity involve removal of most of a particle with many rugosities outside the aqueous phase. For contact angles $< 90°$ this is energetically unfavorable and will therefore have a relatively low probability of occurrence.

D. Particle Size and Kinetics

An essential feature of the bridging mechanism for antifoam action by particles is that the particle size be of the same order as the thickness of the foam film. Foam films drain and stretch during foam generation and after foam generation has ceased. The rate of such processes will clearly play an important role in determining the frequency of foam film collapse if the film must first drain to the dimensions of any particles present.

Dippenaar [52] has developed a simple model to describe foam film collapse as a function of particle size for the case of foam generation by shaking cylinders. With this method of foam generation repeated shaking will often eventually produce a constant foam volume. Here presumably the rate of production of foam equals the rate of destruction due to the presence of hydrophobic particles. The rate of destruction according to Dippenaar [52] will, however, be proportional to the amount of foam present. Therefore the final foam volume is proportional to the volume rate of foam destruction. That volume may be maintained constant for different particle sizes by adjusting the total mass M of particles present. Then M becomes the mass of particles required to reduce the final foam volume to the chosen reference value. All of this implies a relationship between particle size and the total mass of particles required to achieve a constant volume rate of foam destruction (which is proportional to the rate of film rupture).

At the supposedly constant final foam volume the frequency f_r with which a single particle ruptures a film is simply the rate of film destruction/total number of particles. Therefore f_r is given by [52]

$$f_r = \frac{k_1 a^3}{M} \tag{41}$$

where a is the particle diameter and k_1 is a constant.

As we have seen, for particles to rupture thin films they must first thin down to a thickness proportional to the particle dimension (dependent upon particle size, shape, and contact angle). The time taken for this to occur is difficult to assess with certainty. Thus, film drainage is by viscous flow between the air-water surfaces of the film under the influence of gravity or capillary suction at the plateau borders. It may also possibly occur by marginal regeneration [70]. The relative significance of these processes depends upon film orientation and size. However, in the case of large (>a few square centimetres) vertical free nonrigid (i.e., "mobile") films Mysels et al. [70] show that marginal regeneration makes the most significant contribution to film drainage.

The time t taken for a vertical free rigid liquid film to thin to the thickness h by Poiseuille flow under gravity is [70]

$$t = \frac{4\eta y}{\rho g h^2} \tag{42}$$

where η is the viscosity, y is the height above the bulk liquid surface, and g is the acceleration due to gravity. In the case of a horizontal rigid plane-parallel circular film subject only to capillary suction from the plateau border the time taken may be deduced from the Reynolds equation [71], so that

$$t = \frac{3\eta R^2}{2\,\Delta P_f h^2} \tag{43}$$

where R is the radius of the film and ΔP_f is the capillary pressure. Marginal regeneration, on the other hand, is a complex process, and no expression is available for t as a function of h.

In the derivation of both Eqs. 42 and 43 rigidity not only implies absence of marginal regeneration but also that the velocity of flow at the surfactant contaminated air-water surface is zero. Violation of the latter assumption, even supposing absence of marginal regeneration, would mean shorter film drainage times than those predicted using these equations.

Dippenaar [52] ignores any contribution from marginal regeneration. He supposes that the time taken for a film to thin to the dimension required for rupture by a particle is given by Eq. 43 and is therefore simply inversely proportional to the square of that dimension. This would be true if drainage is described by either Eq. 42 or 43, where under the conditions of constant final foam volume either y or $R^2/\Delta P_f$ are constant. From Eq. 16 Dippenaar [52] therefore obtains the relationship

$$f_r = k_2 \left(\frac{\eta}{t_p}\right)\left(\frac{a}{M}\right) \tag{44}$$

where k_2 is another constant and t_p is the time taken for the film to thin to the thickness where rupture by a particle can occur. Using a range of hydrophobic quartz particles differing by more than two orders of magnitude in size, Dippenaar [52] shows that a/M is essentially constant for foams generated from aqueous solutions of triethoxybutane. This implies, if we take account of Eq. 44, that the frequency of rupture of foam films is inversely proportional to the time taken for the films to thin to a thickness where rupture by a particle can occur. Dippenaar [52] then concludes that the "rate-determining step" for film collapse by particles is "natural" thinning of films. This result is, however, obtained by using an expression for that natural thinning process which is only true if marginal regeneration is absent. It seems unlikely that aqueous solutions of triethoxybutane will form rigid films. It is, however, possible that marginal regeneration makes little contribution to film thinning (even for solutions which form mobile films) when the films are of the small size of those formed between bubbles in foam freshly generated by cylinder shaking. Marginal regeneration appears to be an oscillatory phenomenon, which may mean that if the dimensions of the film are less than a certain critical wavelength then it does not occur.

Equation 44 also means that if f_r is proportional to t_p^{-1} then for a constant particle size the ratio η/M must be constant. Dippenaar [52] has verified this, using glycerol to modify solution viscosity. It implies that the mass of antifoam required to achieve a given final volume of foam is proportional to the viscosity of the solution.

If the rate-determining step for film rupture is generally the time taken for films to thin, then the actual process of film rupture, once particle bridging has occurred, must be relatively rapid. Frye and Berg [63] have developed a model of that process which yields estimates of the rupture time. The model assumes an axially symmetrical configuration similar to that shown in Fig. 22. Here the air-water surface in the vicinity of the particle is assumed to have a circular cross section between the contact line on the particle and flat surface of the plane parallel film. The capillary pressure ΔP_c in the vicinity of the particle is crudely simplified by considering only the curvature in the plane of Fig. 22 (i.e., Eq. 26). The rate of flow of fluid Q from the particle under the influence of ΔP_c is then calculated under the assumption that all the displaced liquid is squeezed out of the film so that it remains plane-parallel except in the vicinity of the particle. A rigid circular film is assumed, so that Q is given by

$$Q = \frac{\pi \Delta P_c h^3}{3\eta} \tag{45}$$

This expression is readily obtained by differentiating Eq. 43, replacing ΔP_f by ΔP_c, and noting that $Q = \pi R^2 \, dh/dt$.

The position of the three-phase contact line is adjusted by small increments Δh^* from $h^* = 0$ to $h^* = h/2$, where h is the thickness of the film (see Fig. 22). Here h^* is the distance from the plane of the film surface to

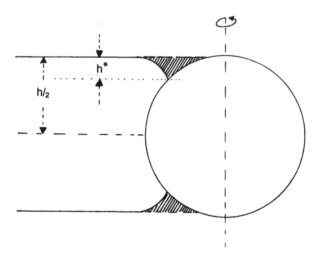

volume fluid
displaced as
contact line moves

FIG. 22 Illustration of procedure for estimation of time of antifoam action. (From Ref. 63, reproduced with permission of Academic Press Inc.)

the plane of the three-phase contact line. The time increment corresponding to each increment Δh^* is calculated from the relationship

$$\Delta t = \frac{\Delta V}{Q} \tag{46}$$

where ΔV is the corresponding incremental change in volume of fluid removed from the shaded area in Fig. 22. The time to rupture is then $N \, \Delta t$, where the number of increments N is given by $N \, \Delta h^* = h/2$.

The assumption that all the displaced fluid is completely removed from the film so that it remains plane parallel facilitates calculation but may not be entirely realistic. In practice, liquid may migrate only a small distance from the particle to form a bulge. Calculated rupture times may therefore be overestimates.

Rupture times calculated with these procedures are dependent upon contact angle and particle size. Thus, as the contact angle declines to the appropriate critical value, the rupture time asymptotically approaches infinity. For example, rupture times $<10^{-3}$ s require contact angles about 15° higher than the critical value of 90° for a rod-shaped particle of 0.1 cm radius. Some experimental evidence is presented by Frye and Berg [63], which indicates that effective antifoam action requires contact angles significantly higher than the critical values due to the effect of rupture time. Thus, in experiments with hydrophobic rods of 0.1 cm radius and free foam films, they show that the probability of perceived instantaneous rupture is significantly less than unity for advancing angles of >100° where the critical angle for the rod is 90°. However, the relevant data are badly scattered. Moreover, the experiments are ambiguous with respect to comparison with measured contact angles. Thus, an advancing angle occurs at the upper surface of the film, and a receding angle should occur at the lower surface. The experiments are also ambiguous with respect to the time allowed for rupture. A perception of instantaneous rupture presumably implies rupture times of < 0.1 s, which would require, according to the calculation of Frye and Berg [63], a contact angle of only 95° for rods of 0.1 cm radius.

The larger the particle the greater the volume of liquid which must be squeezed out of the film before rupture can occur. Rupture times therefore increase with particle size. However, Frye and Berg [63] show that a three-orders-of-magnitude increase in particle size for rod-shaped particles results in a difference of only about 10° in the required contact angle if film rupture is always to occur in less than $\sim 10^{-3}$ s.

Frye and Berg [63] also use Eq. 45, with ΔP_f instead of ΔP_c, to calculate the thinning time which they compare with the rupture time to estimate the rate-determining step. For small particles (≤100 μm) rupture times are minimal and thinning is rate determining in agreement with the conclusion of Dippenaar [52]. However, if the particles are sufficiently large (≥100 μm), then thinning times decline and rupture becomes rate determining. Frye and Berg [63] calculate an optimum particle size for which the total time of antifoam action is minimal. Rates are dominated by rupture times for sizes larger than the optimum and by thinning times for sizes smaller than the optimum. We should note here, however, that particles of optimum size for minimal antifoam action time are not necessarily of optimum size for overall effectiveness. A dominating factor is that the smaller the particle size the greater the number of particles for a given antifoam mass and the higher the probability of film rupture. This is shown empirically by Dippenaar [52], where as we have seen the mass M of antifoam required to reduce the foam volume to a chosen reference value is proportional to the particle size.

Frye and Berg [63] provide no detailed comparison with experiment for these considerations. However, they note that the deviation from constant a/M recorded by Dippenaar [52] for extremely large particles is consistent with rupture times becoming rate determining.

These calculations concerning total time for antifoam action rely essentially upon the relevance of the Reynolds equation [71] and require knowledge of ΔP_f which is not readily accessible by experiment for a real foam. Moreover, it is possible that during foam generation the velocity of flow at the air-water surfaces of films is nonzero so that film-thinning rates are faster than predicted by the Reynolds equation. The calculations also start with a configuration where the particle is already dewetted into one side of the foam film, and implicitly assume that the formation of a second three-phase contact line is instantaneous once the film thins to the relevant thickness. The process by which particles dewet into the air-water surface is ignored. We now address this issue.

E. Capture of Particles by Bubbles

Formation of a three-phase contact line is of course associated with rupture of the aqueous film separating the particle from the air bubble. The process of rupture of that film is usually believed to occur by catastrophic growth of random fluctuations in the air-water surface of the film at a certain critical film thickness h_{crit} in a manner analogous to that described by Vrij and Overbeek [9,72] for free aqueous foam films (see Sec. II.B.). Here we remember two factors influence the growth of those fluctuations. The first is the work done against the air-water surface tension in perturbing the surface, and the second is the work of interaction due to the disjoining force acting across the film.

In the case of hydrophobic particles interacting with air bubbles in an aqueous surfactant solution, both attractive and repulsive disjoining forces will exist. The latter will often be electrostatic in origin and are of relatively long range. They may be due to overlapping electrostatic double layers caused by adsorbed charged surfactant. They may even be a consequence of charge intrinsic to the solid surface in an aqueous environment. Thus, strong electrostatic repulsive forces stabilize films of pure water on silica even when the latter is hydrophobed with methyl silanes [73].

Van der Waals attractive forces tend to be relatively short range. The resultant negative work of interaction exhibits a distance dependence of h^{-2}, which obviously becomes large as $h \rightarrow 0$ [74]. In contrast, the positive work of interaction between overlapping electrical double layers often approximates a distance dependence of $\exp(-Kh)$ (where K is the Debye length), which tends to a constant value as $h \rightarrow 0$ [74]. For hydrophobic solids the van der Waals contribution therefore tends to dominate as $h \rightarrow 0$ leading to negative values of the work of interaction. However, at intermediate distances the two types of interaction often conspire to produce an energy maximum at $h = h_{max}$.

The growth of fluctuations is resisted by the need to do work against the air-water surface tension when perturbing the film (because the surface area increases). If a catastrophic growth in fluctuations is to occur, then this resistance must be overcome by the tendency of the film to thin spontaneously due to the dominance of attractive disjoining interactions.

A complete analysis of the hydrodynamic stability of a thin film between a hydrophobic solid surface and an air-water surface where the interaction potential exhibits a maximum at h_{max} does not appear to be available. It seems probable, however, that in this case $h_{crit} < h_{max}$. If a particle is to adhere to an air bubble, it must first have sufficient kinetic energy to overcome this energy barrier and sufficient time for the film between the particle and the bubble to attain the critical thickness and rupture.

Arguably one of the most careful studies of particle-bubble interactions under dynamic conditions is that due to Anfruns and Kitchener [75]. Here the efficiency of capture by rising bubbles of glass microspheres and rough quartz particles, rendered hydrophobic with methyl silanes, was determined under carefully controlled conditions. Bubbles were in the size range 0.05–0.1 cm, and particles were of size about 30 μm. These experiments enabled the collection efficiency C_e to be obtained as

$$C_e = \frac{\text{number of particles collected}}{\pi R_b^2 L n_p} \tag{47}$$

where R_b is the radius of the bubble, L is the path length of the bubble, and n_p is the number of particles/unit volume suspended in the medium through which the bubble passes. Here $C_e \ll 1$ because most particles ahead of a rising bubble are swept along the fluid streamlines remote from the bubble surface. The process is depicted schematically in Fig. 23.

The conditions of these experiments were such that the relevant streamlines could be calculated so that C_e could in turn be calculated. In Figs. 24A and B the calculated and experimental values of C_e for both smooth glass microspheres and rough quartz particles are compared for various solution conditions. The efficiency of collection of the glass microspheres is seen to be considerably less than that of the quartz particles under all these conditions. Indeed the efficiency of collection of the quartz particles generally approaches theoretical values. This means that every quartz particle following a streamline which takes the particle within a particle radius of the surface of a bubble is captured by the bubble. These differences between the quartz and glass particles occur despite the similarity of both the surface treatment and contact angles measured on representative smooth surfaces (~ 85–90° in distilled water). Similar observations concerning the effectiveness of surface roughness in promoting the efficiency of capture of par-

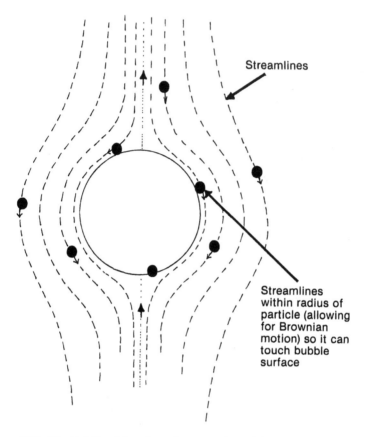

Streamlines

Streamlines
within radius of
particle (allowing
for Brownian
motion) so it can
touch bubble
surface

FIG. 23 Bubble rising through suspension of hydrophobic particles under laminar flow illustrating inefficient interception of particles by bubbles (schematic).

ticles by bubbles are reported by Ducker et al. [76] and Strnad et al. [77].

Interactions due to electrical double layers are essentially suppressed by the addition of 0.1 M KCl. This is seen from Fig. 24B to more than treble the collection efficiency of glass microspheres. Electrical double layers are therefore clearly implicated in contributing to the relatively low collection efficiency of these particles. Further evidence of the importance of electrostatic interactions is provided by the effect of adding sodium dodecylsulfate up to 10^{-3} M. The surfactant adsorbs at the air-water surface to produce an increase in electrostatic repulsion between the particles and the bubbles [75]. This virtually eliminates capture of glass particles despite contact angles of 30° prevailing in these circumstances.

Anfruns and Kitchener [75] deduce that relatively long-range electrostatic repulsion forces are preventing the effective thinning of the aqueous film

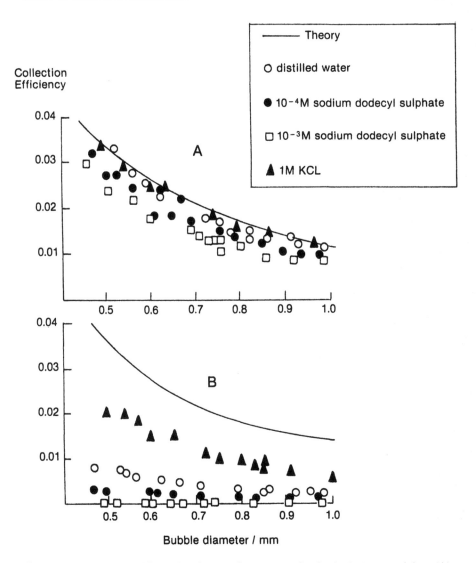

FIG. 24 Collection efficiencies for rough quartz and spherical glass particles: (A) quartz particles of 31 micron Stokes equivalent diameter; (B) spherical glass particles of 32 micron diameter. (From Ref. 75, reproduced by permission of the Institute of Mining and Metallurgy, London).

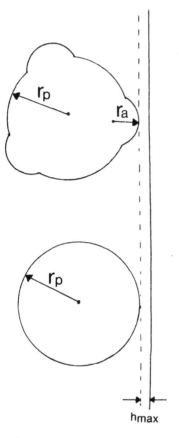

FIG. 25 Approach of smooth spherical particle and spherical particle with asperities to air-water surface (h_{max} exaggerated).

between smooth glass particles and bubbles during the time available when the particle is in near contact with the rising bubble. In contrast, sharp edges and asperities on the surface of the rough particles facilitate rupture of the film despite the presence of the long-range electrostatic forces.

Most antifoam particles are either rough or possess edges. This is also usually true of mineral particles involved in flotation. The effect of edges on capture by bubbles would therefore appear to be of general importance in understanding particle-bubble interactions. Unfortunately no proper theoretical explanation is apparently available. A naive explanation is, however, suggested if we compare a spherical particle of radius r_p having a spherical segment asperity of radius r_a with a smooth spherical particle of the same size. As we have seen, dewetting probably requires these particles to have sufficient kinetic energy to overcome a repulsive energy barrier.

Now imagine the two particles to approach the air-water surface of a bubble in the manner shown in Fig. 25, where deformation of the air-water surface is ignored to a first approximation. The ratio of the height of the energy barrier for the asperity to that for the smooth particle as a whole is r_a/r_p, where $r_a < r_p$ and provided $r_a \gg h_{max}$ (so that we may assume Derjaguin's approximation [78]). Clearly, then, the particle could have sufficient kinetic energy to dewet the asperity and initiate attachment to the bubble without necessarily having sufficient energy to do so in the absence of the asperity. Presumably the asperity could then have a similar effect on the stability of the film still separating most of the particle from the air bubble to that shown by particles bridging free foam films.

Frye and Berg [63] in their study of the effect of particle morphology on antifoam effectiveness were apparently unaware of the work of Anfruns and Kitchener [75]. Clearly, however, they selected similar systems to those selected by the latter. The superior antifoam action of hydrophobed ground-glass particles relative to the effect of spherical glass particles, shown in Fig. 20, may therefore owe something to the more ready attachment of rough particles to bubble surfaces. The importance of roughness in determining dewetting behavior does hint that the contribution to the weak antifoam behavior of the spherical particles by a contaminating proportion of particles with sharp edges is likely to be even more important than Frye and Berg [63] indicate.

F. Melting of Hydrophobic Particles and Antifoam Behavior

Certain hydrophobic materials which exhibit antifoam effects have melting points at temperatures $< 100°C$. Examples include hydrocarbon waxes, triglycerides, and long-chain fatty acids.

Dispersions of hydrophobic particles can be prepared by first emulsifying the molten materials and then cooling the emulsions below the melting point. Observations by Davis and Garrett [79] of the antifoam behavior of the resulting entities reveal some interesting behavior. By way of example in Fig. 26 we plot the ratio F, where

$$F = \frac{\text{volume of air in foam in presence of antifoam}}{\text{volume of air in foam in absence of antifoam}}$$

against temperature for 1.2 g dm^{-3} dispersion of n-docosane, n-eicosane, and paraffin wax in 0.5 g dm^{-3} sodium (C_{10}—C_{14}) alkylbenzene sulfonate solution. Foam measurements were made in a static Ross Miles apparatus. Particle sizes were in the range 0.5–8 μm.

The most striking feature of the behavior of n-docosane is the sharp deterioration of antifoam effect in the region of the melting point at 44°C.

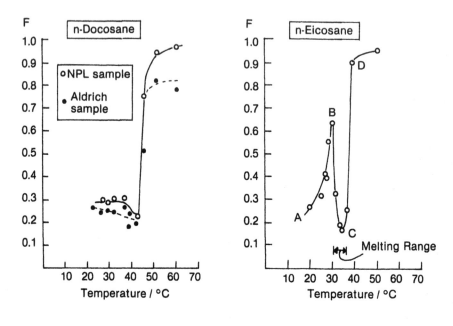

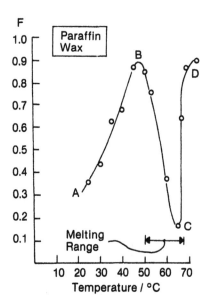

FIG. 26 Effect of temperature on the antifoam behavior of finely divided hydrocarbons (1.2 g dm^{-3} hydrocarbon in 0.5 g dm^{-3} commercial sodium (C$_{10}$—C$_{14}$) alkylbenzene sulfonate (Dobs 055) solution using static Ross Miles). (From Ref. 79.)

An extremely pure sample was used, and therefore a sharp melting transition was observed. Similar observations have been made by Aronson [80] for hydrocarbons and fatty acids in solutions of sodium tridecylbenzenesulfonate.

Neither the n-eicosane nor the paraffin wax used in this work were pure materials. Therefore both exhibit a relatively wide temperature range over which both solid and liquid phases coexist. The antifoam behavior of these materials, shown in Fig. 26, is therefore complex. At low temperatures in region AB an antifoam effect associated with solid hydrocarbon particles is present. Curiously, that effect appears to deteriorate with increase in temperature. However, in region BC, where both solid and liquid hydrocarbon coexist, a pronounced enhancement of the antifoam effect is seen to occur. This type of behavior has apparently also been observed by Kulkarni et al. [81] for impure triglycerides. At sufficiently high temperatures where all solid is converted to liquid the antifoam effect is seen to deteriorate sharply in much the same manner as for pure n-docosane.

That antifoam effects may be associated with solid wax or fatty acid particles is not altogether surprising. Under microscopic observation the particles are observed to exhibit more irregular shapes than the emulsion spheres from which they are derived. Aronson [80] observes particularly marked roughness with many sharp edges for particles of stearic and palmitic acid. These particles are also found to be more effective antifoams than the more symmetrical hydrocarbon particles, for which similar contact angles would be expected to prevail. The equilibrium receding contact angles for the hydrocarbons used to obtain the results shown in Fig. 26 are about 105° against distilled water, which decrease to about 30° against the relevant alkylbenzene sulfonate solutions [79]. It seems probable then that a combination of edges/asperities, finite contact angles, and dynamic effects (where the dynamic contact angle is higher than the equilibrium contact angle due to slow surfactant transport to the relevant surfaces) will mean that antifoam behavior by a bridging mechanism will occur. Both Davis and Garrett [79] and Aronson [80] present evidence which suggests that dynamic effects are important for these systems.

Gradual deterioration of antifoam effectiveness of the solid particles with increase in temperature in region AB for the impure hydrocarbons in Fig. 26 is difficult to explain. No significant change occurs in contact angles over the relevant temperature range [79]. It is, however, possible that the hydrocarbon softens with increasing temperature, so that asperities and edges are gradually removed.

We address the finding that liquid hydrophobic oils are almost completely

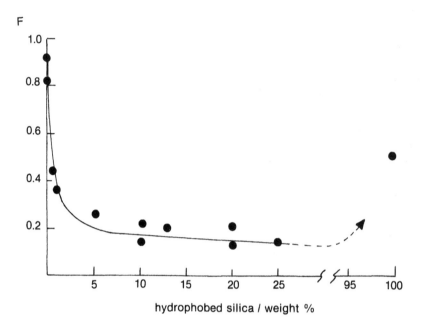

FIG. 27 Antifoam effectiveness F for dispersion of hydrophobed silica/mineral oil antifoam as a function of antifoam composition (silica D17 ex Degussa; 1.2 g dm^{-3} antifoam in 0.5 g dm^{-3} commercial sodium (C_{10}—C_{14}) alkylbenzene sulfonate (Dobs 055) solution; foam generated by cylinder shaking at ambient temperature 22 ± 2°C). (From Ref. 40.)

ineffective as antifoams and that mixtures of hydrophobic liquids and solids are particularly effective in the next section.

V. MIXTURES OF HYDROPHOBIC PARTICLES AND OILS AS ANTIFOAMS FOR AQUEOUS SYSTEMS

A. Antifoam Synergy

One of the most striking aspects of antifoam behavior is the synergy shown by mixtures of insoluble hydrophobic particles and hydrophobic oils when dispersed in aqueous media. We illustrate this behavior in Fig. 27 with a plot of the ratio F against antifoam composition for a mixture of methylsilane hydrophobed silica and a mineral oil. Foam from a solution of 0.5 g dm^{-3} ($\sim 1.4 \times 10^{-3}$ M) commercial sodium (C_{10}—C_{14}) alkylbenzene sulfonate was generated by cylinder shaking [40]. The mineral oil is seen to be virtually without effect on foam behavior, and the silica particles exhibit only a weak effect. Adding only a few percent of hydrophobed silica to the mineral oil is, however, enough to produce a significant enhancement of antifoam performance. Similar observations are reported by others for mixtures of hydrophobed glass with hydrocarbon [39] and hydrophobed silica with polydimethyl siloxanes [30,81–84]. With the latter, silica may be first hydrophobed by use of a suitable agent, such as a silane or alcohol, before mixing with the polydimethylsiloxane. Alternatively the silica may be hydrophobed in situ by heating with polydimethylsiloxane to temperatures in excess of 150°C (possibly in the presence of a catalyst) [85]. The nature of the reaction between silica and polydimethylsiloxane is not well understood. Antifoams produced by this process are often described as silicone "compounds."

We have found that this antifoam synergy appears to be quite general for all manner of hydrophobic particles and oils [40]. Thus, intrinsically hydrophobic organic particulates, such as precipitates of polyvalent metals with long-chain alkyl phosphates and carboxylates, may also be combined with mineral oils [86,87] or polydimethylsiloxanes [88] to produce synergistic antifoam behavior.

Mechanistic studies of this phenomenon did not appear in the scientific literature until the mid-seventies [30,81,83,84]. Indeed only a few references to it appeared before 1970 [29,89]. However, descriptions of many examples of mixtures of hydrophobic particles and oils have been appearing in the patent literature since the early fifties. Examples of those patents are listed in Tables 3–5. Here they are categorized as mixtures of hydrophobed mineral particles and silicone oils (Table 3), mixtures of hydrophobed mineral particles and organic liquids (Table 4), and mixtures of intrinsically hydrophobic organic particles and organic liquids (Table 5).

The finely divided nature of the particles is often stressed in these patents and methods of milling are sometimes described [108]. Whenever particle size is quoted, it usually falls in the range 0.001–1.0 μm. The preferred particle concentration in the oil ranges from 1% to about 30% by weight.

As we have seen, the antifoam performance of both hydrocarbons and polydimethylsiloxane oils may be considerably enhanced by the addition of finely divided silica. This suggests an obvious phenomenological similarity. However, when hydrophobed silica is added to polydimethylsiloxanes, it is often referred to as a "filler" [83,92,94,95] or "activator" [93], implying that the oil plays the active role in the antifoam mechanism. On the other hand, when hydrophobed silica is added to hydrocarbon oils, the oil is sometimes referred to as the "carrier," implying that the silica plays the active role in the antifoam mechanism. Similarly when long-chain organic materials are added to hydrocarbons, the particles are often referred to as the "active" ingredients and the oils as the "carriers" (see, for example, [108]), which once again implies that the particles play the active role in the antifoam mechanism.

TABLE 3 Some Examples of Antifoam Patents Concerning Mixtures of Hydrophobed Mineral Particles and Silicone Oils

Oils claimed to be effective	Particles claimed to be effective	Actual examples given	Preferred concentration of particles/ % by weight
Partially oxidized methyl siloxane polymer. May be diluted with dimethyl siloxane polymer and dispersed in benzene.	Silica aerogel presumably rendered hydrophobic by reaction in situ with the siloxane.	As claimed.	7.5
Methyl polysiloxane.	Finely divided silica presumably rendered hydrophobic by reaction in situ with the siloxane.	Silica + methyl polysiloxane + emulsifiers.	2–10
Silicone oils (product may be diluted by hydrocarbons, ethers, ketones or chlorohydrocarbons. This does not, however, appear as a specific claim).	Aluminium oxides, titanium dioxides, particularly all manner of silicas. These react with silicone oil in situ catalyzed by an acid condensation reagent.	1. Dimethyl polysiloxane + silica + aerogel + $AlCl_3$ diluted by toluene. 2. Dimethyl polysiloxane + alumina + $SnCl_4$ diluted by toluene. 3. Dimethyl polysiloxane + precipitated silica + phosphorous nitrile chloride + polyethylene glycol stearate.	1–30
Silicone oils.	Pyrogenic or precipitated silicas hydrophobed with "chemically bound methyl groups"*. Special reference is made to the relative ineffectiveness of the oil alone (*probably silanized).	As claimed.	2–8 (preferably 5)

Particle size	Any other additives	Application	Date	Ref.
—	—	Aqueous alkaline solns. for paper industry, rubber industry, metalworking industry.	1953	90
"Finely divided".	Patent really concerns emulsifiers which are monostearic acid ester of polyethylene glycol + monostearic acid ester of sorbitol.	Patent particularly concerns emulsifiers for silicone antifoams for cosmetics and drug application.	1958	91
0.01–25 micron.	1. Acid condensation catalyst necessary (e.g., $AlCl_3$, $SnCl_4$ BF_3, etc.) 2. Emulsifiers such as methyl cellulose, polyethylene glycol monostearate, polyethylene glycol trimethyl nonyl ether added to prepare stable aqueous O/W emulsions of the antifoam if desired.	Alkaline aqueous solns. particularly alkaline cleaning products for automatic washing machines.	1966	92
0.015–0.05 micron.	—	Aqueous surfactant solutions.	1968	93

TABLE 3 continued

Oils claimed to be effective	Particles claimed to be effective	Actual examples given	Preferred concentration of particles/ % by weight
Organosiloxane liquid polymers which may be partly or entirely replaced by other nonaqueous fluids such hydrocarbon oils, polyalkylene glycols etc.	Hydroxyl containing inorganic "fillers" TiO_2, Al_2O_3 and preferably SiO_2 reacted with a dialkyl-amino-organosilicone. Special provision of patent is that necessity for catalysts is obviated. Heating and cooling cycles are also obviated.	1. Dialkyl amino silicone treated silica in dimethyl siloxane polymer diluted with polypropylene glycol. 2. Dialkyl amino organo silicone-treated silica in polypropylene glycol. 3. Dialkyl amino organosilicone treated silica in mineral seal oil diluted with polypropylene glycol.	1–30
Dimethyl polysiloxane or about equal proportions of mixture of dimethylpolysiloxane + diorganopolysiloxane (containing silicon-bonded Me, Et, and 2-phenylpropyl groups).	Silica of surface area >50 m^2 gm^{-1}. (List of previous silicone antifoam patents given as examples of the type of materials considered.)	Silica + polydimethyl siloxane.	1–10
Alkylated polysiloxane materials of various types.	Silicas rendered hydrophobic by a variety of methods.	Dimethyl siloxane + trimethyl silanated silica.	10–50

Particle size	Any other additives	Application	Date	Ref.
0.007–0.025 micron.	—	Alkaline aqueous solns, particularly aqueous paints, latex systems, laundry and detergent products.	1971	94
	1. Whole mixed with *sodium tripolyphosphate* to produce free-flowing powder. *This is the particular provision of the patent.* 2. Emulsifying agents, mold inhibitors, etc.	Fabric washing formulations.	1974	95
Not >0.1 micron, preferably 0.01–0.02 micron.	Antifoam incorporated in detergent impermeable water soluble or dispersible carrier material such as gelatin, polyethylene glycol, etc. This material is in turn coated with a water-soluble granular material.	Detergent formulations for automatic clothes and dishwashing machines.	1975	96

TABLE 4 Some Examples of Antifoam Patents Concerning Mixtures of Hydrophobed Mineral Particles and Organic Liquids

Oils claimed to be effective	Particles claimed to be effective	Actual examples given	Preferred concentration of particles/ % by weight
Kerosene, naphthenic mineral oil, paraffinic mineral oil, chlorinated naphthenic mineral oil, chlorinated paraffinic mineral oil, liquid tri-fluorovinyl chloride polymer, fluorinated hydrocarbons.	Aerogel, fume, or precipitated silicas hydrophobed by any suitable method (but only silicone oil or alkyl chlorosilanes mentioned).	Silicas hydrophobed with silicone oil or with alkyl chlorosilanes + most of the oils claimed.	3–20
Aliphatic or aromatic hydrocarbons with at least six carbon atoms.	Precipitated silica having pH from 8–10. Rendered hydrophobic by any suitable method but only polysiloxane or alkyl (aryl or alicyclic) silanes claimed. Type of silica employed is a special provision of the patent.	Precipitated silica hydrophobed with polymethylsiloxane and mixed with various mineral oils.	3–30
Water-insoluble polyalkylene glycol.	Silica subjected to heat and shear treatment in oil. Silica may also be hydrophobed before mixing with oil by treatment with silane or alcohol.	1. Polypropylene glycol + silica. 2. Polybutylene glycol + silica 3. Polypropylene glycol + silica hydrophobed with trimethyl silane. 4. Polypropylene glycol + silica hydrophobed with isobutanol.	1–10

Particle size	Any other additives	Application	Date	Ref.
Most preferred from 0.02–1 micron.	Spreading agent claimed to be essential (anionic, cationic or nonionic surfactant).	Probably mainly intended for paper pulp mills although patent not restrictive in this respect.	1963	97
0.005–0.050 micron.	Need for a spreading agent is eliminated—a special feature of this patent. Use of a spreading agent is claimed to reduce rather than increase the effectiveness of the antifoam.	Especially adapted for paper pulp mill application although patent apparently not restrictive in this respect.	1965	98
Finely divided high surface area (>50 m^2 gm^{-1}).	—	No specific application quoted.	1967	99

TABLE 4 continued

Oils claimed to be effective	Particles claimed to be effective	Actual examples given	Preferred concentration of particles/ % by weight
Aliphatic hydrocarbon or paraffin oil including mineral seal oil, kerosene, various light aliphatic fuel oils, gas oils, paraffin waxes, etc.	Any type of silica possessing reactive surface hydroxyl groups rendered hydrophobic with dialkyl dihalosilane. Reaction made in the oil. Special claim of patent concerns reaction conditions giving rise to dialkyl-substituted cyclic siloxanes on the surface of silica.	Mineral seal oil + fume silica + dimethyl dichlorosilane.	Preferred 1–40 (but reference made to the use of the particles by themselves i.e., 100).
Paraffinic and napthenic mineral oils, cutting oils, kerosene, similar petroleum fractions including food-grade mineral oils and halogenated hydrocarbons. Synthetic oils also claimed to be suitable: aliphatic diesters, silicate esters and polyalkylene glycol or their derivatives (detailed description of viscosity, density, etc., of oils also specified).	Any type of silica possessing reactive surface hydroxyl groups rendered hydrophobic by reacting with alkoxysilicon chloride $SiCl_m(OR)_n$ Reaction made in the oil.	Mineral seal oil with silica treated with various alkoxysilicon chlorides.	7–45
Halocarbon or hydrocarbon fluid.	Synthetic alkali metal or alkaline earth metal silicoaluminate rendered hydrophobic by reaction with halosilane. Reaction made in the oil.	Sodium or calcium silicoaluminates hydrophobed with methyl chlorosilanes in paraffinic mineral oil, chloronaphthalene or kerosene.	5–30

Particle size	Any other additives	Application	Date	Ref.
0.15–0.005 micron (ultimate particle size but states that greatly preferred are nonaggregated particles of silica).	Emulsifiers and dispersants may be added.	Wide variety of industrial applications quoted. A specific claim concerns paper pulp stock.	1968	100
Less than 0.1 micron but preferably less than 0.05 micron (ultimate particle size).	1. Need for spreading agent is eliminated—a special feature of this patent. Surface-active agent claimed to stabilize the foam. 2. Alkylene oxide may be added to react with HCl formed when alkyoxy silicon chloride reacts with silica (in the oil).	Paper pulp mill application.	1969	101
<0.2 micron. This is a particular provision of the patent.	Lewis base may be added to neutralize haloacid formed in reaction of halosilane with silicoaluminate.	Paper pulp mills represent a preferred application.	1970	102

TABLE 4 continued

Oils claimed to be effective	Particles claimed to be effective	Actual examples given	Preferred concentration of particles/ % by weight
Water-insoluble organic liquid selected from vegetable oils, aliphatic hydrocarbons, alicyclic hydrocarbons, halogenated aromatic hydrocarbons, long-chain alcohols, long-chain esters and amines (organic liquid has boiling point >65°C and is liquid at room temperature). A dependent claim states that liquid is hydrophobic.	Aluminium oxide reacted in situ with alkali or alkaline earth hydroxide and fatty acid (with 6 to 24 carbon atoms).	Colloidal aluminium oxide prepared by hydrolysis of aluminium chloride in a flame + paraffinic hydrocarbon oil + Ca(OH)$_2$ + variety of fatty acids.	Most preferred from 6–16.
Water-insoluble organic liquid selected from kerosene, naphthenic mineral oil, paraffinic mineral oil, chlorinated napthenic mineral oil, chlorinated paraffinic mineral oil and liquid difluorovinyl chloride polymer. (Detailed description of viscosity, volatility, etc., also specified.)	Patent claim exclusively concerns the preparation of an alkalized microfine precipitated silica (which may be hydrophobed by any method).	Silica hydrophobed with dimethylpolysiloxane and dispersed in naphthenic, mineral or mineral seal oils.	3–30

Particle size	Any other additives	Application	Date	Ref.
Alumina <15 micron preferred 0.01– 1.3 micron.	(i) Water may be added (forming water in oil emulsion). (ii) Surfactant may be added.	Probably intended for paper pulp mills although reference not made to such an application in the claims.	1972	103
<0.05 micron (ultimate particle size).	Catalyst to promote reaction of silicone with silica (e.g., tin octoate). Addition of spreading agents or other surfactants usually not helpful except occasionally when silicone oil loading of silica low.	Particularly paper pulp mill application.	1973	104

TABLE 5 Some Examples of Antifoam Patents Concerning Mixtures of Intrinsically Hydrophobic Particles and Organic Liquids

Oils claimed to be effective	Particles claimed to be effective	Actual examples given	Preferred concentration of particles/% by weight
Water-immiscible organic liquid boiling above 100°C, e.g., a hydrocarbon.	Polyamide of polymethylane polyamine having 2–12 methylene groups and a carboxylic acid of the group consisting of aliphatic and cycloaliphatic monocarboxylic acids, each acyl group containing 11–18 carbon atoms.	N,N'-distearyl ethylene diamine + white spirit.	0.05–10
Organic liquid is nonsolvent for particle and is immiscible with water. Polar + nonpolar are listed. That is saturated unsaturated aliphatic hydrocarbons. Polar materials include alcohols, esters, ketones, chlorinated aromatic hydrocarbons, fluorinated hydrocarbons.	Finely divided poly-α-olefin polymers such as polypropylene, polyisobutylene, etc. Also claims thermoplastic polyesters such as nylon.	1. Polypropylene + xylene 2. Polypropylene + commercial aromatic hydrocarbon mixture 3. Polypropylene + mineral oil + surfactant 4. Polyethylene tetraphthalate + mineral oil, etc.	2–25

This terminological distinction between polydimethylsiloxane hydrocarbon-based systems suggests a complete mechanistic dissimilarity which seems unlikely. It is more probable that the terminology simply reflects the relative costs of the oil and particulate ingredients!

It is clear from the patent literature that an antifoam mechanism invoking

Particle size	Any other additives	Application	Date	Ref.
"Fine particles" (prepared by heating till dissolved and rapidly cooling or by milling).	—	Anionic detergent solutions. (has no detrimental effect on the wetting or detergent power of the solution or on its content of active material).	1955	105
0.02–50.0 micron but preferred size 0.2–5.0 microns.	Surface active agent may be added in order to provide "an improvement in the property of these dispersions to spread at the air-water surface."	Reference made to paper pulp mill application only.	1972	106

Marangoni spreading of the oil (see Fig. 7) underlies some of the thinking behind the development of these oil-based antifoams. It is, however, equally clear that it represents an inadequate explanation for all reported phenomena. Thus, Boylan [97], in a patent concerning the addition of hydrophobed silica to organic liquids, claimed that the addition of a spreading agent (a soap for

TABLE 5 continued

Oils claimed to be effective	Particles claimed to be effective	Actual examples given	Preferred concentration of particles/% by weight
Hydrophobic organic solvents e.g., aliphatic cycloaliphatic, hydroaromatic, aromatic hydrocarbons, natural fatty oils (e.g., olive oil), silicone oils, phosphoric acid esters.	Hydrates of fatty acid mixed salts of polyvalent metals and/or of lower dibasic amines.	Mineral oil + i. Aluminium-magnesium stearate. ii. Zinc-ethylene diamine stearate. iii. Magnesium-zinc stearate	2–20
Mineral oil of specified viscosity	Alipathic diamide derivatives of polymethylene diamine.	N,N'-distearyle-thylene diamine + mineral oil.	4–12
Hydrocarbon oil	Amide which is the reaction product of a polyamine containing at least one C_2—C_6 alkylene group and a C_6—C_{18} fatty acid.	Stearic diamide of ethylene diamine + paraffin oil.	1–20
Mineral oil or esters of unsaturated fatty acids with mono or polyhydric alcohols, liquid fatty acids or alcohols, terpene hydrocarbons.	Mono- or diester of hydroxystearyl alcohol with saturated fatty acid or hydroxy-fatty acid.	1. Mineral oil + hydroxy-stearyl monobehenate + various other additives, etc.	5–15 (Particles may however dispersed in aqueous phase directly and demonstrate antifoam properties by themselves).
Mineral oils, fatty oils, fatty acids, tetraisobutylene.	Polyethylene having molecular weight from 500 to 25,000.	As claimed.	0.5–15

Particle size	Any other additives	Application	Date	Ref.
—	Nonionic emulsifier	General industrial application.	1972	107
15–20 microns (by Hegman grind gauge)	1. Spreading agent 2. Silicone oil (dependent claims)	Paper pulp mill application	1973	108
Prepared by heating till particles dissolve in oil followed by quick chilling and homogenizing to "obtain smaller particles."	i. Oil-soluble organic polymer ii. a fat iii. silicone oil	General industrial application	1973	109
"Finely divided"	Silicone oils, nonionic ethoxylated surfactants, metal stearates, etc.	Paper pulp mills, manufacture of dispersions of plastics, etc.	1975	110
Finely divided "probably in the range of 0.1 micron." Prepared by dissolving in oil at high temperature and then cooling to obtain optimum particle size.	Emulsifier.	General application for aqueous systems.	1975	111

TABLE 6 Initial Values of Entry and Spreading Coefficients for Polydimethyl Siloxanes

Siloxane (ex DOW Corning)	Viscosity (mPa s)	Average mol. weight[a]	Temperature (°C)	γ_{AO} (mN m^{-1})	γ_{OW} (mN m^{-1})
MS 200	1.0	2.3×10^2	24	16.4	42.7
MS 200	5.0	6.7×10^2	24	19.3	42.8
DC 200	10.0	—	20	20.1	39.5
DC 200	50	—	20	20.8	34.3
DC 200	1,000	2.5×10^4	20	21.2	36.9
DC 200	60,000	7.0×10^4	20	—	—
DC 200	100,000	8.5×10^4	20	21.3	—

[a]Number average
[b]Direct measurement of initial spreading pressure using Wilhelmy plate.

example) to the oil is necessary in order to allow the water-insoluble organic liquid to spread at the air-water surface. In later patents concerning the same system, however, it is claimed that the addition of a spreading agent is unnecessary. Buckman [101], for example, states that Boylan's product is not entirely satisfactory because the presence of a surface-active agent in the system tends to stabilize the foam. In another patent also concerning this system, Miller [104] states that the addition of spreading agents may give an improvement only if the surface treatment of the silica is inadequate. Boylan apparently realized the error of his ways and eventually issued a patent [112] in which the spreading agent was claimed as an optional additive.

B. Spreading Behavior of Antifoam Oils on Water

As we have seen there appears to be a consensus that necessary properties of antifoam oils include insolubility and a tendency to emerge into the air-water surfaces of foam films as marked by $E > 0$ (see Sec. III.B.). However, we have also seen that the spreading behavior of the oil may play a role in the mode of action of the antifoam. Before examining this possibility, we consider the relevant properties of typical oils used in oil-particle antifoams. In the main this means polydimethylsiloxanes and hydrocarbons.

Polydimethylsiloxanes are characterised by low air-oil surface tensions of ≤ 20 mN m^{-1} at ambient temperatures. Values of γ_{AO} and the oil-water surface tension γ_{OW} for typical oils are shown in Table 6. Entry and spreading coefficients are also presented. These are calculated by assuming a value of γ_{AW} equal to that of distilled water. They therefore represent nonequilibrium "initial" values of E and S. Equilibrium values of E'' and S'' should

Water		8 × 10⁻³ M SDS Solution (γ_{AW} = 39.0 mN m⁻¹)			
E (mN m⁻¹)	S (mN m⁻¹)	γ_{OW} (mN m⁻¹)	E (mN m⁻¹)	S (mN m⁻¹)	References
+98.2	+12.8	9.8	+32.4	+12.8	113
+95.4	+ 9.8	12.0	+31.7	+ 7.7	113
+90.9	+11.9	—	—	—	48
+85.0	+16.4	—	—	—	48
+87.2	+13.4	—	—	+ 7.3[b]	48
—	+10.7[b]	—	—	—	
—	+11.0[b]	—	—	—	

be calculated with the values of γ_{AW} obtained after contamination with po-lydimethylsiloxane (which forms a monolayer at the air-water surface). Un-der those circumstances it is possible that $S'' < 0$ and the oil will not spread but will form a lens (where we still have $E'' > 0$) [32,33].

If a small droplet of polydimethylsiloxane oil is inserted into the air-water surface of distilled water in a narrow through, it spreads as a duplex film at a rate given by Eq. 19. This equation implies that the rate is dependent upon S but is independent of the viscosity of the polydimethylsiloxane. In contrast to viscosity, S is only weakly dependent upon molecular weight for high molecular weights (see Table 6). Therefore the spreading rate is es-sentially independent of molecular weight provided Eq. 19 is valid. We have, however, observed that for number average molecular weights $> 2.5 \times 10^4$ (and viscosities > 1000 mPa s) Eq. 19 tends to overestimate spreading rates which become dependent upon polydimethylsiloxane viscosities at high mo-lecular weights [114]. After spreading, the film appears to very slowly (de-pending on the viscosity of the oil) break up into oil lenses. A second droplet of oil introduced to any free air-water surface formed as the film breaks up does not spread but forms a lens.

The effect of adding polydimethylsiloxane oils to the surfaces of surfac-tant solutions is somewhat less well documented. Again, however, the oil usually spreads as a duplex film and subsequently slowly breaks up to form lenses. A second droplet added to the siloxane-contaminated surface does not spread but forms a lens, so that equilibrium values of the spreading coefficient $S'' < 0$. This is illustrated in Fig. 28 for a 5×10^{-3} M solution of 6-phenyldodecylbenzenesulfonate in 4×10^{-2} M NaCl to which two drops of 1000 mPa s polydimethylsiloxane were added in sequence.

Estimates of the initial values of E and S for polydimethylsiloxanes on sodium lauryl sulfate solutions calculated from the surface tension results of

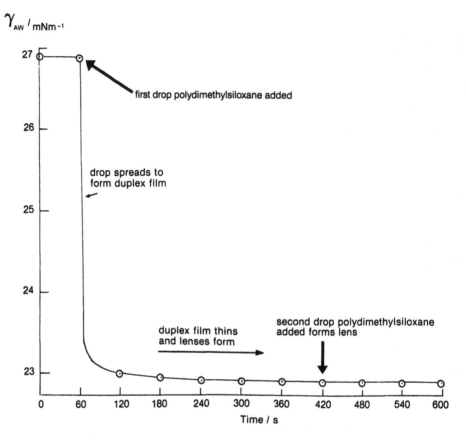

FIG. 28 Spreading behavior of polydimethylsiloxane on $5 \times 10^{-3}M$ sodium 6-phenyldodecylbenzenesulfonate in $4 \times 10^{-2}M$ NaCl at 25°C (polydimethylsiloxane of 1000 mPa s (DC200 ex Dow Corning); drops of ~0.01 cm^3 added to dish of 10 cm diameter). (From Ref. 114.)

Kannelopoulos and Owen [113] are given in Table 6. Spreading rates may well be lower than values predicted by Eq. 19 because γ_{OW} increases as the oil-water interface expands and because compression of the surfactant monolayer ahead of the spreading front may also occur. There is, however, a paucity of data concerning the spreading behavior of polydimethylsiloxane oils on surfactant solutions. A detailed study of the spreading behavior of polydimethylsiloxane on surfactant solution would serve to educate the debate concerning the relevance of that property for antifoam behavior.

The relevant surface properties of hydrocarbons are shown in Table 7. It is seen that homologues higher than n-heptane exhibit initial spreading coef-

TABLE 7 Initial Values of Entry and Spreading Coefficients for Hydrocarbons on Water

Hydrocarbon	Temperature (°C)	γ_{AO}(mN m^{-1})	γ_{OW}(mN m^{-1})	E^a(mN m^{-1})	S^a(mN m^{-1})	Ref.
n-pentane	20	16.0	50.2	+107.2	+6.8	115
n-hexane	20	18.4	50.8	+105.4	+3.8	115
n-heptane	20	20.1	51.2	+104.1	+1.7	115
n-octane	20	21.7	51.7	+103.0	−0.4	116
n-decane	20	23.8	52.3	+101.5	−3.1	116
n-dodecane	20	25.4	52.8	+100.4	−5.2	116
n-tetradecane	20	26.5	53.8	+99.8	−6.8	116
n-hexadecane	20	27.4	53.8	+99.4	−8.2	116
Typical mineral oil	25	29.7	44.9^b	+87.5	−2.3	40

^{a}Taking γ_{AW} = 73.0 mN m^{-1} at 20°C and 72.3 mN m^{-1} at 25°C.
^{b}Low, presumably due to polar impurities.

ficients $S < 0$ on distilled water at near ambient temperatures. For these compounds $E > 0$ and lens formation occurs. Hydrocarbons used for antifoams tend to be mixtures of high molecular weight such as mineral oils (or "white" oils, etc.). These, too, do not spread on distilled water at ambient temperatures.

However, spreading on distilled water of all hydrocarbons including mineral oils can be induced by dissolution of oil-soluble surfactants in the oil to reduce γ_{OW}. Such surfactants are often present as impurities in hydrocarbon oils.

Hydrocarbons which do not spread on distilled water can spread on surfactant solution surfaces even when impurities are absent from the hydrocarbons. Thus, certain hydrocarbons will exhibit a positive initial spreading coefficient on surfactant solutions. Such hydrocarbons first spread as a duplex film, which subsequently breaks up to form lenses in equilibrium with surfactant monolayer contaminated with hydrocarbon. However, the results of Aveyard et al. [117,118] suggest that if the hydrocarbon has a sufficiently long chain length relative to that of the surfactant, then this phenomenon does not occur and the initial spreading coefficient becomes <0. We have found, for example, that the mineral oil used to obtain the foamability results shown in Fig. 19 does not detectably (using a Wilhelmy plate) reduce the surface tension of the relevant sodium (C_{10}—C_{14}) alkylbenzene sulfonate solutions [40].

The effect of surfactant in the aqueous phase on the initial dewetting and spreading behavior of hydrocarbons can be predicted if certain assumptions are valid. Thus, suppose that the surfactant is ionic with a single charge, does not dissolve in the oil, and is not adsorbed at the oil-air surface (reasonable assumptions for anionic and cationic surfactants in the absence of added electrolyte). Then if we use the Gibbs equation, we may write for the entry coefficient

$$\frac{dE}{d \ln a\pm} = \frac{d(\gamma_{AW} + \gamma_{OW} - \gamma_{OA})}{d \ln a\pm} = 2RT(-\Gamma_{AW} - \Gamma_{OW}) \qquad (48)$$

and for the spreading coefficient

$$\frac{dS}{d \ln a\pm} = \frac{d(\gamma_{AW} - \gamma_{OW} - \gamma_{OA})}{d \ln a\pm} = 2RT(-\Gamma_{AW} + \Gamma_{OW}) \qquad (49)$$

where Γ_{AW} and Γ_{OW} are the surface excesses at the air-water and oil-water surfaces of the surfactant solution, and $a\pm$ is the mean ionic surfactant activity. Contact angle measurements have shown that adsorption at the hydrocarbon solid-water surface is equal to adsorption at the air-water surface for anionic surfactants [119]. This is a consequence of the dominance of the hydrophobic effect in determining adsorption behavior. It seems reasonable to expect that adsorption at the air-water and the liquid hydrocarbon-water surfaces will show the same behavior provided that the chain length of the hydrocarbon is long so that it does not contaminate the air-water surface. Therefore we may substitute $\Gamma_{AW} = \Gamma_{OW}$ into Eqs. 48 and (49) to find

$$\frac{dE}{d \ln a\pm} = -4RT\Gamma_{AW} \tag{50}$$

and

$$\frac{dS}{d \ln a\pm} = 0 \tag{51}$$

Addition of surfactant to the aqueous phase may reduce the entry coefficient (and may even "wet-in" the hydrocarbon so that $E < 0$). It will, however, only induce spreading in a hydrocarbon which does not spread on distilled water if the assumption $\Gamma_{AW} = \Gamma_{OW}$ is violated (which is presumably the case for shorter-chain-length hydrocarbons, for which we must have $\Gamma_{OW} > \Gamma_{AW}$ before hydrocarbon contamination of the surfactant monolayer, so that $dS/d \ln a\pm > 0$ from Eq. 49). We have observed that the spreading coefficient for a mineral oil on solutions of anionic surfactant is essentially unaffected by surfactant concentration and is similar to that calculated for distilled water where $S < 0$, so that $\Gamma_{AW} = \Gamma_{OW}$ and Eq. 51 is satisfied [40].

C. Role of the Oil in Synergistic Oil/Particle Antifoams

Mixtures of hydrophobed silica and polydimethylsiloxane are effective antifoams for aqueous solutions of anionic surfactants. However, the oil alone is not. Such oils can, as we have seen, spread at both the air-water surface and air-surfactant solution surfaces.

Ross and Nishioka [84] have examined the stability of air bubbles (of ~0.8 cm radius) released under monolayers of polydimethylsiloxane spread on both distilled water and surfactant solution. With a distilled water substrate bubbles were more stable than in the absence of polydimethylsiloxane (presumably as a result of reduced film draining rates due to lack of mobility of the polydimethylsiloxane-contaminated surface). Similar findings are reported by Trapeznikov and Chasovnikova [120]. With a surfactant solution substrate the presence of a spread polydimethylsiloxane monolayer did not diminish bubble stability. All of this suggests that displacement or contamination of a surfactant monolayer with polymethylsiloxane to produce unstable foam films cannot be the role of this material.

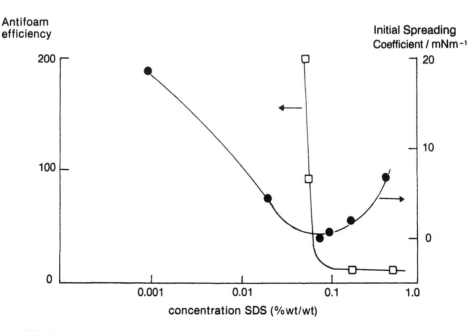

FIG. 29 Comparison of initial spreading coefficient S and antifoam efficiency for polydimethylsiloxane/hydrophobed silica antifoam on sodium dodecylsulfate solution. (From Ref. 30.)

Another possibility must be foam film rupture induced by spreading (see Sec. III.C.). However, Kulkarni et al. [81] describe an experiment where a drop of polydimethylsiloxane is added to foam formed from sodium lauryl sulfate in a tray. This caused no foam destruction despite the obvious inference that spreading from the droplet would occur, which would in turn cause collapse if the spreading mechanism is effective in this context. Moreover, although Kulkarni et al. [30,81] claim that antifoam action usually requires a positive value of the initial spreading coefficient, they show that antifoam efficiency does not correlate with the magnitude of S [30]. A comparison of antifoam efficiency with S for mixtures of hydrophobed silica/polydimethylsiloxane is presented in Fig. 29. Here the increase in initial spreading coefficient at high sodium lauryl sulfate concentration (where presumably S increases due to solubilization of dodecanol impurity in micelles) contrasts with declining antifoam efficiency. However, as suggested by Ewers and Sutherland [26], spreading would be expected to become more significant for a greater spreading coefficient. Thus, the shear force applied to the intralamellar liquid is $dS/d\chi$, where χ is the distance of the spreading layer from its source in the plane of the air-water surface. Higher initial values of S should therefore mean higher shear forces and a greater probability of foam film rupture.

It is possible, however, to query the relevance of the initial spreading coefficients used by Kulkarni et al. [30] in Fig. 29. The method of foam generation used release of gas through a glass frit. Any foam volume collapse due to the effect of antifoam must involve the continuous air-water surface which exists at the top of the resulting foam column. It seems reasonable to suppose that accumulation of polydimethylsiloxane will occur on that surface even before the experiment begins. Moreover, it seems probable that silicone contamination of bubbles may have occurred by the time they have reached the upper surface of the foam. On the whole, then, use of initial spreading coefficients rather than equilibrium spreading coefficients would appear to have a dubious basis. Equilibrium spreading coefficients will be ≤ 0.

As we have seen, hydrocarbons also exhibit antifoam behavior when mixed with hydrophobic particles. Here, however (provided precautions are taken to eliminate contamination by oil-soluble surfactant), it is possible to achieve antifoam effects for oil-particle mixtures where the initial spreading coefficient for the oil is unequivocally <0. Thus, a detailed study of the spreading behavior of the mineral-oil-based antifoam, used to obtain the foam results shown in Fig. 27, on solutions of sodium ($C_{10} - C_{14}$) alkylbenzene sulfonate revealed no evidence of spreading under either initial or dynamic conditions (where the air-water surface is expanded in a controlled manner) [40].

We therefore arrive at the conclusion that the role of the oil, if it is hydrocarbon, need not concern spreading at the air-water surface. If it is polydimethylsiloxane, then, although the oil may spread, the evidence for a role for that spreading is at best equivocal and does not support a mechanism of film rupture by surface tension gradient induced shear. It is tempting then to infer that spreading of the antifoam oil, irrespective of type, is not a necessary condition for antifoam mechanism with mixtures of oils and particles. In this we differ from Kulkarni et al. [30,81], who state that a positive initial spreading coefficient is a necessary, but not a sufficient, property of such oils.

As we have seen, oils for which $E > 0$ and which do not spread at the air-water surface form lenses. If the oil lenses are small, then gravitational forces may be neglected and the configuration adopted is that shown in Fig. 30a. Here the air-water surface is everywhere planar, and both the oil-water and oil-air surfaces form spherical segments. The angles formed by the three fluid-fluid surfaces at the three-phase contact line are determined by Neumann's triangle of forces as shown in the figure.

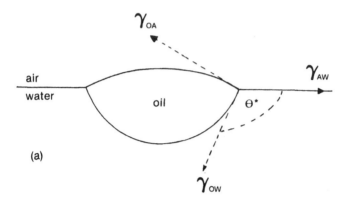

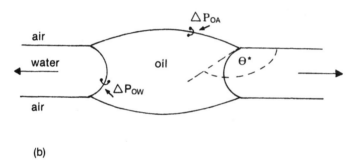

FIG. 30 Oil lens and definition of θ^*: (a) oil lens at air-water surface illustrating Neumann's triangle of surface forces; (b) oil lens bridging aqueous foam film. (From Ref. 121.)

Garrett [121] has shown that if such an oil lens bridges a plane-parallel foam film by emerging into both air-water surfaces then no configuration of mechanical stability is available to the lens if the angle θ^*, formed by the tangents to the air-water and oil-water surfaces, is $>90°$. The condition $\theta^* > 90°$ is satisfied provided we have

$$\gamma_{AW}^2 + \gamma_{OW}^2 - \gamma_{OA}^2 > 0 \tag{52}$$

Condition 52 gives rise to unstable bridging configurations even when the air-water surface is everywhere planar as shown in Fig. 30b. Here it can be shown that the Laplace pressure drop across the oil-air surface ΔP_{OA} cannot equal that across the oil-water surface ΔP_{OW}. In this situation $\Delta P_{OA} > \Delta P_{OW}$, and an unbalanced capillary force will cause enhanced rates of drainage in the aqueous film away from the lens.

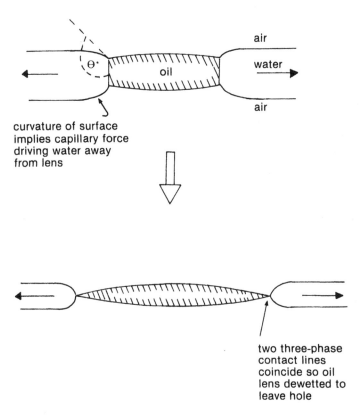

curvature of surface
implies capillary force
driving water away
from lens

two three-phase
contact lines
coincide so oil
lens dewetted to
leave hole

FIG. 31 Foam film collapse induced by bridging oil lens. (From Ref. 39.)

If, on the other hand, the air-water surface is nonplanar, then a configuration similar to that shown in Fig. 31 occurs when condition 52 is satisfied. Here again there is a capillary pressure enhancing the rate of drainage
away from the droplet. In this case, however, it is possible, as suggested
by Frye and Berg [39], to envisage a process where the oil lens is completely
dewetted from the foam film to form a hole. The process is shown schematically in Fig. 31. Here thinning of the aqueous film causes the lens to
expand and ultimately results in coincidence of the upper and lower three-
phase contact lines. It is immediately obvious that this process is analogous
to the dewetting of a cylindrical particle in a foam film and therefore requires
$\theta^* > 90°$.

The process depicted in Fig. 31 would, however, be only possible for
comparatively low values of θ^*. As $\theta^* \rightarrow 180°$, then the curvature of the
oil-air surfaces will of necessity become concave if the oil-water surface is

to remain cylindrical. In this case the point of rupture could be at the center of the expanding oil lens as the two concave air-oil surfaces approach one another.

Condition 52 for formation of destabilizing bridging configurations by oil lenses is strongly dependent upon γ_{AW} and γ_{OA} because, in general, $\gamma_{OW} \ll \gamma_{OA}$ and $\gamma_{OW} \ll \gamma_{AW}$. Indeed, essentially all that is required for such behavior is that the oil-air surface tension be less than the air-water surface tension of the foam film surfaces. Such a condition is readily satisfied for polydimethylsiloxane oils but will not always be satisfied for hydrocarbon oils.

Some evidence that the behavior of hydrocarbon oil-based antifoams follows that predicted by condition 52 has been presented by Garrett et al. [122]. This study concerned the effect of a mineral oil/hydrophobed silica antifoam on the foamability and foam stability of a homologous series of sodium alkylbenzene sulfonate isomer blends. Of particular interest is the effect of that antifoam on the stability of the foam several minutes after foam generation ceased. The ambiguities associated with dynamic effects are then minimized, and the air-water surface tension of foam films is reasonably supposed to approximate the equilibrium surface tension of a free air-water surface. In Table 8 we present the entry coefficient E, the spreading coefficient S, and the bridging coefficient B, where

$$B = \gamma_{AW}^2 + \gamma_{OW}^2 - \gamma_{OA}^2 \tag{53}$$

for each of the chain lengths of alkylbenzene sulfonates calculated by using the equilibrium values of γ_{AW} and γ_{OW}. Here it is clear that everywhere $E > 0$ and $S < 0$. However, for chain lengths $> C_{12}$ the bridging coefficient $B > 0$ and therefore $\theta^* > 90°$.

In the absence of antifoam the foam volume of each of these solutions of alkylbenzene sulfonate was stable for up to 1800 s after foam generation ceased (although bubble disproportionation and drainage significantly affected the appearance of the foam). A measure of the effectiveness of the antifoam after foam generation has ceased is the ratio $F(t = 960 \text{ s})/F(t = 0)$, where t is the foam age and where we remember F = volume of air in foam in presence of antifoam/volume of air in foam in the absence of antifoam. In effect, the ratio $F(t = 960 \text{ s})/F(t = 0)$ is the ratio of the volume of air in the foam in the presence of antifoam after 960 s to the volume of air in the foam in the presence of antifoam immediately after foam generation has ceased. This follows because the foam volume is stable in the absence of antifoam. The ratio $F(t = 960)/F(t = 0)$ has the value of unity if F does not change with the age of the foam because the antifoam does not function under the near-equilibrium conditions then prevailing. It is compared with S and B in Table 8. Here we see that $F(t = 960)/F(t = 0) \ll 1$ for chain lengths $<C_{12}$ where $B > 0$. Therefore, significant foam film col-

TABLE 8 Comparison of Entry E, Spreading S, and Bridging B Coefficients with Mineral-Oil-Based Antifoam[a] Effectiveness after Foam Generation[b] Has Ceased for Sodium Alkyl Benzene Sulfonates[c]

Chain Length	γ_{AW} (mN m^{-1})	γ_{OW} (mN m^{-1})	E (mN m^{-1})	S (mN m^{-1})	B (mN m^{-1})^2	Antifoam effectiveness $F(t = 960\ s)/F(t = 0\ s)$
C_9	33.0	6.5	+8.5	−4.5	+170.3	0.50
C_{10}	34.0	6.8	+9.8	−3.8	+241.2	0.33
C_{11}	31.7	6.75	+7.5	−6.1	+89.4	0.62
C_{12}	30.2	4.9	+4.1	−5.7	−24.9	0.94
C_{13}	28.8	4.0	+1.8	−6.2	−115.6	1.00
C_{14}	27.7	3.3	0.0	−6.6	−182.8	0.90

[a] Antifoam is a mixture of mineral oil $\gamma_{AO} = 31$ mN m^{-1} and hydrophobed silica.
[b] Foam measured using Ross Miles apparatus.
[c] A mixture of all possible isomers except the 1-phenyl isomer.
Source: Ref. 122.

93

lapse due to the antifoam after foam generation has ceased only occurs for solutions where $B > 0$, and therefore $\theta^* > 90°$. This could represent some evidence that foam film collapse can be induced by a bridging oil lens which does not spread at the air-water surface. It does, however, ignore the role of the particle.

The bridging oil lens then provides an explanation for the role of the antifoam oil, which does not require that the oil spread at the air-water surface. It fails, however, to address the central question of the role of the particle.

D. Hypotheses Concerning the Role of the Particles in Synergistic Oil/Particle Antifoams

In the past decade or so a number of different hypotheses have been advanced in explanation for the role of particles in synergistic oil-particle antifoams. Much of the relevant work has concerned mixtures of oil with silica where the requirement that the silica be surface treated to render it hydrophobic has been amply demonstrated.

Some of these hypotheses are essentially naive and are easily disposed of. Thus, Sinka and Lichtman [123] attribute the role of the particle to inhibition of solubilization of the oil. However, it is easily shown that the presence of concentrations of antifoam oil well in excess of the (usually extremely low) amounts which are solubilized may still produce negligible antifoam effect with aqueous solutions of, say, anionic surfactants [40]. Povich [83] shows that the increase in bulk shear viscosity of polydimethylsiloxane due to the presence of the silica is not responsible for the sevenfold increase in antifoam effectiveness accompanying presence of that material in the oil. Moreover, Ross and Nishioka [85] show that the presence of silica has no effect on the surface shear viscosity of polydimethylsiloxane. Garrett et al. [40] have shown that although the presence of hydrophobed silica particles can facilitate dispersal of the antifoam oil that is not the principal role of the particles. Thus, mineral oil dispersions of essentially the same size distribution as hydrophobed silica/mineral oil dispersions reveal a markedly different antifoam effectiveness. This is exemplified in Fig. 32, where the relevant size distributions are compared with the respective antifoam effectiveness as measured by F for a solution of a commercial sodium alkylbenzene sulfonate.

The possibility that the spreading coefficient of polydimethylsiloxane oils is modified by the addition of hydrophobic particles has been explored by Povich [83]. Here the initial spreading coefficient for polydimethylsiloxane has been shown to slightly *decrease* upon addition of hydrophobed silica (despite the effectiveness of this silica in promoting the antifoam behavior

Fraction of Particles of
given Size Range

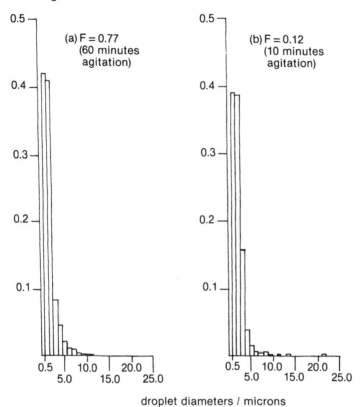

droplet diameters / microns

FIG. 32 Comparison of antifoam effectiveness F with size distribution of (a) mineral oil, (b) mineral oil/hydrophobed silica (90/10 by weight). Here the size distributions were adjusted by altering the agitation time so that they are approximately equal. (From Ref. 40.)

of the oil). Povich [83] attributes this to adsorption of oil-soluble surface-active impurities on the silica. Hydrophobed silica has also been shown to be without significant effect on the spreading behavior of a mineral oil [40].

An often-quoted mechanism for mixtures of hydrophobic oils and particles is that due to Kulkarni et al. [30,81]. These authors claim that the oil spreads over the air-water surface exposing the particles to the aqueous solution. Adsorption of surfactant onto the surface of the particles is then supposed to occur, rendering the particle hydrophilic so that particles are progressively extracted from the oil into the aqueous phase. Rapid local depletion of surfactant in the aqueous film is then in turn supposed to produce a "surface stress" which renders the foam film unstable so that rupture occurs.

There are a number of problems with this mechanism. The first problem clearly concerns the requirement that the oil spread at the air-water surface. We have already established that this is not a necessary property of antifoam oils. A second problem is conceptual—depletion of surfactant in the foam film by adsorption on particles will give rise to an increase in surface tension, which will in turn produce a Marangoni flow in the direction of the foam film. This will tend to enhance film stability. Finally it is clear that an aspect of the mechanism is an intrinsic tendency for the particle to be removed from the oil phase to the aqueous phase. However, it has been shown [40] that addition of mineral oil to a dispersion of hydrophobed silica in sodium alkylbenzene sulfonate solution produces an *enhancement* of antifoam effectiveness. The results are given in Table 9. This is clearly not consistent with a process of removal of particles from oil having a central role in the synergistic mode of action of the antifoam. On the whole, then, we are forced to conclude that the mechanism proposed by Kulkarni et al. [30,81] is probably wrong.

Dippenaar [52] has suggested that the particle may have a central role in the mode of action of the antifoam. He supposes that the particle functions by a dewetting mechanism similar to that outlined in Sec. IV.B. The oil is supposed to adhere to the particle to yield a higher contact angle than would otherwise prevail. Dippenaar [52] illustrates this mechanism by comparing the contact angle and effectiveness at film rupture of sulfur particles before and after contamination with liquid paraffin. This mechanism would, however, be most effective in the case of rough particles where the oil may adhere in the rugosities to increase the contact angle at the air-water surface. It is, however, difficult to see how this mechanism could function where the oil forms the overwhelming proportion in the antifoam and where the particle is preferentially wetted by the oil. In this case, entities where a rough particle is embedded in an oil droplet with only minimal exposure of the particles to the aqueous phase will be favored. This issue is addressed in more detail below (Sec. V.E.).

Frye and Berg [39] have proposed that if the oil completely wets the particles then the size of those oil droplets containing particles will be determined by the size of the particles. Sufficiently large particles should therefore reduce the time required for foam films to thin to the point where bridging and foam film collapse can occur according to the mechanism suggested by condition 52 and shown in Fig. 31. However, the presence of particles which produce synergistic foam behavior does not necessarily mean larger oil droplets, as we have shown in Fig. 32. Moreover, reduced thinning time does not necessarily imply increased frequency of foam film collapse be-

TABLE 9 Effect of Adding Mineral oil to a Dispersion of Hydrophobic Silica in 0.5 g dm^{-3} Sodium Alkyl Benzene Sulfonate[a] Solution (0.03 g of hydrophobed silica[b] in 25 cm^3 of surfactant solution. Foam generated by shaking 100 cm^3 cylinder)

After 15 s shaking	After 2 h standing	After 15 s shaking for second time	After 15 s shaking for third time	After 1.5 h standing	After 15 s shaking for fourth time
0.41	0.36	0.55	0.18	0.0	0.15
0.43	0.40	0.52	0.10	0.0	0.11

Approximately 0.03 ± 0.01g mineral oil added

[a]A commercial alkylbenzene sulfonate of C_{10}—C_{14} chain length (Dobs 055).
[b]A silanized Aerosil 200.
Source: Ref. 40.

97

cause larger antifoam entities means fewer entities. Thus, we remember the finding of Dippenaar [52] that the mass of antifoam required to remove a given proportion of foam is proportional to the size of the antifoam—smaller entities therefore mean a higher overall antifoam efficiency. The hypothesis of Frye and Berg [39] also ignores the usual observations [40,123] that the hydrophobic particles are not completely wetted by the oil but adopt a finite contact angle θ_{OW} at the oil-water surface so that

$$90° < \theta_{OW} < 180° \tag{54}$$

where θ_{OW} is measured through the aqueous phase.

On the whole, then, we are left with the conclusion that no published hypothesis concerning the role of the particles in synergistic oil-particle antifoams is entirely satisfactory.

E. Capture of Oil Droplets by Bubbles and the Role of the Particles

1. Experimental Observation

As we have seen, the particulate component of an oil-particle antifoam has a contact angle at the oil-water surface so that condition 54 is satisfied. This is in fact the condition required for water-in-oil Pickering emulsion formation where particles adhering to the oil-water surface ensure high stability. Indeed a characteristic of oil-particle antifoams is that if equivolume amounts of the antifoam and the solution to be defoamed are shaken together a water-in-oil emulsion invariably forms [40,123]. If, however, the oil alone is shaken with the same solution, an oil-in-water emulsion usually results. Clearly, then, there is strong evidence to suggest that the particles in the oil-particle antifoam adhere to the oil-water surface. Measurements of contact angles on compressed disks of particles or on smooth plates of representative materials seem to confirm this conclusion and imply that condition 54 is satisfied [40,123]. A finding by Frye and Berg [39] that the receding contact angle θ_{OW} is 180° for solutions against hexadecane on hydrophobed glass plates supposedly representative of the hydrophobed glass antifoam ingredient used by these authors would appear to contradict these findings. However, Frye and Berg [39] do not present emulsion evidence where the presence of their antifoam particles should not invert the emulsion behavior of the solution-hexadecane system (which should be oil-in-water) if their contact angle measurements are correct.

The balance of evidence then suggests that the particulate component of the antifoam adheres to the oil-water surface with a contact angle which satisfies condition 54. A schematic illustration of such an entity is shown in Fig. 33. It seems probable then that the location of the particles at the

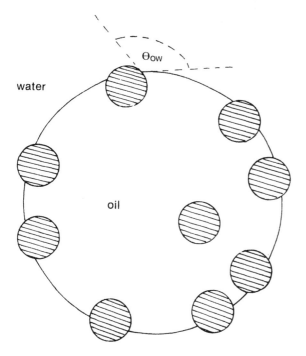

FIG. 33 Composite antifoam entity comprising mixture of oil and spherical hydrophobic particles with $90° < \theta_{OW} < 180°$.

oil-water interface is the key to their role. Kulkarni et al. [124] have emphasised the importance of long-range electrostatic repulsion forces between antifoam entities and bubbles in contributing to antifoam behavior with hydrophobed silica/polydimethylsiloxane antifoams. These authors present some evidence of diminished antifoam effectiveness accompanying increasing zeta potentials in sodium lauryl sulfate solution with increasing concentration. It seems probable then that the role of particles at the oil-water surface will concern the (often electrostatic) forces between bubbles and antifoam entities.

Further evidence concerning this proposition is given if we note that hydrophobic particles are often not added to polydimethylsiloxanes when these materials are used as antifoams for nonaqueous systems. In a study of polydimethylsiloxane antifoam behavior in lube oils, Shearer and Akers [17] have shown that electrostatic interactions are essentially absent in that system.

Anfruns and Kitchener [75], as we have seen, have shown that the capture of particles by bubbles is greatly facilitated by the presence of asperities. This has been attributed to the penetration, by an asperity, of the energy barrier associated with long-range (electrostatic) repulsion forces between

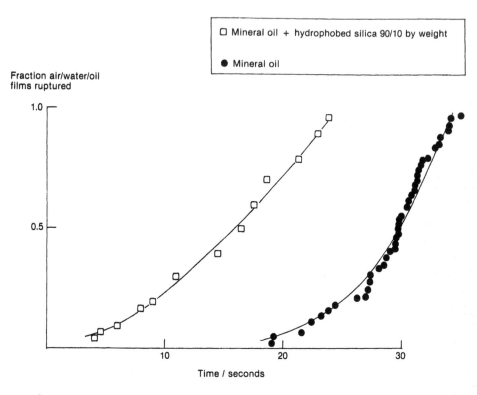

Fraction air/water/oil
films ruptured

FIG. 34 Effect of hydrophobed silica on time of emergence of mineral oil droplets into air-water surface of 0.5 g dm^{-3} sodium (C$_{10}$—C$_{14}$) alkylbenzene sulfonate solution (Dobs 055). Drop volumes 0.01 cm^3, hydrophobed silica D17 ex Degussa. (From Ref. 40.)

particles and bubbles (see Sec. IV.E.). It seems reasonable to attribute a similar role to particles at the oil-water surface. Thus, in effect, the particle should facilitate rupture of the aqueous film separating an oil droplet from the air-water surface. This hypothesis has been tested by examining the effect of hydrophobed silica on the time of emergence of mineral oil droplets into the air-water surface of the solution of a sodium (C$_{10}$—C$_{14}$) alkylbenzene sulfonate [40] used to obtain the foam results shown in Fig. 27. The droplets were of volume −0.01 cm^3. Results are presented in Fig. 34. Here it is clear that the particles significantly reduce the emergence time for the oil droplets. A similar result has been obtained for oil droplets coalescing with a layer of oil on the surface of a solution [40]. This then implies that the emulsion behavior found with these particle-oil mixtures is determined by a tendency of the particles to rupture the aqueous film between oil droplets. Essentially the same explanation for the effect of particles on emulsion behavior has been offered by van Boekel and Walstra [125] and by Mizrahi and Barnea [126].

It would seem that the role of the particles in oil-particle antifoams is to facilitate the emergence of oil droplets into the air-water surface. Thus, capture of an oil droplet by a bubble will not occur if the time taken for the oil droplet to be swept along streamlines around the bubble is less than the time for emergence. The presence of particles will therefore increase the probability of capture. Here, however, we would expect the time scales to be different under such dynamic conditions than those revealed by Fig. 34.

2. Spherical Particles

The process of emergence of an oil droplet into the air-water surface will clearly be initiated by penetration of the energy barrier due to long-range forces. However, ultimately the particle must facilitate hole formation in the oil-water-air film in much the same manner as particles rupture air-water-air films. The process is illustrated in Fig. 35 for a spherical particle satisfying condition 54. Here we suppose that the particle is small so that the effect of gravity on the shape of the oil-water surface may be neglected. It is easily seen that for the particle to rupture the film we must have

$$\theta_{AW} > 180° - \theta_{OW} \tag{55}$$

If the particle also satisfies condition 54, then condition 55 implies that rupture of an oil-water-air film by a spherical particle can occur if $\theta_{AW} < 90°$. Such a particle will not, as we have seen, rupture symmetrical air-water-air films (for which we require $\theta_{AW} > 90°$). We therefore find that spherical particles which satisfy conditions 54 and 55 will promote the emergence of an oil droplet into the air-water surface without exhibiting any antifoam behavior when used alone. If the oil satisfies condition 52, then we have a clear explanation for the antifoam synergy which Frye and Berg [39] report for mixtures of hydrophobed glass spheres and hexadecane. Thus, the particle alone will not function as an antifoam if the contact angle at the air-water surface is <90° and the oil will not function because of low probability of emergence into the air-water surface. Particles in the oil facilitate the latter process, enabling the resultant oil lens to bridge an aqueous film and participate in the foam film rupture mechanism shown in Fig. 31.

Inversion of the tendency of the oil to form oil-in-water emulsions by the presence of spherical particles is also easily understood. Thus, if the particles are preferentially wetted by the oil so that they satisfy condition 54 then they may rupture aqueous films between oil droplets in much the same

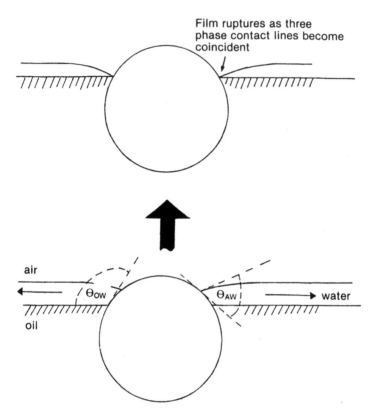

FIG. 35 Rupture of air-water-oil film by spherical particle with $90° < \theta_{OW} < 180°$ and $\theta_{AW} > 180° - \theta_{OW}$ so that $\theta_{AW} < 90°$.

manner as spherical particles, with $\theta_{AW} > 90°$, rupture aqueous films between air bubbles (Fig. 12 and 14).

3. Smooth Particles with Edges

In the case of smooth particles with edges then, as with air-water-air films, a more complex set of conditions for rupture of oil-water-air films arises. With an axially symmetrical particle defined by the angle θ_p and shown in Fig. 18 we have only one configuration which unambiguously yields film rupture if $\theta_{AW} < 90°$. Thus, if we have

$$\theta_p < \theta_{OW} < 180° - \theta_p \tag{56}$$

then a configuration similar to that of B in Fig. 18 will occur where the oil-water surface replaces the air-water surface and the particle is half immersed in oil. The configuration is depicted in Fig. 36. A particle adopting such a

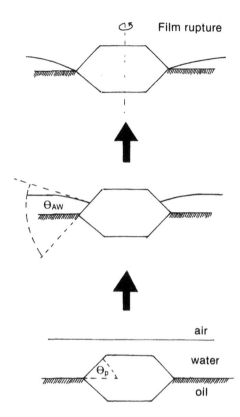

FIG. 36 Effect of axially symmetrical particles with edges on stability of oil-water-air film with $\theta_p < \theta_{OW} < 180° - \theta_p$ and θ_{AW} θ_p.

configuration at the oil-water surface will rupture any oil-water-air film provided that $\theta_{AW} > \theta_p$. This condition is of course the same as that (i.e., condition 29) required for the particle alone to cause air-water-air foam film rupture. Oil-particle synergy would therefore not be apparent with particles of this geometry. Such particles are not readily prepared. However, similar conclusions are relevant for cubic crystalline particles where the same film rupture configuration as at the air-water surface (see Sec. IV.B.) is possible at the oil-water surface. An example of a cubic particle is a galena crystal rendered hydrophobic by suitable surfactants. Assuming perfect crystal geometry, such particles should be equally effective at rupturing oil-water-air films as for rupturing air-water-air foam films (provided $\theta_{AW} > 45°$). Synergy would therefore be absent. However, the probability of achieving the required diagonal orientation (see Fig. 16b) at the relevant interfaces will in general be different, and a significant incidence of crystalline imperfections would mean additional edges and asperities. All of this could mean that synergy may still be apparent.

As we have seen, more complex particles with straight edges can be modeled by projection of the cross section shown in Fig. 18 in a direction normal to the plane of the paper (see Fig. 19). In Table 10 we present the configurations such a particle could adopt at the oil-water interface if condition 54 is satisfied and if $\theta_p = 30°$. The conditions required for θ_{AW} if the particle is to cause rupture of oil-water-air films and facilitate emergence of the oil droplet into the air-water surface are also deduced. Here we see that three configurations give rise to unambiguous rupture of oil-water-air films with $\theta_{AW} < 90°$. The remaining configurations will only affect rupture of such films if the critical thickness of rupture of the aqueous film on the solid is greater than that of the aqueous film on the oil.

Synergy between an oil which satisfies condition 52 and a particle which satisfies condition 54 requires that the air-water contact angle conditions for the particle to rupture air-water-air films be more severe than those for rupture of air-water-oil films. This situation can, however, only occur for the particle considered in Table 10 if the configuration $j = 4$ occurs because the conditions for θ_{AW} for all other configurations which yield rupture of air-water-oil films would also mean rupture of air-water-air films (see Table 1). It is clear from Table 10 that the configuration $j = 4$ can coexist with both $j = 5$ and 6, where θ_{OW} lies between 135° and 150°. In these circumstances the work of emergence of the particle into the oil-water surface from the oil phase will largely determine the relative probabilities of the relevant configurations (ignoring the effect of "degeneracy"). The work of emergence W_j^0 of each of the configurations at the oil-water surface is given (by analogy with the corresponding Eq. 33 for the air-water surface)

$$W_j^0 = \gamma_{OW}(\Delta A_{SW} \cos(180° - \theta_{OW}) - \Delta A_{OW}) \tag{57}$$

where ΔA_{SW} and ΔA_{OW} are respectively the changes in solid-water and oil-water surface areas accompanying emergence of the particles into the oil-water surface from the oil phase. If we consider the emergence of a particle into the oil-water surface of an oil droplet where the radius of the oil droplet is large compared to the size of the particle, then we may neglect the reduction in surface area of the oil droplet due to changes in radius as the particle emerges into the oil-water surface. The geometry of the problem of estimating W_j^0 then becomes exactly the same as that involved in estimating the work W_i of emergence of the same particle from the aqueous phase into the air-water surface (where $180° - \theta_{OW}$ is substituted for θ_{AW} and γ_{OW} for γ_{AW} in the relevant expressions given in Table 1).

Whether configuration $j = 4$ has the largest negative work of emergence depends upon θ_{OW}. Thus, it is easy to deduce from the data presented in

Table 10 that synergy may occur for these hypothetical particles if $\theta_{OW} = 145°$ and $15° < \theta_{AW} < 30°$. This follows because under those conditions configuration $j = 4$ is the most probable configuration (because it has the largest negative work of emergence when coexistence with other configurations is possible) and the particle cannot function as an antifoam alone (because this requires $\theta_{AW} > 30°$; see Table 1). In the case of $\theta_{OW} = 140°$, however, configuration $j = 5$ at the oil-water surface is most probable and collapse of the air-water-oil film will require $\theta_{AW} > 30°$. This latter is, however, the same condition as that required for the particle to collapse air-water-air films, so synergy will be absent in this case. Clearly, then, for this hypothetical particle we have potentially complex behavior where synergy may or may not occur, depending upon the magnitudes of θ_{OW} and θ_{AW}, which will in turn be influenced by the type and concentration of surfactant present.

No experimental study of the antifoam effect of mixing hydrophobed crystals of well-characterized habit with oils has been reported. Experimental confirmation of the expectation of synergy only under restricted circumstances is therefore not available. However, Frye and Berg [39] have shown evidence of only weak antifoam synergy when hydrophobed ground-glass particles, which possess sharp edges (and are therefore relatively effective antifoams when used alone), are mixed with hexadecane.

4. Rough Particles with Many Edges

The case of rough particles with many edges can be modeled by selecting the axially symmetrical particle with regular rugosities shown in Fig. 21. In the interests of clarity the particle is again depicted in Fig. 37 for the case of interaction with an air-water-oil film. At the oil-water surface such a particle may adopt configurations $jj = 1$ to 15 where that surface hinges at the rugosities provided

$$\theta_r < \theta_{OW} < 180° - \theta_r \tag{58}$$

where $\theta_r < 90°$ and is defined in Fig. 37. This particle can only cause rupture of oil-water-air films when $\theta_{AW} < 90°$ if configuration $jj = 8$ is adopted (where most of the particle is immersed in the oil) and $\theta_{AW} > \theta_r$. A particle with such a configuration would also cause rupture of oil-water-oil films and would therefore cause inversion of emulsion behavior so that water-in-oil emulsions would be found.

The work of emergence from the oil phase W_{jj}^{+0} of each of the configurations of this rough particle at the oil-water surface may be calculated in a similar manner to W_j^0 by using Eq. 57 if due allowance is made for the relevant geometry. If again we assume that the particles are small relative

TABLE 10 Air-Water-Oil Film Collapse and the Configurations at the Oil-Water Surface of a Particle with Several Edges

Possible configurations at oil-water surface if $90° < \theta_{ow} < 180°$	Conditions for realization of configurations	"Degeneracy"	Air-water-oil film collapse if $\theta_{AW} < 90°$	Work of emergence[a] of particle into oil-water surface from oil = $W_j^0/\gamma_{ow}X^2$ for $Z = 5X$	
				$\theta_{ow} = 145°$	$\theta_{ow} = 140°$
$j = 1$ water oil	$150° < \theta_{ow} < 180°$	2	—	—	—
$j = 2$	$150° < \theta_{ow} < 180°$	4	—	—	—
$j = 3$	$90° < \theta_{ow} < 120°$	2	Yes provided $\theta_{AW} > 60°$	—	—

106

$j = 4$	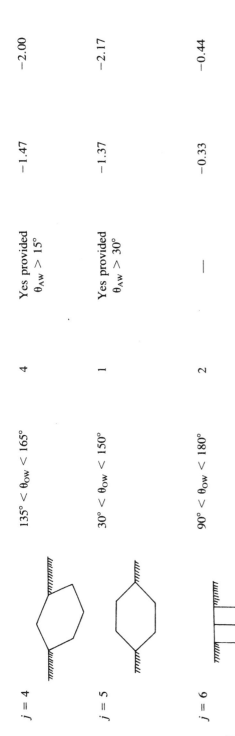	$135° < \theta_{OW} < 165°$	4	Yes provided $\theta_{AW} > 15°$	-1.47	-2.00
$j = 5$		$30° < \theta_{OW} < 150°$	1	Yes provided $\theta_{AW} > 30°$	-1.37	-2.17
$j = 6$		$90° < \theta_{OW} < 180°$	2	—	-0.33	-0.44

Particle geometry as shown in Fig. 19 with $\theta_p = 30°$ and $Y = X$. Z is the length of the particle defined in the figure.
[a]Calculated using expressions shown in Table 1 for emergence into the air-water surface from water, which is exactly analogous except that γ_{OW} replaces γ_{AW} and $180° - \theta_{OW}$ replaces θ_{AW}.

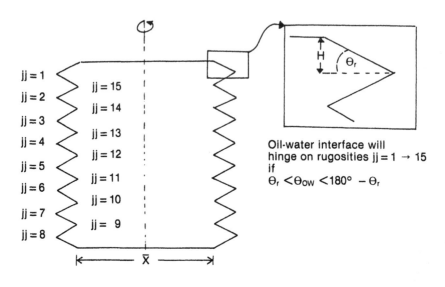

Oil-water interface will
hinge on rugosities jj = 1 → 15
if
$\Theta_r < \Theta_{OW} < 180° - \Theta_r$

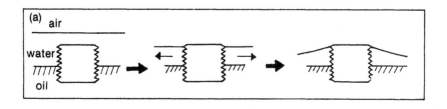

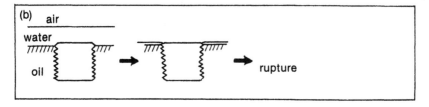

FIG. 37 Axially symmetrical particle with many edges at oil-water interface: (a) example of configuration which stabilizes air-water-oil film where oil-water interface hinges on any rugosity $jj \neq 8$; (b) configuration where oil-water interface hinges on rugosity $jj = 8$ will cause air-water-oil film rupture provided $\theta_{AW} > \theta_r$.

TABLE 11 Work of Emergence into the Oil-Water Surface W_{jj}^{+0} of an Axially Symmetrical Particle with Regular Rugosities

Configuration with oil-water surface hinging on rugosity jj	$W_{jj}^{+0}/\gamma_{ow}\bar{X}^2$	
	$\theta_{ow} = 135°$	$\theta_{ow} = 95°$
$jj = 1$	6.951	−0.391
2	5.908	−0.520
3	4.866	−0.648
4	3.823	−0.777
5	2.781	−0.905
6	1.738	−1.034
7	0.695	−1.163
8	−0.347	−1.291

$\theta_r = 30°$; and $H/X = 0.1$; see Fig. 37 for definition of θ_r, H, and $\bar{X}$.

to the size of the oil droplets, then we find

$$\frac{W_{jj}^{+0}}{\gamma_{ow}\bar{X}^2} = \pi \left[p_{jj}\left(\frac{H/\bar{X}}{\sin \theta_r}\right)\left(1 + \frac{H/\bar{X}}{\tan \theta_r}\right) + 0.25 \right]\cos\left(180 - \theta_{ow}\right)$$
$$- \pi \left[0.5 + \frac{H/\bar{X}}{\tan \theta_r}\right]^2$$
(59)

where $\bar{X}$ is defined in Fig. 37, and p_{jj} is the number of conical segments of height H (see Fig. 37) transferred from the oil phase to the aqueous phase.

Calculations of $W_{jj}^{+0}/\gamma_{ow}\bar{X}^2$ for a particle which satisfies both conditions 54 and 58, so that

$$90° < \theta_{ow} < 180° - \theta_r$$
(60)

are given in Table 11. From this table it is clear that the configuration $jj = 8$, with the particle mainly in the oil phase, yields the largest negative value of W_{jj}^{+0}. That is therefore the most probable configuration.

An alternative configuration may occur if the particle is completely removed from the bulk of the oil so that the oil just fills the rugosities. This is illustrated in Fig. 38. It is the configuration considered by Dippenaar [52], where the effective contact angle of the oil-filled particle at the air-water surface is supposed higher than that of the uncontaminated particle. This configuration can occur if the oil-water contact angle satisfies the condition

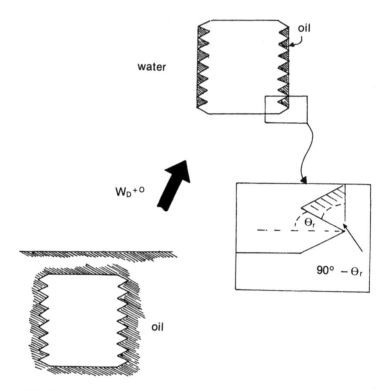

FIG. 38 Formation of hydrophobic particle with oil-filled rugosities after dispersal of oil-particle mixture in aqueous phase.

$$90° + \theta_r < \theta_{ow} < 180° \tag{61}$$

Conditions 60 and 61 overlap so that the configuration depicted in Fig. 38 can coexist with the configurations $jj = 1$ to 15 if

$$90° + \theta_r < \theta_{ow} < 180° - \theta_r \tag{62}$$

However, the configuration where the rugosities are just filled with oil (shown in Fig. 38) will have a higher work of formation W_D^{+0} than that of the most probable configuration at the oil-water surface where $jj = 8$ and where most of the particle is immersed in the oil. This will arise because of both the greater oil-water surface area and the greater particle-water surface area. Thus, if we assume the particles are small compared to any oil droplet in which they are initially dispersed then the difference in work of formation of the two configurations is

$$\frac{W_D^{+0} - W_8^{+0}}{\pi \gamma_{ow} \bar{X}^2} = \left[\left(\frac{H/\bar{X}}{\sin \theta_r} \right) \left(1 + \frac{H/\bar{X}}{\tan \theta_r} \right) + 0.25 \right] \cos (180° - \theta_{ow})$$

$$+ \left[0.5 + \frac{H/\bar{X}}{\tan \theta_r} \right]^2 \tag{63}$$

$$+ \left[14 \left(\frac{H}{\bar{X}} \right) \left(1 + \frac{H/\bar{X}}{\tan \theta_r} \right) \right]$$

so that we always have $W_D^{+0} > W_8^{+0}$ provided $\theta_{ow} > 90°$ since $\theta_r < 90°$.

We therefore find for a rough particle, which satisfies conditions 54 and 58 that the most probable configuration is one where the particle is mostly immersed in the oil phase. The oil-water surface hinges on an edge so that exposure of particle surface to the aqueous phase is minimized. As we have seen, this is also the only configuration which can cause rupture of the oil-water-air film (provided condition 39 is also satisfied so that $\theta_{AW} > \theta_r$). By contrast, a particle of the same geometry which satisfies condition 39 and for which $\theta_{AW} < 90°$ will have a low probability of rupturing air-water-air foam films. This follows because the only configuration which such a particle may adopt at the air-water surface which can cause rupture has the lowest probability of occurrence (see Sec. IV.C.). Therefore, a rough particle of geometry shown in Fig. 37 (or 21) has a high probability of rupturing an oil-water-air film if $\theta_{ow} > 90°$ and $\theta_r < \theta_{AW} < 90°$ but a low probability of rupturing an air-water-air foam film. Again, then, we would expect synergy if such particles are mixed with oils of low γ_{AO} so that the bridging condition 52 is satisfied. Since most particles used as antifoam promoters are best considered to be rough with many edges, then the ubiquitous occurrence of such synergy is clearly consistent with the argument outlined here. The overall process of antifoam action is depicted schematically in Fig. 39 for the case of an oil for which $S'' < 0$ and $\theta^* > 90°$ (i.e., condition 52).

F. Antifoam Dimensions and Kinetics

There have been no systematic studies of the effect of particle size on the efficiency of particle-oil antifoams. However, if part of the role of the particle is to penetrate any electrical double layer between oil droplets and air bubbles, then clearly the particle size should be at least of the same order as the double layer thickness, i.e., >0.1 μm. If the particle size is appreciably smaller than this, then the oil droplet will in effect be smooth so that the presence of particles will be irrelevant (except in that they modify the overall interaction forces between bubbles and antifoam entities).

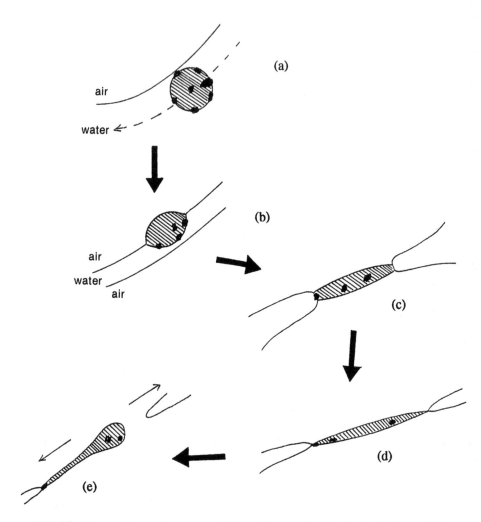

FIG. 39 Proposed overall process of foam film rupture by hydrophobic particle-oil mixtures for oils where $S'' < 0$ and $\theta^* > 90°$ (i.e., condition 52). (a) Particle ruptures air-water-oil film permitting oil droplet to adhere to bubble surface (b) Oil lens formed. (c) Oil lens bridges air-water-air foam film. (d) Capillary pressure in vicinity of bridging lens increases rate of drainage of foam film. (e) Film rupture.

Particle-oil antifoams form composite entities. Increase in particle size at constant weight fraction of necessity increases the size of those entities, otherwise the system will subdivide into oil droplets and oil-coated particles. If these composites are to cause foam film rupture, then they must first form a lens at the air-water surface at one side of the foam film and subsequently emerge into the other side to bridge the film. Drainage of the foam film to dimensions of the same order as that of the antifoam may therefore be expected to play the same role with oil-particle antifoams as with particle antifoams. We would therefore expect that the considerations of Dippenaar [52] and Frye and Berg [63] to be relevant for particle-oil antifoams so that foam film draining is the rate-determining step for antifoam action. Sizes of the antifoam particle-oil composites should therefore be small enough to ensure a high probability of presence in a given foam film but not so small that foam film drainage to the dimensions of the antifoam entity is too slow. Observations with both mineral oil/hydrophobed silica [40] and silicone [127] antifoams suggest that improvements in antifoam effectiveness of a given weight concentration of antifoam may be achieved by decreasing antifoam entity sizes down to at least 1–2 μm.

ACKNOWLEDGMENTS

The author wishes to acknowledge many discussions with Prof. D.G. Hall concerning the calculation of the probabilities of particles adopting various configurations at the fluid-fluid surface. The author also wishes to thank Dr. R. Aveyard and Dr. J. Lucassen for reading through the manuscript and making helpful suggestions.

REFERENCES

1. J. A. Kitchener in Recent Progress in Surface Science (J. F. Danielli, K. G. A. Pankhurst, and A. C. Riddiford, eds.), Vol. 1, Academic Press, New York, 1964, p.51.
2. J. Lucassen in Anionic Surfactants—Physical Chemistry of Surfactant Action (E. H. Lucassen-Reijnders, ed.), Marcel Dekker, New York, 1981, p.217.
3. A. Prins in Advances in Food Emulsions and Foams (E. Dickinson, G. Stainsby, eds.), Elsevier Applied Science, New York, 1988, p.91.
4. P. Walstra in Foams: Physics, Chemistry and Structure A.J. Wilson, ed.), Springer-Verlag, Berlin, 1989, p.1.
5. R. S. Hansen in Foams (R. J. Akers, ed.), Academic Press, London, 1976, p.1.
6. J. W. Gibbs in The Scientific Papers, Vol.1, Dover, New York, 1961.
7. K. Malysa, R. Miller, and K. Lunkenheimer, *Colloids Surfaces 53*: 47 (1991).

8. A. Prins, In Foams (R. J. Akers, ed.), Academic Press, London, 1976, p.51.
9. A. Vrij and J. Th. G. Overbeek, *J. Am. Chem. Soc. 90*: 3074 (1968).
10. B. P. Radoev, A. D. Scheludko, and E.D. Manev, *J. Colloid Interface Sci. 95*: 254 (1983).
11. D. R. Exerowa, *Izv. Khim 11*(3–4): 739 (1978).
12. E. D. Goddard and G.C. Benson, *Can. J. Chem. 35*:986 (1957).
13. J. M. Corkill, J.F. Goodman, and S.P. Harrold, *Trans. Faraday Soc. 60*: 202 (1964).
14. C. H. Fiske, *J. Biol. Chem. 35*: 411 (1918).
15. T. Sasaki, *Bull Chem. Soc. Japan 11*: 797 (1936).
16. T. Sasaki, *Bull Chem. Soc. Japan 13*: 517 (1938).
17. L. T. Shearer and W. W. Akers, *J. Phys. Chem. 62*: 1264, 1269 (1958).
18. J. W. McBain, S. Ross, A. P. Brady, J. V. Robinson, I. M. Abrams, R. C. Thorburn, and C. G. Lindquist, National Advisory Committee for Aeronautics A.R.R. No. 4105, 1944.
19. J. V. Robinson and W. W. Woods, *J. Soc. Chem. Ind. 67*: 361 (1948).
20. S. Ross, *J. Phys. Colloid Chem. 54*: 429 (1950).
21. S. Ross, R. M. Haak, *J. Phys. Chem. 62*: 1260 (1958).
22. S. Okazaki, K. Hayashi, and T. Sasaki, Proceedings of the IV International Congress on SAS, V3, Brussels, 67, 1964.
23. T. A. Koretskaya and P. M. Kruglyakov, *Izv. Sib. Otd. An SSSR, Ser. Khim. Nauk. 7*(3):129 (1976).
24. S. Ross and T. H. Bramfitt, *J. Phys. Chem. 61*: 1261 (1957).
25. E. J. Burcik, *J. Colloid Sci. 5*: 421 (1950).
26. W. E. Ewers and K. L. Sutherland, *Aust. J. Sci. Res. 5*: 697 (1952).
27. J. A. Kitchener and C. F. Cooper, *Quart. Rev. 13*: 71 (1959).
28. A. Leviton and A. Leighton, *J. Dairy Sci. 18*: 105 (1935).
29. S. Ross and G. J. Young, *Ind. Eng. Chem. 43* (11): 2520 (1951).
30. R. D. Kulkarni and E. D. Goddard, *Croatica Chem. Acta 50*(1–4): 163 (1977).
31. P. M. Kruglyakov and P. R. Taube, *Zh. Prikl. Khim. 44*(1): 129 (1971).
32. W. D. Harkins, *J. Chem. Phys. 9*: 552 (1941).
33. J. S. Rowlinson and B. Widom in Molecular Theory of Capillarity, Oxford University Press, Oxford, 1982.
34. P. M. Kruglyakov, in Thin Liquid Films, Fundamentals and Applications (I. B. Ivanov, ed.), Marcel Dekker, New York, 1988, p. 767.
35. P. M. Kruglyakov and T. A. Koretskaya, *Kolloid. Zh. 36*(4): 682 (1974).
36. A. P. Koretsky, A. V. Smirnova, T. A. Koretskaya, and P. M. Kruglyakov, *Zh. Prikl. Khim. 50*: 84 (1977).
37. T. A. Koretskaya, *Kolloid. Zh. 39*(3): 571 (1977).
38. M. N. Fineman, G. L. Brown, and R. J. Myers *J. Phys. Chem. 56*: 963 (1952).
39. G. C. Frye and J.C. Berg, *J. Colloid Interface Sci. 130*(1): 54 (1989).
40. P. R. Garrett, J. Davis, and H. M. Rendall, unpublished work.
41. Y. Abe and S. Matsumura, *Tenside Detergents 20*(5): 218 (1983).
42. I. C. Callaghan, C. M. Gould, R. J. Hamilton, and E. L. Neustadter, *Colloids Surfaces 8*: 17 (1983).

43. C. B. McKendrick, S. J. Smith, and P. A. Stevenson, *Colloids Surfaces 52*: 47 (1991).
44. P. R. Garrett, *J. Colloid Interface Sci. 69*(1): 107 (1979).
45. A. Prins in Food Emulsions and Foams (E. Dickinson, ed.), Royal Society of Chemistry Special Publication 58, 1986, p.30.
46. J. A. Fay in Oil on the Sea (D. P. Hoult, ed.), Plenum Press, New York, 1969, p.53.
47. D. P. Hoult, *Ann. Rev. Fluid Mech. 4*: 341 (1972).
48. C. Huh, M. Inoue, and S. G. Mason, *Can. J. Chem. Eng. 53*: 367 (1975).
49. P. Joos and J. Pintens, *J. Colloid Interface Sci. 60*: 507 (1977).
50. S. U. Pickering, *J. Chem. Soc.*, 2001 (1907).
51. O. Bartsch, *Kolloidchem. Beihefte 20*: 1 (1924).
52. A. Dippenaar, *Int. J. Mineral Process. 9*: 1–22 (1982).
53. V. M. Lovell in Flotation; A. M. Gaudin Memorial Volume (M. C. Feurstenau, ed.), Vol. 1, American Institute of Mining, Metallurgical, and Petroleum Engineers, Inc., 1976, p. 597.
54. J. R. Tate and A. C. McRitchie (assigned to Procter and Gamble), GB 1,492,938, November 23, 1977, filed January 1, 1974.
55. S. G. Mokrushin, *Kolloidn. Zh. 12*: 448 (1950).
56. N. Dombrowski and R. P. Fraser, *Phil. Trans. Royal Soc. London, Ser.A 247*: 13 (1954).
57. A. K. Livshitz and S. V. Dudenkov, *Tsvet. Metally 30*(1): 14 (1954).
58. J. H. Schulman and J. Leja, *Trans. Faraday Soc. 50*: 598 (1954).
59. A. K. Livshitz and S. V. Dudenkov, Proceedings of 7th IMPC New York, (1965), p. 367.
60. S. V. Dudenkov, *Tsvet. Metally 40*: 18 (1967).
61. R. R. Irani and C. F. Callis, *J. Phys. Chem. 64*: 1741 (1960).
62. H. Peper, *J. Colloid Sci. 13*: 199 (1958).
63. G. C. Frye and J. C. Berg, *J. Colloid Interface Sci. 127*(1): 222 (1989).
64. A. J. De Vries, *Rec. Trav. Chim. 77*: 383 (1958).
65. A. K. Livshitz and S. V. Dudenkov, *Tsvet. Metally 33*: 24 (1960).
66. F. Tang, A. Xiao, J. Tang, and L. Jiang, *J. Colloid Interface Sci. 131*(2): 498 (1989).
67. J. F. Oliver, C. Huh, and S. G. Mason; *J. Colloid Interface Sci. 59*(3): 568 (1977).
68. A. J. Hopkins, and L. V. Woodcock, *J. Chem. Soc. Faraday Trans. 86*(12): 2121 (1990).
69. R. Vacher, T. Woignier, J. Pelous, and E. Courtens, *Phys. Rev. B. 37*(11): 6500 (1988).
70. K. J. Mysels, K. Shinoda, and S. Frankel in Soap Films—Studies of Their Thinning, Pergamon Press, London 1959.
71. O. Reynolds, *Phil. Trans. Royal Soc. London, Ser. A 177*: 157 (1886).
72. A. Vrij, *Disc. Faraday Soc. 42*: 23 (1966).
73. T. D. Blake and J. A. Kitchener, *J. Chem. Soc. Faraday Trans. 1, 68*: 1435 (1972).

74. J. N. Israelachvili in Intermolecular and Surface Forces with Applications to Colloidal and Biological Systems, Academic Press, London, 1985.

75. J. F. Anfruns and J. A. Kitchener, *Trans Inst. Min. Metall., London C 86*: 9 (1977).

76. W. A. Ducker, R. M. Pashley and B. Ninham, *J. Colloid Interface Sci.* *128*(1), 66 (1989).

77. J. Strnad, H. Kohler, K. Heckmann, and M. Pitsch, *J. Colloid Interface Sci.* *132*(1): 283 (1989).

78. B. Derjaguin, *Trans. Faraday Soc. 36*: 203 (1940).

79. J. Davis and P. R. Garrett, unpublished work.

80. M. Aronson, *Langmuir 2*, 653 (1986).

81. R. D. Kulkarni, E. D. Goddard, and B. Kanner, *Ind. Eng. Chem., Fundam.* *16*(4): 472 (1977).

82. R. Birtley, J. Burton, D. Kellett, B. Oswald, and J. Pennington, *J. Pharm. Pharmac. 25*: 859 (1973).

83. M. J. Povich, *A. I. Ch. E. J. 21*(5): 1016 (1975).

84. S. Ross and G. Nishioka in Emulsions, Lattices and Dispersions (P. Becker and M. Yudenfreud, eds.), Marcel Dekker, New York, 1978, p. 237.

85. S. Ross and G. Nishioka, *J. Colloid Interface Sci. 65*(2): 216 (1978).

86. M. N. A. Carter and P. R. Garrett (assigned to Unilever Ltd.), GB 1,571,502, July 16, 1980, filed January 23, 1976.

87. O. F. Schweigl and G. P. Best (assigned to Unilever Ltd.), GB 1,099,502, January 17, 1968, filed July 8, 1965.

88. P. R. Garrett (assigned to Unilever Ltd.), EP 75,433, March 30, 1983, filed September 16, 1981.

89. S. Ross, *Chem. Eng. Prog. 63*(9): 41 (1967).

90. C. C. Currie (assigned to Dow Chemical Co.), US 2,632,736; March 24, 1953, filed August 22, 1946.

91. M. M. Solomon (assigned to General Electric Co.), US 2,829,112; April 1, 1958, filed September 22, 1955.

92. S. Nitzsche and E. Pirson (assigned to Wacker-Chemie GmbH), US 3,235,509; February 15, 1966, filed October 3, 1962.

93. (Assigned to Degussa), Fr 1,533,825; July 19, 1968, filed August 8, 1967.

94. M. J. O'Hara and D. R. Rink (assigned to Union Carbide Corp.), GB 1,247,690, September 29, 1971, filed August 11, 1987.

95. K. W. Farminer and C. M. Brooke (assigned to Dow Corning Ltd.), US 3,843,558; October 22, 1974, filed June 16, 1972.

96. G. Bartolotta, N. T. de Oude, and A. A. Gunkel (assigned to Procter and Gamble Co.), GB 1,407,997; October 1, 1975, filed August 1, 1972.

97. F. J. Boylan (assigned to Hercules Powder Co.), US 3,076,768; February 5, 1963, filed Apri 5, 1960.

98. R. Leibling and N. M. Canaris (assigned to Nopco Chemical Co.), US 3,207,698, September 21, 1965, filed February 13, 1963.

99. R. E. Sullivan (assigned to Dow Corning Corp.), US 3,304,266, February 14, 1967, filed May 6, 1963.

100. E. Domba (assigned to Nalco Chemical Co.), US 3,388,073, June 11, 1968, filed December 16, 1963.
101. H. Buckman (assigned to Buckman Laboratories Ltd.), GB 1,166,877, October 15, 1969, filed December 8, 1967.
102. G. C. Harrison and A. J. Stumpo (assigned to Pennsalt Chemicals Corp.), GB 1,195,589; June 17, 1970, filed August 10, 1967.
103. H. Lieberman, C. A. Duharte-Francia, and J. W. Henderson (assigned to Betz Laboratories Inc.), GB 1,267,479; March 23, 1972, filed May 26, 1969.
104. J. R. Miller, R. H. Pierce, R. W. Linton, and J. H. Wills (assigned to Philadelphia Quartz Co.), US 3,714,068; January 30, 1973, filed December 28, 1970.
105. M. Caviet (assigned to Shell Development Co.), Canada 508,856; April 4, 1955, filed November 28, 1949.
106. F. J. Boylan (assigned to Hercules Inc.), US 3,705,859; December 12, 1972, filed December 30, 1970.
107. G. Boehmke, M. Quaedrlieg, and G. Kolla (assigned to Farbenfabriken Bayer Aktiengesellschaft), GB 1,267,482; March 22, 1972, filed July 14, 1970.
108. H. J. Shane, J. E. Schill and J. W. Lilley (assigned to Hart Chemical Ltd.), Canada 922,456; March 13, 1973, filed March 31, 1971.
109. I. A. Lichtman and A. M. Rosengart (assigned to Diamond Shamrock Corp.), Canada 927,707; June 5, 1973, filed September 3, 1971.
110. (Assigned to Henkel and Cie GmbH), GB 1,386,042; March 5, 1975, filed January 25, 1973.
111. F. M. Ernst (assigned to Mobil Oil Corp.), US 3,909,445; September 30, 1975, filed September 21, 1972.
112. F. J. Boylan (assigned to Hercules Inc.), US 3,408,306; October 29, 1968, filed July 7, 1961.
113. A. G. Kanellopoulos and M. J. Owen, *Trans. Faraday Soc. 67*: 3127 (1971).
114. P. R. Garrett and P. Gratton, unpublished work.
115. J. Timmermans, Physico-Chemical Constants of Pure Organic Compounds, Vols. 1, 2, Elsevier, Amsterdam, 1950.
116. R. Aveyard and D. A. Haydon, *Trans. Faraday Soc. 61*: 2255 (1965).
117. R. Aveyard, P. Cooper, and P. D. I. Fletcher, *J. Chem. Soc. Faraday Trans. 86*(1): 211 (1990).
118. R. Aveyard, B. P. Binks, P. Cooper, and P. D. I. Fletcher, *Adv. Colloid Interface Sci. 33*: 59 (1990).
119. F. Van Voorst Vader, *Chem. Ing. Techn. 49*(6): 488 (1977).
120. A. A. Trapeznikov and L. V. Chasovnikova, *Colloid J. USSR 35*: 926 (1973).
121. P. R. Garrett, *J. Colloid Interface Sci. 76*(2): 587 (1980).
122. P. R. Garrett and P. Moore, unpublished work.
123. J. Sinka and I. Lichtman, *Int. Dyer Textile Printer*: (May), 489 (1976).
124. R. D. Kulkarni, E. D. Goddard, and B. Kanner, *J. Colloid Interface Sci. 59*(3): 468 (1977).
125. M. A. J. S. van Boekel and P. Walstra, *Colloids Surfaces 3*: 109 (1981).
126. J. Mizrahi and E. Barnea, *Br. Chem. Eng. 15*(4): 497 (1970).
127. V. Veber and M. Paucek, *Acta Fac. Pharm. Univ. Comenianae 26*: 221 (1974).

2

Antifoams for Nonaqueous Systems in the Oil Industry

IAN C. CALLAGHAN   BP Research, Sunbury-on-Thames, Middlesex, United Kingdom

I.	Introduction	119
II.	Occurrence of Foams in the Oil Industry	121
	A. Crude oil production	121
	B. Oil refining and oil products	125
III.	Characterization of Nonaqueous Foam Systems	127
	A. Foam structures	128
	B. Methods of assessing the foaming characteristics of nonaqueous systems	128
IV.	The Relationship between Physical Properties and Foam Stability	138
	A. Bulk rheological properties	138
	B. Nature of the filler gas	139
	C. Composition of the foaming system	140
	D. Interfacial rheological properties	140
V.	Antifoam Types and Methods of Selection	143
	A. Antifoam types	143
	B. Selection of antifoams	146
VI.	Future Directions	147
	Acknowledgments	147
	References	147

I. INTRODUCTION

The stability and breaking of nonaqueous foams is a subject of great importance to the oil industry, as such foams occur, for example, in the production and refining of crude oil, as well as during the use cycle of many products such as lubricants. Foam, whenever it occurs, leads to a reduction in process/product efficiency with consequent loss of revenue, if the problem remains untreated. Consequently, the reliable assessment of the foamability of oils and oil products is of prime importance as it allows potential foaming problems to be anticipated and allowed for at an early stage in either the design of a production/refining process or in the development of an oil product. However, despite the obvious importance of such nonaqueous foams, much of the available literature on foam stability and breaking relates to aqueous systems, with no clear distinction being made between them and nonaqueous systems. Indeed, Ross [1] having reviewed aqueous foams in some detail, states that "the same principles can be applied mutatis mutandis to the problem of foam in non-aqueous media."

A number of authors [2–11] have worked specifically on nonaqueous foams and reported that foam stability is due to a number of factors such as bulk viscosity, surface rheology, and the presence of surface active species. Brady and Ross [2] studied a series of engine oils and medicinal-grade paraffin oils and showed that foam stability increased linearly with the kinematic viscosity of the oil. McBain and Robinson [3] confirmed this view and showed that a high viscosity, both in the bulk and at the surface, was often associated with high foam stability, but cautioned that viscosity alone was not the sole cause. They attributed the high surface viscosities they had observed to the formation of plastic films at the gas-liquid interface giving rise to a semi-immobilized oil layer underneath. Ross [1] reported that in aqueous foams x-ray diffraction studies had shown that such structures did indeed exist. He has also described experimental evidence for the existence of surface plasticity in crude and fuel oils [5]. However, in contrast, Mannheimer and Schecter [6] did not find any evidence for surface viscosity being an important factor in determining the stability of mineral oil/calcium sulphonate foam systems. Furthermore, Scheludko and Manev [7] have shown that the thickness at which foam films rupture occurs spontaneously in the aniline/lauryl alcohol system is independent of viscosity.

For crude oils, the presence of specific acids and phenols has been shown [8] to be important for foam formation and stability. The presence of surface active materials has also been shown by other authors [4,9,10,12] to be necessary for foaming to occur in nonaqueous media. However, although surface tension [7] has been shown to play a major role in determining nonaqueous foam stability, it does not appear to be as important as the preceding factors [3,11]. Dynamic surface tension effects, on the other hand, have been postulated [13] as being particularly important to foam stability. Although a number of such studies, e.g., [14,15], have been reported for aqueous systems, only a relatively small number [8,16,17] have been reported for nonaqueous systems.

The main method recommended for breaking nonaqueous foams is the use of foam inhibitors (antifoams). Normally, these are added to the foaming system in a suitable diluent. It has been shown experimentally [9] that many effective antifoams have low solubility in the foaming medium. This was confirmed by Robinson and Woods [18], who also studied the effects of the surface tensions of the foaming medium and antifoam, as well as that of the interfacial tension between them. They concluded that the main requirement for an antifoam was that a droplet of the antifoam should penetrate the gas-liquid interface of the foam on contact. They defined a rupture or entering coefficient, E, as follows:

$$E = \gamma_F + \gamma_{DF} - \gamma_D \qquad (1)$$

where

γ_F = surface tension of the foaming medium
γ_D = surface tension of the antifoam
γ_{DF} = interfacial tension between the antifoam and foaming medium

Using this parameter, they argued that for foam breaking to occur E must have a high positive value. Ross [1,19] considered that the main requirement for an antifoam was its ability to spread over the foaming liquid. He suggested using Harkins' [20] spreading coefficient, S, to characterize foam inhibitors:

$$S = \gamma_F - \gamma_{DF} - \gamma_D \qquad (2)$$

In either case, it appeared that an effective antifoam must be readily dispersible in the foaming system, and must have a lower surface tension than that of the foaming liquid. This latter point was confirmed by Shearer and Akers [21] in a study on the use of silicone oils as foam inhibitors for lubricating oils. These authors showed that silicone oils were only effective as antifoams if they were dispersed as droplets of less than 100 μm in diameter.

In this chapter, many of the points brought out above will be highlighted further when we discuss the specific problems encountered with foams in the oil industry.

II. OCCURRENCE OF FOAMS IN THE OIL INDUSTRY

A. Crude Oil Production

When crude oil is produced, it is forced from the reservoir by pressure up the production tubing to the surface (see Fig. 1). As the oil rises up the tubing, the pressure becomes less and the associated gas is progressively

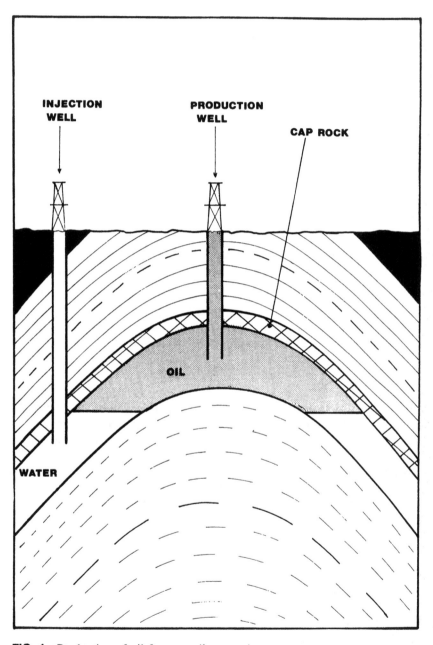

FIG. 1 Production of oil from an oil reservoir.

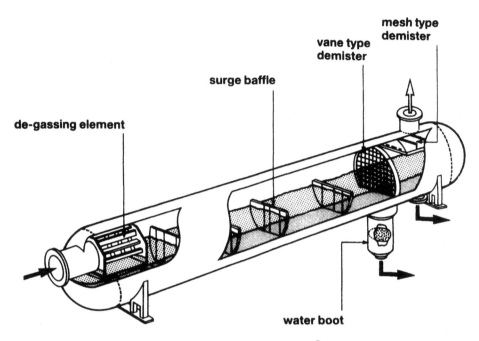

FIG. 2 A gas-oil separator. (From Ref. 24, copyright © 1985, SPE-AIME.)

released from solution. On emerging from the well, it is usually necessary to process the oil-gas mixture to remove both free and dissolved gas, which is likely to come out of solution when the oil is maintained at about atmospheric pressure, as, for example, during transport in a tanker or storage. Sometimes the separation of the gas is effected near the wellhead or, alternatively, the "live" crude oil may be conveyed under high pressure, e.g., 30 bar, in a pipeline to a distant location whereupon the separation of the oil and gas is effected. Occasionally, during this process, the crude oil forms a stable foam, with the result that liquid oil is carried over into the gas stream, creating serious problems in downstream compression plant. Such carryover situations can, if not treated, result in severe production losses and long equipment downtimes. To overcome this type of problem various procedures can be used: they may be totally physical in nature or a combination of both physical and chemical methods.

1. Physical Methods of Foam Control

It is common to describe the device used to separate gas from oil as a gas-oil or three-phase separator (see Fig. 2) when the bulk component and principal product are liquid hydrocarbon, at a nominally "steady" production rate. The term *three-phase separator* is used in this context since water is also separated from the gas and oil phases during this process.

The factors which affect separator design and performance are as follows:

a. the properties of the reservoir fluids, such as fluid densities and compositions, solids loading, and viscosities
b. the process conditions upstream of the separator
c. the process conditions downstream
d. external influences such as that of imposed motion, as for example, in floating production systems

The liquid-handling capacity of a separator is generally determined by a nominal mean residence time, typically 2 to 5 min. The selection of this residence time is based partly on an assessment of foam lifetime (gas disengagement time), partly on surge control requirements. The assessment of gas disengagement time has been based, in the past, on design experience and not on quantitative data. However, much effort is now being expended on improving this situation through the modeling of foam collapse under separator conditions, so as to allow separators to be designed with specific crudes in mind. The recent improvements in the laboratory assessment of foam characteristics under simulated separator conditions is proving useful in this latter respect.

In many gas-oil separators, internal devices are employed to help improve performance. The most common types of internal device are outlined briefly below.

(a) Transverse Baffles (see Fig. 2). These are located perpendicular to the direction of flow and act as skimmers to hold back the foam in the upper part of the separator and, thereby, allow time for collapse to occur.

(b) Parallel Plates. These devices are very common and are variously known as Dixon (Natco) or Arch (B.S.&B.) plates. The basic principle on which these plates operate is that the separating fluids are forced to flow in thin layers, giving rise to increased separation area, and some mechanical enhancement of foam breaking. It is believed that the orientation of these plates has some effect on their efficiency. However, as the bulk flow of fluid is normally along the separator axis, this claim is by no means obvious. There is some evidence from the author's studies that the wetting characteristics of the plate materials used to manufacture these devices may have some bearing on their efficacy as foam breakers.

(c) Random Packings. These are similar to those used in mass transfer columns, and they can assist in the breaking of crude oil foams or in preventing their formation. Again the wetting characteristics of the materials used to make these packings has an influence on their overall performance.

(d) Inlet Devices. These may be cyclonic, dished, or flat-plate constructions. They are claimed to assist gas-liquid separation, but may in fact promote foam formation and hence worsen the overall situation in some instances.

(e) Wire Mesh Pads. These are commonly installed at the gas outlet of a separator to act as demisters, removing any entrained droplets from the gas stream. Special meshes of lower bulk density are used to promote foam collapse. These latter devices are positioned in the separator headspace, upstream of the demister pads, and are thought to function by stressing the foam bubbles as they pass through the mesh, causing them to rupture.

2. Chemical Methods of Foam Control

The use of chemicals to prevent foam formation on entry into a gas-oil separator is well established. The principal types of antifoam used in this application are polydimethylsiloxanes and, for particularly difficult or aggressive crude oils, fluorosilicones. These liquids are ideally added sufficiently far upstream of the separator to allow thorough dispersal throughout the crude oil: this process can be aided by the proper choice of carrier solvent. In general, the dose rates required for these two classes of chemical fall within the range of 0.1 to 20 ppm. It is important to ensure that the correct dosage is predetermined in the laboratory, since profoaming can occur if the antifoam is applied (in the field) at a concentration below its solubility limit.

B. Oil Refining and Oil Products

1. Crude Oil Distillation Units

Foaming in crude oil distillation columns can lead to flooding of the trays, causing foam to pass down the downcomers (see Fig. 3), resulting in a loss of separation efficiency due to the recycling of vapour. It has been suggested [22] that this problem can be solved by careful tray design to produce jetting rather than bubbling at the perforations. Alternatively, one can use antifoam chemicals to reduce the foaming tendency of the liquid being distilled: the preferred materials are again the high molecular weight polydimethylsiloxanes.

2. Thermal Cracking Processes

When hydrocarbons are heated to temperatures in excess of 450°C they begin to decompose or "crack" into smaller molecular units. Paraffinic hydrocarbons are the most easily cracked materials followed by naphthenes, aromatic hydrocarbons being extremely refractory. As more heavy feedstocks have become available, the importance of thermal cracking processes has increased. One such process which is currently in use is visbreaking. Visbreaking is the thermal cracking of viscous residues to reduce their vis-

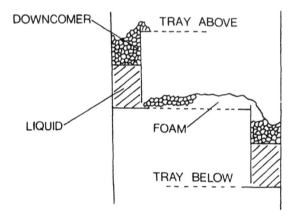

FIG. 3 Foam in a distillation column downcomer.

cosity by breaking down the large complex molecules in the feedstock to smaller ones. Using this process a satisfactory fuel oil can be made with minimum recourse to the blending of gas-oil or kerosine with the viscous residue.

Another thermal cracking process is delayed coking. This process is normally applied to atmospheric or vacuum residues from low-sulfur crudes for the production of electrode-grade coke. Also, it is increasingly being used for the disposal of heavy residues from high-sulfur crudes. In both these processes, the release of volatile products as the feed is heated can lead to foam formation with a consequent reduction in capacity, and fouling of overhead lines. The use of silicone-based antifoams controls the foam buildup and allows faster feed rates, greater capacity, and a reduction in fouling of the overhead lines. High molecular weight silicone fluids are again the preferred materials in these applications.

3. Lubricating Oils

Lubricating oils cover a diverse range of products, specially blended to meet specific in-use requirements in such machinery as compressors, steam turbines, and internal combustion engines. Lubricating oils are classified according to end use and include automotive lubricants, metalworking fluids, and compressor oils. The formation of foam during the use of these oils is deleterious to their performance and needs to be avoided at all costs. This problem is usually solved through careful formulation of the product and involves the use of foam inhibitors such as alkyl polyacrylates and low molecular weight silicone oils. Of these two classes of antifoam, the alkyl polyacrylates are to be preferred since they have much better air-release properties than silicone oils.

4. Automotive Fuels

The foaming of automotive fuels can be problematic, leading to such problems as difficulty in completely filling a car's fuel tank at a service station. This problem is particularly pronounced with diesel fuel and can result in overflow and early filler pump cutout. As in the case of lubricating oils, the solution to the problem lies in proper formulation of the fuel's additive package. Silicone-polyether-type antifoams appear to be the preferred materials for this application, although both dimethyl silicone oils and fluorosilicones also have utility if formulated correctly.

5. Propane Deasphalting

During the removal of asphaltenes from residue by precipitation with deasphalting agents, foam may be produced in the tower mixer during flash distillation. This foaming causes crude oil concentrate to overflow with consequent loss of process efficiency. The addition of a silicone antifoam to the feed to the deasphalting tower prevents foaming and the escape of residual oil in the spray condenser and asphalt stripper. However, the main potential problem with foam in this process is during the steam stripping of the asphalt. Normal practice in this case would be to add a silicone antifoam agent to the feed to the stripper rather than to the deasphalting tower.

III. CHARACTERIZATION OF NONAQUEOUS FOAM SYSTEMS

Foams are colloidal systems in which a gaseous phase is dispersed throughout a continuous liquid phase in such a way that the gas cavities are separated by thin liquid films. It is the stability of these liquid films that determines the overall stability of the foam. A fundamental characteristic common to all foams is their very large interfacial area. As a consequence, they are not stable in the thermodynamic sense, although some aqueous foams, e.g. shaving foam, can exist in a metastable state for very long periods in the absence of any external disturbances. Nonaqueous foams, on the other hand, are usually (though not always) incapable of existing in such a state and undergo complete collapse on removal of the motivation to foam, e.g., removing the source of filler gas. Thus, the stability of nonaqueous foams is usually discussed in terms of their collapse kinetics: stable foams being those that undergo collapse at a slow rate.

A. Foam Structures

There are two extreme structures which foams can assume; these are the classical polyhedral structure and the round-bubbled or gas dispersion structure. Both structures can be experienced in nonaqueous foam systems.

Polyhedral foams comprise large pockets of gas separated by thin liquid films. These liquid films form a three-dimensional network down which liquid drains from the foam under the influence of capillary forces. A typical crude oil polyhedral foam is shown in Fig. 4. Round-bubbled foams, on the other hand, consist of a dispersion of nearly spherical gas bubbles separated by relatively thick films of the bulk liquid. This type of foam has been termed *kugelschaum* by Manegold [23]. An example of a round-bubbled crude oil foam is shown in Fig. 5. In crude oil systems, both extremes of foam can be found and the particular type encountered often depends on the prevailing conditions of temperature and pressure under which the foam is formed. Since foam type can influence the choice of antifoam employed, it is clearly important to ascertain whether a round-bubbled or polyhedral foam forms under a given set of process conditions. For example, round-bubbled foam requires a silicone antifoam of low polydispersity [24], whereas polyhedral foam can be treated with a conventional polydisperse silicone oil. In the case of crude oils, polyhedral foams tend to form in high gas-oil ratio (GOR) systems at low pressures, whereas round-bubbled foams tend to form in either high GOR systems at moderate to high pressures or in low GOR systems. These latter structures eventually undergo synersis to the polyhedral state as liquid drains from between the gas bubbles (see Fig. 6).

B. Methods of Assessing the Foaming Characteristics of Nonaqueous Systems

1. General Methods

The simplest apparatus in which foam can be studied is that due to Bikerman [25] (see Fig. 7). Bikerman's gas sparging tube comprises a glass column with a glass sinter at one end onto which a known volume of the test liquid is pipetted. The flow of gas through the sinter causes the liquid to foam: different filler gases can be used to simulate different situations. Callaghan and Neustadter [26], in a study of crude oil foaming, used a modified Bikerman foaming column that was 30 cm in length with two fine (no. 2 porosity) sintered glass disks placed 1 cm apart, situated at the base of the tube just above the gas inlet. The gas used to create the foam was admitted to the column via a pressure reduction and flowmeter assembly after being saturated with the crude oil by passage through a series of Dreschel bottles

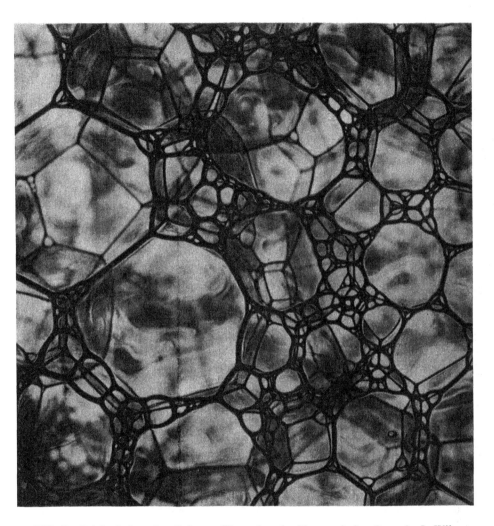

FIG. 4 Polyhedral crude oil foam. (Reproduced with permission from A. J. Wilson, Foams: Physics, Chemistry, and Structure, Chapter 7, Springer-Verlag, London, 1989.)

containing the oil. These workers used a fixed gas flowrate of 40 mL/min and allowed the aliquot of oil to be completely taken up into the foam before extinguishing the gas supply and recording the decrease of foam height with time. Using this method, two parameters which characterize the foam can be derived: these are the foaminess index (Σ) and the average foam lifetime

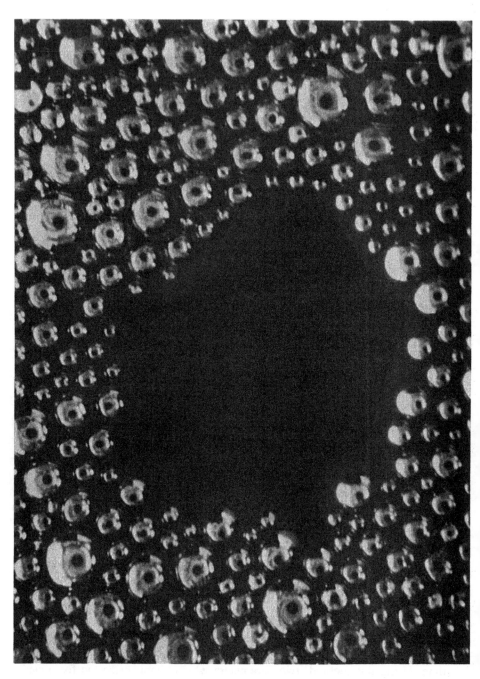

FIG. 5 Round-bubbled crude oil foam. (Reproduced with permission from A. J. Wilson, Foams: Physics, Chemistry, and Structure, Chapter 7, Springer-Verlag, London, 1989.)

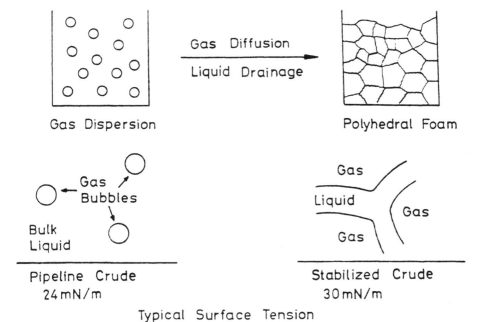

FIG. 6 Foam syneresis. (From Ref. 24, copyright © 1985, SPE-AIME.)

(L_F). The foaminess index is the ratio of the foam volume generated by a given gas flowrate to that flowrate:

$$\text{Foaminess index } \Sigma = \frac{\text{foam volume}}{\text{gas flowrate}} \tag{3}$$

This parameter has dimensions of time: it is a measure of foam-forming tendency. The average foam lifetime is determined from the collapse curve (see Fig. 8) and is a measure of foam stability. It too has the dimensions of time.

$$\text{Average foam lifetime } L_F = \frac{\displaystyle\int_0^T H(t)\, dt}{H(0)} \tag{4}$$

where

$H(0)$ = steady foam height at time $t = 0$
$H(t)$ = foam height at time t

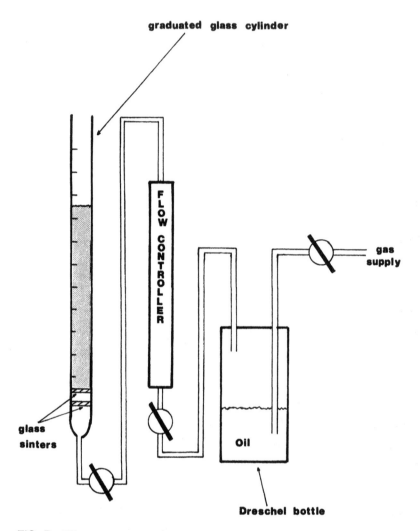

FIG. 7 Bikerman sparge tube apparatus.

T = time at which the foam height has reached zero; i.e., all foam
        bubbles have disappeared

The choice between foaminess index and average foam lifetime as the
measured parameter is very much dependent upon the system under investigation. If that system is a gas-oil separator then the average lifetime is
generally measured, since it gives information that can be readily related to
residence time in the separator. However, where the system is a distillation

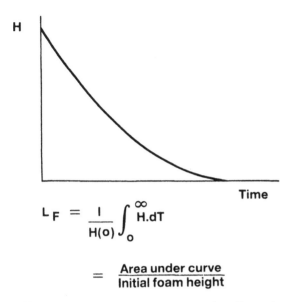

$$L_F = \frac{1}{H(o)} \int_0^\infty H.dT$$

$$= \frac{\text{Area under curve}}{\text{Initial foam height}}$$

FIG. 8 Foam collapse curve (schematic). (Reproduced with permission from A. J. Wilson, Foams: Physics, Chemistry, and Structure, Chapter 7, Springer-Verlag, London, 1989.)

column, the foaminess index is more appropriate since, in distillation processes, the liquid (downcoming) phase is continuously sparged with the gaseous (rising) phase. In general, however, it is likely that the choice is arbitrary since the two parameters are closely related.

2. Methods Specific to Crude Oil Systems

A major disadvantage of the Bikerman sparge tube method as applied to crude oil foaming is that it does not reflect the real situation in, say, a gas-oil separator. Consequently, a method of assessing foam under high pressure/temperature conditions is required. Such an apparatus is shown [27] in Fig. 9. The foam is created by allowing a sample of "live" crude oil to flow from an injection vessel at, say, pipeline pressure into a sample cell held at gas-oil separator pressure and temperature. This sudden depressurisation causes the crude oil to degas and form a foam. Once the foam has been fully formed the sample is isolated from the reference cell, and the subsequent foam collapse monitored by a differential pressure transducer placed across the two cells. As the foam in the sample cell collapses, an increase in the pressure of this cell relative to that in the reference cell occurs as a result of the release of the excess pressure within the foam bubbles to the headspace as the bubbles rupture. This method of foam analysis was devised by Ross and

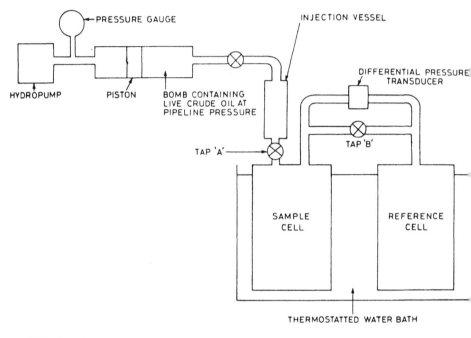

FIG. 9 High-pressure foam cell apparatus.

Nishioka [28]. A particular attraction of their method is that, in principle, it allows the gas and liquid holdup fractions of the collapsing foam to be determined as a function of time, as opposed to the foam height measurement of the Bikerman method. These more fundamental foam parameters can be related to the pressure above a foam collapsing in a closed vessel by Ross's equation of state for a foam [29]. This technique can also be used to study foams formed from stabilised crude oils. In practice, the method is difficult to apply as strict control of both background pressure and temperature are necessary if one is to achieve reliable results. To overcome these difficulties, an endoscope system can be incorporated into the apparatus so as to allow visual observation of the foam under test. In this way, the foam type and collapse characteristics can be observed under different conditions of temperature and pressure.

An alternative method of foam height analysis has been recently introduced which enables more reliable data to be obtained using a sparging tube. This method, due to Camp and Lawrence [30], is an adaptation of that first proposed by Ross and Cutillas [31]. It makes use of an optical transmission foam meter [30] (see Fig. 10) in which a conventional (but ungraduated) gas sparging tube is placed between a uniform light source and a light-sen-

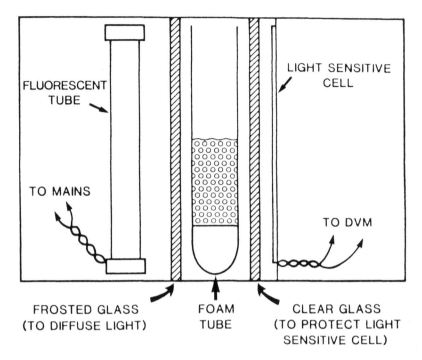

FLUORESCENT
TUBE

TO MAINS

LIGHT SENSITIVE
CELL

TO DVM

FROSTED GLASS
(TO DIFFUSE LIGHT)

FOAM
TUBE

CLEAR GLASS
(TO PROTECT LIGHT
SENSITIVE CELL)

FIG. 10 Optical transmission foam cell. (From Ref. 72).

sitive cell that has good linearity in terms of voltage output against incident light intensity. In this method the foam, created either by the conventional sparging technique or by shaking, collapses and allows more light to reach the light-sensitive cell. This results in a proportionate increase in voltage output from the photocell which is coupled via an analog-to-digital converter to a desktop computer. The voltage-time data so obtained is then analyzed and a foam height-time curve produced from which an average foam lifetime can be calculated. This technique has the advantage of being easier to use than either the conventional sparge tube or the high-pressure foam cell. It is particularly useful for the study of rapidly collapsing foams where the conventional Bikerman technique is both difficult to use and prone to large errors. Its principal disadvantage is that it is limited to use with stabilized crude oils, although efforts are currently being made to develop a high-pressure version suitable for use with live oils.

An intermediate method of assessing crude oil foam stability has been recently developed by the Goldschmidt Company [32]. Their method allows foam measurements to be made on simulated live crude oils at pressures up to about 10 bar. The apparatus is shown schematically in Fig. 11 and com-

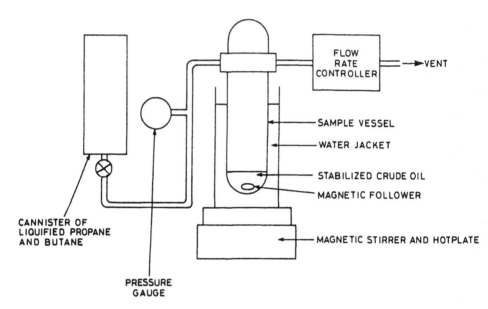

FIG. 11   Portable foam cell. (From Ref. 72.)

prises a quartz cell coupled to a liquified hydrocarbon gas canister and a flow control valve. In this apparatus, the simulated live crude oil is created by mixing liquified hydrocarbon gas (butane, propane) with stabilized oil. The pressure in the system is then slowly released via a flow control valve until it reaches the pressure at which the liquified hydrocarbon boils; a foam is then created in a similar manner to that occurring in practice. The nature and collapse rate of the foam in the absence and presence of antifoams can be observed through the walls of the quartz vessel. The anti-foams are introduced via a purpose-built injection cap [33].

Humphries and co-workers [34] have patented an alternative method of determining the foaming characteristics of crude oils under simulated process conditions. Their method uses reconstituted live crude oil obtained by saturating stabilized oil with gas under high pressure and comprises passing this gas-saturated oil into a flash separator. The amount of foam formed is then determined with respect to the amount of gas withdrawn from the separator and the amount of oil contained therein. A full description of this technique is given in the patent specification [34].

3.   Methods Applicable to Fuel and Lubricating Oils

The standard method for assessing the foaming characteristics of lubricating oils is the ASTM D892-IP 146 test. In this test, an apparatus of the type

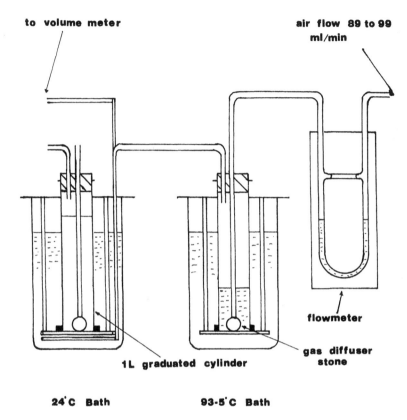

to volume meter

air flow 89 to 99 ml/min

1 L graduated cylinder

flowmeter

gas diffuser stone

24°C Bath

93-5°C Bath

FIG. 12 Apparatus for ASTM D892-IP 146 foam test.

shown in Fig. 12 is employed. It comprises a graduated glass cylinder and an air inlet tube to the bottom of which is fastened a spherical, fused crystalline alumina diffuser stone of specified maximum pore diameter (>80 μm) and air permeability (3 to 6 L/min at 250 mm water pressure). Air is blown through the inlet tube into the oil under test for a period of 5 min, the oil being maintained at 24°C. At the end of this period, the air supply is shut off and the foam volume recorded. The sample is then allowed to stand for 10 min and the foam volume again recorded. The test is then repeated on a second sample at 93.5°C, and then, after collapsing the foam, at 24°C. The foam volume measured after a specified time is reported as the foaming tendency (at end of 5 min blowing period) or foam stability (at end of 10 min settling period) in milliliters. However, this test has been shown by Watkins [35] to be intrinsically prone to a lack of reproducibility. To overcome this problem, Watkins has devised an improved foam test [35] based on the use of a conical, as opposed to a cylindrical, foam vessel. His results showed that the conical geometry gave good repeatability. Recently, Ross and Suzin have reexamined Watkins' hypothesis and concluded that his results are sound. Furthermore, they have highlighted two points of immense practical importance; namely, that the shape of the vessel used has a positive influence on foam flooding and that the propensity toward flooding is high for foam formed in a cylindrical vessel. Thus, it would seem that cylindrical vessels are a poor choice for foaming tests if repeatability of the test is a key factor.

4. Other Test Methods

In addition to the methods already discussed, numerous other attempts [37–47] have been made to devise a reliable method of assessing foaming characteristics. While each of these alternative methods has its own specific advantages, no single method has yet been found which can be regarded as ideal for all situations. This is perhaps not too surprising since, according to Bikerman [25], "an ideal measurement of the foaminess would result in a number independent of the apparatus and the procedure being employed and being characteristic for the solution tested (at a given temperature, etc.), as say, viscosity or surface tension." However, if Bikerman's ideal could be realized then the problem of characterizing and comparing foaming systems would be greatly simplified.

IV. THE RELATIONSHIP BETWEEN PHYSICAL PROPERTIES AND FOAM STABILITY

The stability of nonaqueous foams has been shown by numerous authors [2–5,8,26] to be related to the following physicochemical properties:

1. the bulk rheological properties of the foaming liquid
2. the nature of the filler gas
3. the composition of the foaming system
4. the shear and dilatational properties of the gas-liquid interface

Of these four factors, the last three have been found to play the most significant roles in antifoam selection.

A. Bulk Rheological Properties

The drainage of liquid from a single foam film can be likened to that of liquid from between parallel plates. Using this analogy, it can readily be seen that the approach toward a critical film thickness will be governed by the bulk viscosity of the foaming liquid. Obviously, highly viscous liquids will take longer to drain down to a given thickness than less viscous liquids. Consequently, highly viscous liquids will give rise to the longer-lived foams, all other factors being equal. Thus, it is perhaps not too surprising that for stabilized crude oil foams, it has been found [26] that the average foam lifetime L_F is linearly related to the crude oil bulk viscosity in the following manner:

$$L_F = a\eta_B + b \qquad (5)$$

where

$L_F =$ the average foam lifetime in s
$\eta_B =$ the bulk viscosity of the crude oil in mPa·s

and a,b are the constants characteristic of the filler gas used to create the foam; e.g., for natural gas, $a = 1.8$ and $b = 7.7$.

Furthermore, it has been found that this relationship holds for all temperatures up to at least 65° C [48]. Since the bulk viscosity of crude oils (and indeed most other liquids) decreases with an increase in temperature, it is expected that the foam stability of such oils should follow a similar pattern. This is, in fact, found to be the case for stabilized crude oils. However, for live crude oils, the situation is more complex since the volume of gas evolved as the live oil is depressurized increases with increasing temperature. This enhanced evolution of gas causes the production of not only a larger quantity of foam but also, in many instances, of a different type of foam to that produced at lower temperatures. Hence, in assessing crude oil foam stability one needs to ensure that both stabilized and live oil foam tests are carried out at the relevant operating conditions.

B. Nature of the Filler Gas

Although bulk viscosity is a key factor in determining foam lifetime, it alone does not control foam collapse rate. Lawrence and co-workers [49] have investigated the mechanism of foam collapse and shown the relative importance of both liquid drainage and interbubble gas diffusion. Their work clearly shows that interbubble gas diffusion plays an important part in determining foam collapse rates and that the nature of the filler gas is of prime importance in this process. Sharovarnikov and co-workers [50] have also investigated the role of filler gas type in determining foam stability. These authors [50] found that for aqueous foams the stability decreased with increasing filler gas solubility in the foaming liquid. Callaghan and Neustadter [26] found the reverse effect for crude oil foams, in that stability increased on going from air as the filler gas to natural gas. This latter finding seems to be counterintuitive since the relative solubilities of air and natural gas in crude oil suggest that air-filled foams should have the greater stability. It has been suggested [26] that this difference may be due to wax particles being produced more readily at the gas-liquid interface in the case of natural gas foams, leading to enhanced stabilization. Attempts to test this hypothesis have proved to be inconclusive: undoubtedly this anomaly needs to be investigated in more detail.

C. Composition of the Foaming System

The composition of a crude oil has been shown [8] to have a marked influence on its foamability. Callaghan and co-workers [8] in a series of experiments with 16 different crude oils showed that the major foam-stabilizing agents were heavy metal salts of short-chain carboxylic acids and alkyl phenols of molecular weight <400. Furthermore, they showed that this same suite of chemicals was responsible for the foaming characteristics of a wide range of crude oils from many different sources. In many instances, this fraction of the crude oil responsible for its foam behavior represented no more than 0.02% w/w of the whole crude.

D. Interfacial Rheological Properties

The rheological properties of interfacial films plays a major role in the stabilization of foams. Prins and van den Tempel [51] have shown that a stable foam possesses both a high surface dilatational viscosity and elasticity. This has also been shown to be true for a limited number of stable crude oil foams [15,27]. Furthermore, it has been shown [15] that the dilatational properties of the crude oil-gas interface can be used to distinguish between profoaming and antifoaming chemicals. The surface rheological relaxation spectra derived from such interfacial dilatational measurements are useful in fingerprinting antifoam chemicals, thereby enabling the profoaming components of an antifoam composition to be identified.

1. Techniques Used to Measure Interfacial Rheological Properties

a. Interfacial Shear Rheology Measurements.   Numerous techniques have been suggested for the measurement of shear rheological properties of gas-liquid interfaces (see, for example, Joly [52]). In the case of nonaqueous systems, the technique described by Mannheimer [11] has been shown to be useful [26]. Basically, Mannheimer's rheometer consists of an annular canal with a moving floor. The surface shear viscosity is determined by comparing the flow times of a tracer particle (e.g., talc or PTFE) placed on the surface of a pure liquid at the centerline of the canal (see Fig. 13) to that of a similar particle placed on the surface of a crude oil in a similar

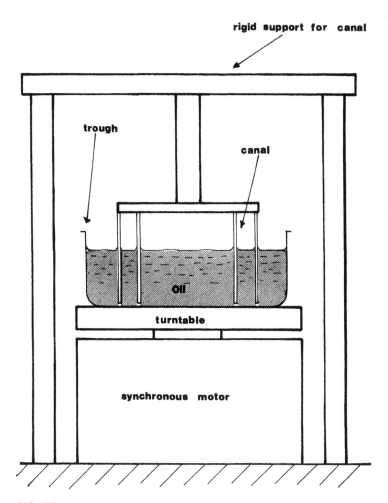

FIG. 13 Canal viscometer.

experiment. The glass trough containing the liquid (reference or crude oil) is rotated at a constant speed, and the flow times determined for complete revolutions of the particle in each case. The dimensions of the canal are chosen so that the simplified equation of Burton and Mannheimer [13] can be used, namely

$$\eta_s = \frac{Y_0 \eta_B}{\pi} \left(\frac{t_1}{t_i^* - 1} \right) \tag{6}$$

where

Y_0 = canal width in cm
η_B = bulk viscosity of the oil in mPa·s
t_1 = time in seconds for the tracer particle to complete one revolution on crude oil
t_1^* = time taken for the same tracer particle to complete one revolution on a pure liquid with the same bulk viscosity as the crude oil
η_S = surface shear viscosity of the crude oil in g/s

Using this method, it has been shown [26] that crude oil foam stability increases with increasing surface shear viscosity.

Alternative methods of measuring surface shear viscosities have been proposed by Davies [53], Oh and Slattery [54], and de Feijter and Benjamins [55].

b. *Interfacial Dilatational Measurements.* The dilatational modulus, ε^* is defined as the surface tension increment $(d\gamma)$ per unit fractional surface area change (dA/A), namely,

$$\epsilon^* = A \frac{d\gamma}{dA} \tag{7}$$

This parameter is of both fundamental and applied interest in foam/colloid science. Normally, data on this parameter are obtained from the properties of longitudinal surface waves or, more directly, from the periodic surface tension variation resulting from an imposed areal variation. Since ϵ^* is a complex function of frequency, a scanning of the system at a number of frequencies is usually involved in its determination [56]. However, Loglio and co-workers [57] have developed a single pulsed Fourier transform (FT) technique to obtain ϵ^* as a function of frequency. In Loglio's method, measurements are made on a Langmuir trough (see Fig. 14) by expanding (or contracting) rapidly the interfacial area by a known amount and recording the recovery of the surface pressure with time. The resulting surface pressure/time curve is then analysed using the FT method to give both the surface dilatational elasticity and viscosity over a range of frequencies. An alternative method of obtaining such dilatational data is the pulsed-bubble (drop) technique as used by Lunkenheimer and co-workers [58] and Clint and co-workers [59].

Callaghan and co-workers [16,26] have used the Langmuir technique of Loglio to obtain surface dilatational data on both crude oil and crude oil/antifoam systems. Using such data, these workers were able to differentiate between profoamers and antifoams, and were also able to optimize antifoam dose rates. They found that profoamers increased the dilatational elasticity, whereas antifoams decreased it markedly. Furthermore, they found that the surface dilatational relaxation spectrum of a crude oil/antifoam system was a sensitive tool in dose optimization and could be used to define profoaming limits.

V. ANTIFOAM TYPES AND METHODS OF SELECTION

A. Antifoam Types

The principal antifoam chemicals used in the treatment of nonaqueous foams in the oil industry (see Pape [60]) are polydimethylsiloxanes (PDMS), fluorosilicones, and silicone glycols. The basic structures of these materials are given in Fig. 15.

The polydimethylsiloxane antifoams are available as 100% active odorless fluids; as dispersions in a range of nonaqueous carrier solvents; as compounded materials consisting of silica dispersed in PDMS fluid; and as aqueous emulsions. Of these various types of PDMS antifoam, the 100% fluids and dispersions are the preferred forms for oil industry use, since they have low solubility in hydrocarbon liquids and are effective at low concentrations. In general terms, the 12,500 and 60,000 mm^2/s grades of PDMS fluid are most useful in oil production and refining operations, whereas the lower-viscosity (ca. 100 mm^2/s) fluids have some utility in oil products. The compounded materials do not appear to be as useful in crude oil operations as they are in aqueous systems. Indeed, it is questionable whether or not the presence of particles is necessary for effective nonaqueous foam control. This latter point may be worthy of further investigation. For further information on the PDMS antifoams the reader is referred to the patent literature (see, for example, Refs. 61–64).

The fluorosilicone materials are also available as 100% fluids and dispersions. These products are useful in aggressive environments, such as sour crude treatment systems, where their greater resistance to chemical attack and solubilization makes them highly effective. They are also useful in situations where low dosage rates are desirable. In use, these materials appear to be less sensitive to the proportion of light end fractions in crude oil than the PDMS antifoams. Examples of this class of antifoam can be found in Refs. 65–68.

Silicone glycol antifoams [69] tend to be provided in 100% fluid form, although they are also available as dispersions in carrier solvents. This class of antifoams is proving to have utility in the control of diesel fuel foaming [70].

Another class of antifoam used in the oil industry are the alkyl polyacrylates [71]. These materials are usually supplied in a carrier solvent such

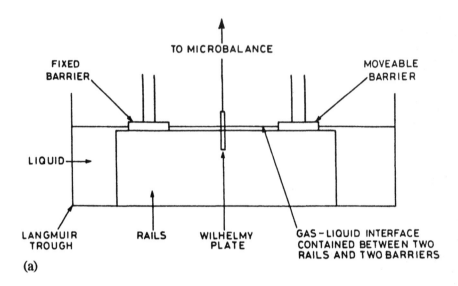

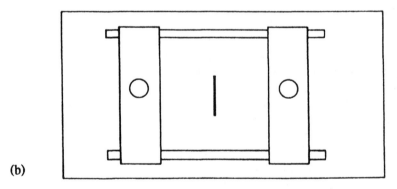

FIG. 14 Langmuir trough for dilatational rheology measurements: (a) viewed from the side; (b) viewed from above.

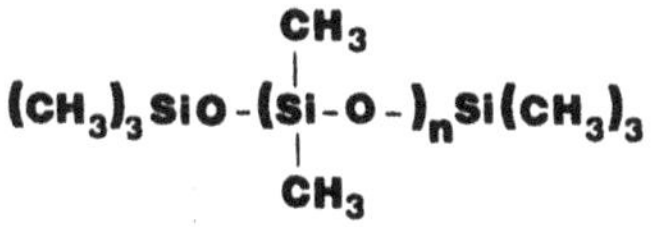

$$(CH_3)_3 SiO - (\underset{\underset{\displaystyle CH_3}{|}}{\overset{\overset{\displaystyle CH_3}{|}}{Si}} - O -)_n Si(CH_3)_3$$

POLYDIMETHYLSILOXANES

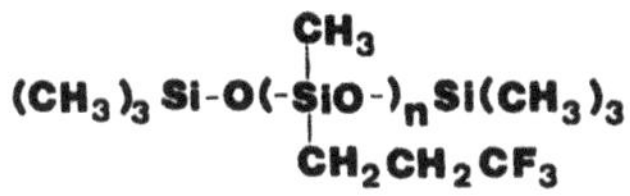

$$(CH_3)_3 Si-O(-\underset{\underset{\displaystyle CH_2CH_2CF_3}{|}}{\overset{\overset{\displaystyle CH_3}{|}}{Si}}O-)_n Si(CH_3)_3$$

FLUOROSILICONES

$$(CH_3)_3 SiO(\underset{\underset{\displaystyle CH_3}{|}}{\overset{\overset{\displaystyle CH_3}{|}}{Si}}-O)_m (\underset{\underset{\displaystyle R}{|}}{\overset{\overset{\displaystyle CH_3}{|}}{Si}}-O)_n Si(CH_3)_3$$

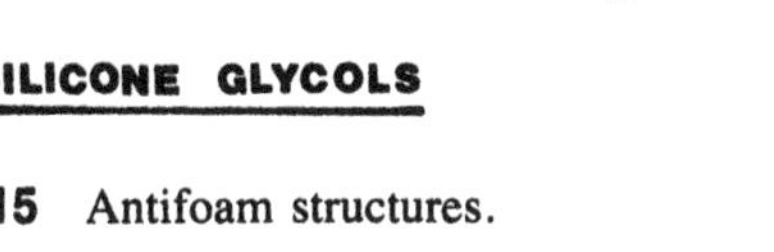

where $R = -CH_2CH_2CH_2O(C_2H_4O)_x H$, for example.

SILICONE GLYCOLS

FIG. 15 Antifoam structures.

as a petroleum distillate. Their particular area of use is in oil products such as lubricating oils where their good air-release characteristics make them more attractive than PDMS materials. It is possible that the better air release observed with these materials is a direct consequence of their poorer antifoam performance when compared to silicone oils. The polyacrylates have a much less pronounced effect on surface elasticity than the PDMS materials and, as a result, are less likely to reduce the beneficial circulatory flows in the rising bubble surface which aid air release.

TABLE 1 Antifoams and Diluents Commonly Used in Oil Industry Processes

Antifoam type	Application(s)	Diluent(s)
Polydimethylsilicone (1,000, 12,500, or 60,000 mm²/s grades)	Gas-oil separation Vacuum distillation Delayed coking Crude distillation Asphalt extraction Visbreaking	White spirit Kerosene Diesel oil Toluene Xylene MEK pet. ether
Polydimethylsilicone (<1,000 mm²/s grades)	Lube oils	As above
Compound silicone (silica dispersed in silicone oil)	Propane deasphalting	Toluene Naphtha Amyl acetate
Fluorosilicone	Diesel fuel Sour crude processing	2-Ethoxyethyl acetate Acetone MEK White spirit
Silicone glycol	Diesel fuel	Toluene Gas oil
Alkyl polyacrylate	Lube oil	Petroleum distillate

Table 1 gives an outline of the antifoam and diluent types used for a number of oil industry applications. The influence of added solvents on antifoam performance has been discussed by Callaghan and co-workers [72].

B. Selection of Antifoams

In order to be an effective antifoam, a material must meet four basic requirements:

1. It must be insoluble in the foaming medium under the relevant process conditions.
2. It must have a lower surface tension than the foaming medium.
3. It must be rapidly dispersed in the foaming medium.
4. It must be chemically inert.

Insolubility of the antifoam compound in the foaming medium is very important, since an antifoam with its low surface tension will actually become a profoamer if soluble in the foaming medium. Silicones with their low surface tension and low interfacial tension against hydrocarbon oils, are ideally suited to enter and spread in nonaqueous systems and, as such, are ideal antifoams. However, the selection of the precise silicone required needs to be carried out carefully so as to avoid any possible profoaming problems.

The preferred method for selecting an antifoam for a specific process is as follows:

a. Determine the foam type in a process simulator.
b. Measure the surface tension of the foaming liquid, preferably under simulated operating conditions.
c. Rapidly screen a number of potential antifoam candidates, using surface tension as an initial sifting tool, using the Bikerman sparge tube described earlier.
d. Determine the surface rheological spectra of the best candidates selected in (c).
e. Optimize dosage, and define profoaming limits (if any), using both surface rheological and Bikerman foamability measurements.
f. Finally, check the performance and profoaming limit of the chosen antifoam in a process simulator prior to recommendation.

The effect of both profoamers and antifoams on the surface rheological relaxation spectra of crude oil-gas interfaces has been described in Ref. 16.

VI. FUTURE DIRECTIONS

In general terms, the antifoams currently available can tackle most oil industry problems satisfactorily. However, two areas where improvements could still be made are sour crude oil defoaming and inhibiting foam formation in oil products, especially diesel fuel and lubricating oils. In the latter case, the problem is to produce an antifoam that is effective at low (<5 ppm) concentration and yet does not adversely affect the oil's air-release characteristics.

ACKNOWLEDGMENTS

I would like to thank The British Petroleum Co. plc for permission to publish this chapter, and my colleagues, Messrs. B. J. Oswald and G. J. J. Jayne, for their counsel during its preparation.

REFERENCES

1. S. Ross, *Chem. Eng. Prog. 63*: 41–47 (1967).
2. A. P. Brady and S. Ross, *J. Am. Chem. Soc. 66*: 1348–1356 (1944).
3. J. W. McBain and J. V. Robinson, Natl. Advisory Comm. Aeronaut., Tech. Note No. 1844, 1949.

4. J. W. McBain, S. Ross, A. P. Brady, J. V. Robinson, I. M. Abrams, R. C. Thorburn, and C. G. Lindquist, Natl. Advisory Comm. Aeronaut., War time Report, ARR no. 4105, 1944.

5. S. Ross and R. M. Haak, J. Phys. Chem., 62: 1260–1264 (1958).

6. R. J. Mannheimer, and R. S. Schecter, J. Colloid Interface Sci. 32: 212–224 (1970).

7. A. Scheludko and E. Manev, Trans. Faraday Soc. 64: 1123–1134 (1968).

8. I. C. Callaghan, A. L. McKechnie, J. E. Ray, and J. C. Wainwright, Soc. Petrol. Eng. J., 171–175 (1985).

9. S. Ross and J. W. McBain, Ind. Eng. Chem. 36: 570–573 (1944).

10. I. B. Ivanov, B. Radoev, E. Manev, and A. Scheludko, Trans. Faraday Soc. 66: 1262–1273 (1970).

11. R. J. Mannheimer, A. I. Ch. Eng. J. 15: 88–93 (1969).

12. J. W. McBain, J. V. Robinson, W. W. Woods, and I. M. Abrams, Natl. Advisory Comm. Aeronaut., Tech. Note No. 1845, 1949.

13. R. A. Burton and R. J. Mannheimer, Adv. Chem. Ser. 63: 315–328 (1967).

14. J. Lucassen and M. van den Tempel, Chem. Eng. Sci. 27: 1283–1291 (1972).

15. G. Loglio, U. Tesei, and R. Cini, J. Colloid Interface Sci. 71: 316–320 (1979).

16. I. C. Callaghan, C. M. Gould, R. J. Hamilton, and E. L. Neustadter, Colloids Surf. 8: 17–28 (1983).

17. G. Loglio, U. Tesei, and R. Cini, J. Colloid Interface Sci. 100: 393–396 (1984).

18. J. V. Robinson, and W. W. Woods, J. Soc. Chem. Ind. 67: 361–365 (1948).

19. S. Ross, J. Phys. Colloid Chem. 54: 429–436 (1950).

20. W. D. Harkins and E. Boyd, J. Phys. Chem. 45: 20–43 (1941).

21. L. T. Shearer and W. W. Akers, J. Phys. Chem. 62: 1269–1270 (1958).

22. A. D. Barber and E. F. Wijn, Proc. 3rd International Symposium on Distillation, 2: 3.1/15–3.1/35, 1979.

23. E. Manegold, Schaum, strassenbau, chemie und technik, 83, Heidelberg (1953).

24. I. C. Callaghan, C. M. Gould, A. Reid, and D. H. Seaton, J. Petrol. Tech. 37: 2211–2218 (1985).

25. J. J. Bikerman, Foams, Springer-Verlag, New York, 1973.

26. I. C. Callaghan and E. L. Neustadter, Chem. Ind. 53–57 (1981).

27. I. C. Callaghan and C. M. Gould (assigned to The British Petroleum Co. plc), GB 2149127A; June 5, 1985; filed October 29, 1984.

28. S. Ross and G. M. Nishioka, J. Colloid Interface Sci. 81: 1–7 (1981).

29. S. Ross Am. J. Phys. 46: 513–516 (1978).

30. M. A. Camp and F. T. Lawrence (assigned to The British Petroleum Co. plc), GB 2158574A; November 13, 1985; filed May 2, 1985.

31. S. Ross and M. J. Cutillas, J. Phys. Chem. 59: 863–866 (1955).

32. H. Lattek, H.-F. Fink, and G. Koerner (assigned to Th. Goldschmidt AG), GB 2154738A; September 11, 1985; filed December 14, 1985.

33. F. T. Lawrence (assigned to The British Petroleum Co. plc), GB 2176602A; December 31, 1986; filed June 12, 1986.

34. C. L. Humphries, E. F. Schultz, and A. M. Winkelman (assigned to Mobil Oil Corporation), US 4426879; January 24, 1984; June 21, 1982.

35. R. C. Watkins, *J. Inst. Petrol. 59*: 106–113 (1973).
36. S. Ross and Y. Suzin, *Langmuir 1*: 145–149 (1985).
37. E. G. King, *J. Phys. Chem. 48*: 141–154 (1944).
38. G. L. Clark and S. Ross, *Ind. Eng. Chem. 32*: 1594–1598 (1940).
39. M. A. Amerine and L. P. Martin, *Ind. Eng. Chem. 34*: 152–157 (1942).
40. P. P. Gray and I. Stone, *Wallerstein Labs. Commun. 3*: 159–171 (1940).
41. E. J. Burcik, *J. Colloid Interface Sci. 5*: 421–436 (1950).
42. W. B. Hardy, *Proc. Roy. Soc. (London) 86A*: 610–635 (1912).
43. W. E. Glausser, *Chem. Eng. Prog. 60*: 67–68 (1964).
44. S. Hartland, and A. D. Barber, *Trans. Inst. Chem. Eng. 52*: 43–52 (1974).
45. H. Tsuge, J. Ushida, and S. Hibino, *J. Colloid Interface Sci. 100*: 175–184 (1984).
46. Deutsche Industrie Norm (DIN) Test no. 53902, Teil 1, 1981.
47. S. Okazaki and S. Sasaki; *Tenside 3*: 115–118 (1966).
48. I. C. Callaghan and F. T. Lawrence, unpublished results.
49. I. C. Callaghan, F. T. Lawrence, and P. M. Melton, *Colloid Polym. Sci. 264*: 423–434 (1986).
50. A. F. Sharovarnikov, V. N. Tsap, A. Ya. Korol'chenko, and A. V. Ivanov, *Kolloid. Zh. 43*: 808–812 (1981).
51. A. Prins and M. van den Tempel, Proc. IVth Int. Congress Surface Active Subst., 2: 1119–1131, 1964.
52. M. Joly, *Surface Colloid Sci. 5*: 1–77 (1972).
53. J. T. Davies, Proc. Int. Congress Surface Activity, 2nd., London, 1: 220–224 (1957).
54. S.-G. Oh and J. C. Slattery, *J. Colloid Interface Sci. 67*: 516–525 (1978).
55. J. A. de Feijter and J. Benjamins, *J. Colloid Interface Sci. 70*: 375–382 (1979).
56. J. Lucassen and M. van den Tempel, *Chem. Eng. Sci. 27*: 1283–1291 (1972).
57. G. Loglio, U. Tesei, and R. Cini, *Ber. Bunsenges. Phys. Chem. 81*: 1154–1156 (1977).
58. K. Lunkenheimer, K.-D. Wantke, and R. Miller, *ABH Akad. Wiss. DDR*, 445–453 (1976).
59. J. H. Clint, E. L. Neustadter, and T. J. Jones, *Dev. Petrol. Sci. 13*: 135–148 (1981).
60. P. G. Pape, *J. Petrol. Tech 35*: 1197–1204 (1983).
61. I. C. Callaghan, H. F. Fink, C. M. Gould, G. Koerner, H-J. Patzke, and C. Weitmeyer (assigned to The British Petroleum Co. plc and Th. Goldschmidt AG), US 4564665; January 14, 1986; filed April 5, 1983.
62. R. E. Moeller (assigned to General Electric Co.), GB 1523654A; September 6, 1978; filed October 21, 1975.
63. A. Guillaume and F. Sagi (assigned to Rhone-Poulenc Industries), GB 1519541A; August 2, 1978; filed June 6, 1975.
64. K. W. Farminer (assigned to Dow Corning Corp.), FR 2379307; September 1, 1978; filed July 4, 1977.
65. J. W. Keil (assigned to Dow Corning Corp.), US 4537677; August 27, 1985; filed November 5, 1984.

66. H. F. Fink, G. Koerner, R. Berger, and C. Weitmeyer (assigned to Th. Gold-schmidt AG), DE 3635093C; March 10, 1988; filed October 15, 1986.

67. D. Boerner, H. F. Fink, G. Koerner, and G. Rossmy (assigned to Th. Gold-schmidt AG), DE 2444073B; August 19, 1976; filed March 25, 1976.

68. E. R. Evans (assigned to General Electric Co.), US 4329528; May 11, 1982; filed December 1, 1980.

69. I. C. Callaghan, C. M. Gould, and W. Grabowski (assigned to The British Petroleum Co. plc), US 4711714; December 8, 1987; June 21, 1985.

70. Dow Corning Corp., US 4690688; September 1, 1987; filed July 7, 1986.

71. J. E. Fields, (assigned to the Monsanto Co.), US 3166508; January 19, 1965; filed January 16, 1963.

72. I. C. Callaghan, S. A. Hickman, F. T. Lawrence, and P. M. Melton, RSC Special Publication No. 59, 48–57 (1986).

3
Defoaming in the Pulp and Paper Industry

S. LEE ALLEN   McGill University, Montreal, Quebec, Canada

LAWRENCE H. ALLEN   Pulp and Paper Research Institute of Canada, Pointe Claire, Quebec, Canada

TED H. FLAHERTY   Dorset Industrial Chemicals, Ltd., Chateauguay, Quebec, Canada

I.	Overview of the Industry	151
II.	Sources of Foam	155
III.	Problems Caused by Foam	159
IV.	Physical Control of Foam	161
V.	Chemical Control of Foam	164
	A. Early antifoams	164
	B. Development of brownstock washer antifoams	164
	C. Other pulp and paper antifoams	168
	D. Disadvantages of chemical defoaming	169
VI.	Measurement of Foam	170
VII.	Testing of Antifoams	171
VIII.	Handling and Use of Antifoams	171
IX.	Future Research and Development	173
	References	173

I. OVERVIEW OF THE INDUSTRY

The pulp and paper industry is the world's biggest single user of defoaming agents. Forecasts of pulp and paper capacities [1] and typical antifoam use rates for various processes [2] suggest that a conservative estimate of world-wide antifoam consumption in 1990 is 160,000 tonnes. Indeed, antifoams are the largest single class of process control chemicals in the industry [2].

Theoretically, the production of paper from wood is simple. Wood fibers are dispersed chemically or mechanically and suspended in water to form a

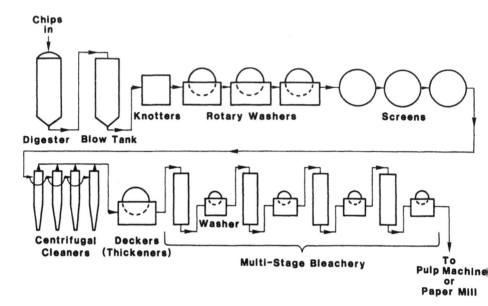

FIG. 1 Flowchart of a typical bleached kraft pulp mill. Only pulp flow is indicated.

pulp slurry. The slurry is deposited on a fine screen, creating a fiber web which is drained and dried, resulting in paper. In practice, the process is considerably more complicated. The manufacture of a single type of paper, in a specific pulp and paper mill, may require dozens of chemicals for pulping, bleaching, process control, and modification of end properties; each different grade of paper will require a different chemical recipe and process conditions.

In chemical pulp production, wood chips are cooked at elevated temperatures in solutions of various chemicals in pressurized vessels called digesters. In sulfite pulping the cooking solution is composed of sulfur dioxide and a metal ion hydroxide (Ca^{2+}, Mg^{2+}, Na^+; also NH_4^+); depending on the process, the conditions range from pH 1 and 120°C to pH 9 and 180°C. The kraft process uses sodium hydroxide and sodium sulfide solutions, at 180°C and pH above 13. A typical kraft pulp mill is outlined in Fig. 1.

In both processes, the chemicals soften and degrade the lignin (Fig. 2) (the material that binds the fibers together in the wood), liberating the individual fibers. After pulping, the cooked wood chips are usually "blown" from the digester under pressure into a blowtank. From there, the pulp proceeds to the knotters, which remove the largest pieces of undispersed material. The pulp then typically goes to a series of rotating drum washers, where it forms a thick mat on the screen which forms the surface of each

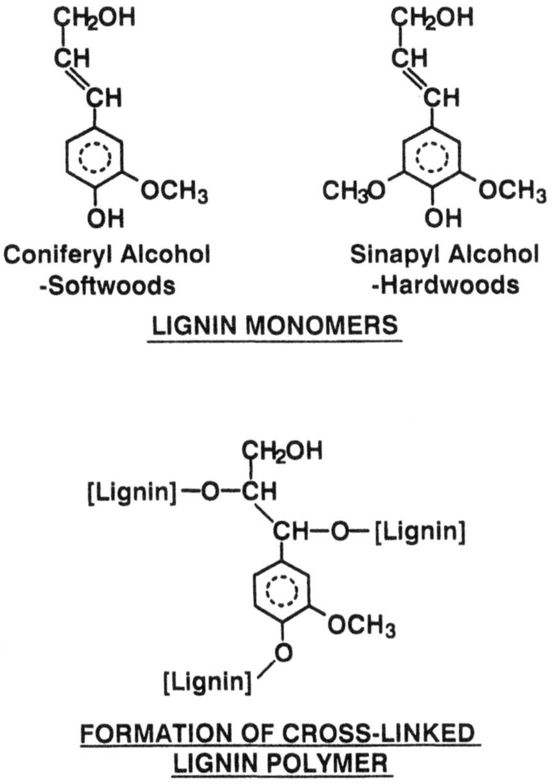
Coniferyl Alcohol
-Softwoods
Sinapyl Alcohol
-Hardwoods
LIGNIN MONOMERS
[Lignin]—O—CH
CH—O—[Lignin]
[Lignin]
FORMATION OF CROSS-LINKED
LIGNIN POLYMER

FIG. 2 Representative structural formulae of lignin monomers and lignin.

drum. The spent cooking liquor (black liquor) is drained off (usually by suction applied from inside the washer drum), as the pulp mat is sprayed with recycled process water.

It is here in the washing of unbleached pulp that over 60% of the defoaming agents are used [2]. To make the kraft process economically viable, spent pulping chemicals must be recovered as efficiently as possible; hence, thorough washing is essential. In addition, in pulp that will be bleached, good washing is necessary to remove undesirable components that would otherwise consume bleaching chemicals. Foam hampers the washing process, and must be controlled to ensure efficient washing and good chemical recovery. Ironically, it is usually during the washing of the kraft brownstock (freshly cooked pulp) that the worst foam problems are encountered. Antifoam may be added immediately before the washers, at the knotters, or at any other appropriate point in the washing sequence. Antifoam can also be

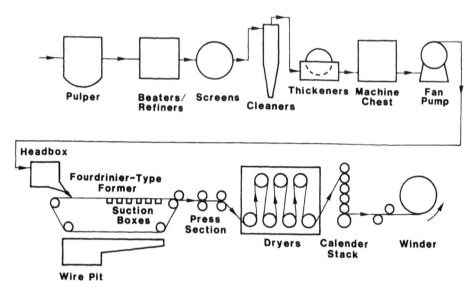

FIG. 3 Flow chart of a typical fine paper mill. Only pulp/paper flow is indicated.

sprayed across the pulp mat on the washers. A typical use rate is about 1 kg/t dry pulp.

In the screen room, screens permit individual fibers to pass through, but fiber bundles (shives) are retained. Antifoam is sometimes required. Extra makeup water is added at the screens, and the screen room filtrate is customarily recycled back to the washers. The pulp then goes to the cleaners, where small dirt specks are centrifugally separated from the fibers.

If a white pulp is desired, the pulp proceeds to the bleach plant, where residual lignin and color are removed by sequential treatments with chemicals such as oxygen, chlorine, sodium hydroxide, hydrogen peroxide, chlorine dioxide, and sodium hypochlorite. After each stage of bleaching, the pulp is usually washed to remove residual chemicals. Foam problems may be encountered here too.

The bleached pulp may be further screened and cleaned, and sent directly to the paper mill; alternatively, it may go to a pulp-drying machine, if it is to be sold as pulp. On the pulp machine the pulp slurry is distributed continuously onto a mesh, either a horizontal mesh belt or a drum resembling a washer, where it forms a thick fiber mat. The pulp mat is then dewatered, dried, and baled for shipment or storage.

In the paper mill (Fig. 3), the pulp may be beaten or refined to increase strength in the final product. Then additives are mixed into the pulp slurry.

These vary with the type of paper being made; a list of the more common ones is given in Sec. II. Such additives frequently cause, or stabilize, foam; also, air may be entrained when they are mixed into the pulp suspension.

On the paper machine, the pulp enters the headbox, where it is agitated for optimum dispersion. The pulp is dispensed through a slit ("slice") in the headbox onto a moving screen ("wire") on which the web is formed and partly dewatered. The wet web is further dewatered in the nips of the press section, and is dried on the steam-heated cylinders of the dryer section. The paper may then be calendered (to increase surface smoothness) and/or coated. Foam may be generated in any of the wet stages of papermaking and coating, and must be controlled to maintain uniform quality in the finished product.

In contrast to chemical pulping, mechanical pulping separates the wood into its component fibers by pressing logs against rotating grindstones, or by refining wood chips with disc refiners (which damages the fibers less than grinding). The lignin is not dissolved, remaining largely with the fibers. Mechanical pulps are not usually washed, but are screened and cleaned. Although foam problems can be severe, they are generally less than with chemical pulps.

To reduce fiber damage during mechanical pulping, the lignin in the wood may be softened before refining. In thermomechanical pulping, the chips are heated with pressurized steam, then immediately sent hot to the refiners. In chemimechanical pulping, a partial chemical cooking precedes refining.

II. SOURCES OF FOAM

The three materials necessary for foam production are a liquid, a gas, and a surfactant. In the pulp and paper industry, the liquid is water, the gas is air, and the surfactants are numerous.

The major source of surfactants in the pulp mill is the wood itself. A typical softwood supplied to a mill might consist of approximately 45–50% fiber, 20–30% lignin, 20–35% miscellaneous carbohydrates, and 1–8% wood resin, based on oven-dried weight [3,4]. The worst foaming generally occurs after chemical pulping, where the spent pulping liquors must be washed from the fibers. Table 1 gives examples of compositions of spent pulping liquors; the surface active materials are found predominantly in the lignin and wood resin fractions.

Chemical additives also add to the surfactant load, chiefly in the paper mill. Table 2 summarizes the main classes of additives and the types of chemicals used in each class. No single papermaking furnish (the mixture of pulp and chemicals supplied to the paper machine) will contain all of these additives; for example, wet strength resins are used only for paper that must hold up under damp or humid conditions, such as tissue and towel.

TABLE 1 Compositions of Spent Pulping Liquors

Type of liquor[a]	Class of material	Composition	% Solids[b]
Kraft black liquor 12–15% solids	Lignin	Monomeric and polymeric lignin with a broad range of molecular weights; some substitution; lower molecular weight fraction dissolved	33
	Organic acids	>90% monocarboxylic acids Most polyhydroxy acids	27
	Inorganic components	Spent pulping chemicals, plus small amounts from wood: NaOH, Na_2S, $NaHSO_4$, Na_2SO_4, Na_2CO_3 some K^+, Ca^{2+}, Cl^-	23
	Extractives	Wood resin (60–90% saponified): fatty and resin acids, di- and triglycerides, steryl esters, sterols, flavonoids, etc.	6
	Bound sodium	Sodium bound to lignin, organic acids, and extractives	11
Spent sulfite liquor 11–14% solids	Lignin	Monomeric and polymeric lignosulfonates with a broad range of molecular weights	44
	Carbohydrates	>80% monosaccharides; mainly xylose, mannose, galactose, glucose	23
	Inorganic components	Spent pulping chemicals, plus small amounts from wood (usually sodium salts): $NaHSO_3$, Na_2SO_3, Na_2SO_4, $Na_2S_2O_3$, Na_2CO_3 depending on process, Mg^{2+} or Ca^{2+} salts K^+, Cl^-	16
	Organic acids	Acetic and aldonic acids	7
	Extractives	Wood resin (not saponified)	7
	Bound sodium (calcium, magnesium)	Sodium (calcium, magnesium) bound to lignosulfonates	3

[a]Material exiting from digester = 7–10% fiber, balance spent liquor.
[b]Percent composition for specific examples only. Actual percentages will vary greatly depending on the wood and the pulping conditions.
Source: Ref. 5.

TABLE 2 Paper Mill Additives

Class	Typical chemicals used
Fillers	TiO$_2$ Clay (aluminosilicate platelets with negative faces and positive edges; usually kaolin)
Retention aids	Polyacrylamide Polyethyleneimine Polyethyleneoxide Polyacrylic acid

Wet strength resins	Cationic resins
	  urea formaldehyde
	  melamine formaldehyde
	  polyamideamine epichlorhydrin

Starch
(dry strength)
	Polymers of glucose
	  amylose (linear)
	  amylopectin (branched)
	  may be anionic or cationic

Internal size (water repellency for good printing)	Rosin size (mixture of resin acids) Fortified rosin size (maleic anhydride or fumaric acid added)
Alkaline size	Alkyl ketene dimer Alkenyl succinic anhydride
Slimicides	Organic biocides Cl$_2$, ClO$_2$, NaClO

Resin deposition
(pitch) control additives
	Alum
	Sodium aluminate
	Talc
	Dispersants:
	  pleuronics (ethyleneoxide/
	    propylene-oxide copolymers)
	  alkylphenol/ethyleneoxide
	    condensates
	  aliphatic or aromatic sulfated or sulfonated esters, oils, and phenolic derivatives
	  polyacrylic acid

Dyes and pigments	Acid dyes (require alum)
	Basic dyes
	Direct dyes
	Pigments
	  dispersions of colored particles
	  TiO$_2$ (white)

TABLE 3 Example of Paper Mill Formulation: Coated Publication-Grade Paper from Groundwood and Kraft Pulps

Additive	Use	kg/tonne of paper
Clay	Filler; opacifier	225
Cationic starch	Retention; strength	43
Talc	Control of deposits (pitch control)	20
Alum	pH control; pitch control	5
Sodium hydroxide	pH control	1
Cationic polyacrylamide	Retention	0.3

An example of a papermaking furnish is shown in Table 3; actual formulations vary widely, and depend greatly on the type and grade of paper being made. Many of the additives attach to the fibers and end up in the paper, but certain quantities remain in the white water (the liquid which drains off the fiber web on the paper machine).

Even when the pulp is largely free of additives (e.g., newsprint) the white water contains considerable amounts of dissolved and suspended solids. Table 4 presents analyses of three newsprint white waters, showing substantial quantities of organic and inorganic material dispersed in the water.

Particulates in the process liquid, such as fines (fiber fragments and ray cells that can pass through a paper machine wire) and fillers, although not actually surface active, can stabilize foam under some conditions, or contribute to the residue that dry foam leaves on equipment. Temperature and pH can also influence foam formation and stability, but it is difficult to generalize about their effects because of the complexity and variability of the process liquids.

Cascades and vigorous agitation are among the most obvious ways of introducing air into the stock (process liquid plus pulp). Rapid expulsion of cooked pulp from pressurized digesters causes foam, both by the release of dissolved gases when pressure is reduced, and by uptake of air in the ensuing turbulence. The problem at this point is aggravated by the high concentrations of resin and lignin dissolved in the cooking liquor. Leaking pumps, pipe fittings, and washer seals permit air to enter the system. Addition of cold water to warm stock and heating of cold stock release dissolved gases. Vacuum drainage of process water through screens can also release dissolved air by pressure reduction, and can introduce additional air, if air is drawn through the screen with the water. Dry material mixed directly with the process water can carry air into the stock. The amount of air entrained at a given point will depend on the nature of the process equipment and the

TABLE 4 Compositions of Newsprint White Waters

	Mill A	Mill B	Mill C
Type of pulp	Thermo-mechanical	Thermo-mechanical	72% Stone groundwood 28% Chemi-mechanical
pH	4.8	4.6	4.3
Lignin (g/L)	0.90	0.95	0.15
Dissolved solids (g/L)	3.58	7.00	1.09
Suspended solids (mg/L)	80	198	33
Ca^{2+} (mg/L)	21.8	54.0	14.8
Al^{3+} (mg/L)	1.10	1.82	0.88
Mg^{2+} (mg/L)	5.68	20.9	5.03
Total S (mg/L)	282	370	70
SO_3^- (mg/L)	144	36	nil
SO_4^{2-} (mg/l)	672	954	207
Free sugars (mg/L)	79	74	nil
Hemicellulose (mg/L) (polymeric sugars from degraded cellulose)	879	3852	266
Colloidal wood resin (particles/mL)	2.0×10^6	2.3×10^6	0.13×10^6

Source: Ref. 6.

manner in which it is operated, and cannot be predicted unless one is familiar with on-site conditions.

III. PROBLEMS CAUSED BY FOAM

The most noticeable type of foam is that floating on the surface of the stock. In some respects, this foam is easier to handle, as the problem and its treatment can be monitored readily. Less obvious, and harder to monitor, is foam in the form of small bubbles (1–100 μm in diameter) entrained in the bulk of the liquid or adhering to the solid components of the stock.

Some of the problems caused by foam are obvious. The volume of the foam reduces the effective capacity of pumps and tanks, and thus either reduces output, or cuts down on the residence time in washers, refiners, mixing tanks, bleaching towers, etc. Solids, such as fibers, fines, or fillers, or surface active residues or additives, can be concentrated in surface foam; overflow can then result in loss of material and changes in relative proportions of components, as well as a mess and a possible safety hazard (Fig.

FIG. 4 Overflow of surface foam in a pulp mill.

4). If the level of the surface foam drops, foam left on the walls of a vessel may leave behind deposits that can break off and contaminate the pulp.

As mentioned previously, one of the biggest foam-related problems in chemical pulp (especially kraft) mills is maintaining rate and efficiency of washing on rotary drum washers. As wash water is drawn through the pulp by partial vacuum in the interior of the drum, air bubbles in the stock can give rise to channeling. These channels reduce washing efficiency in two ways: if air passes through the channels, the vacuum is reduced and less black liquor is removed; if wash water from the showers is channeled through the mat, it is less effective at displacing the black liquor from the pulp.

Physical problems are encountered when handling foaming liquids. Pumps can cavitate when attempting to pass large volumes of foam. Metering devices are rendered less accurate. Liquid held in surface foam may not be blended thoroughly into the bulk of the stock during mixing.

In the paper mill, foam may reduce beater and refiner efficiency by cushioning the mechanical action on the pulp [7]. Drainage efficiency on the paper machine is reduced by channeling. Bubbles breaking on the fiber web leave behind traces of solids (fibers, fines, fillers, pigments, dirt, pitch, etc.)

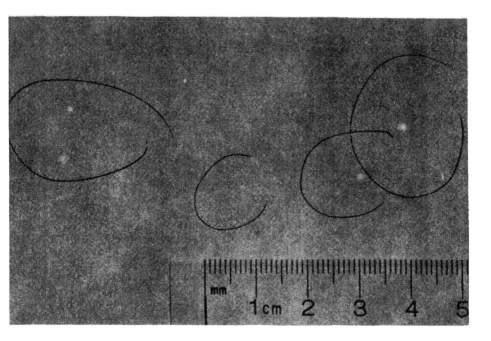

FIG. 5 Foam flaws in fine paper. Small air bubbles in the stock on the paper machine wire result in the appearance of circular thin spots in the final product. The spots were encircled by the quality control technician who rejected the paper.

that mark the sheet or cause it to stick to rolls and rip. If the foam bubbles displace fiber significantly thin spots result, leading to the speckled appearance of the paper shown in Fig. 5; in extreme cases, the web breaks. Foam in paper coating media may result in heterogeneous application and bubble spots (Fig. 6).

Surface foam in effluent is aesthetically unacceptable, and may hamper effluent treatment by impeding settling of particles held in the foam, and by reducing the rate of oxygen uptake by the effluent. Occasionally, foam can be used to advantage by mill personnel; the foam with its high concentration of surfactants (lignin, lignosulfonates, and/or resin [8,9]) can be skimmed off and treated separately.

IV. PHYSICAL CONTROL OF FOAM

The first steps in physical prevention of foam are relatively inexpensive. Elimination of cascades may require only simple repiping; closing up leaks in pumps, fittings, and seals is standard maintenance. Further equipment

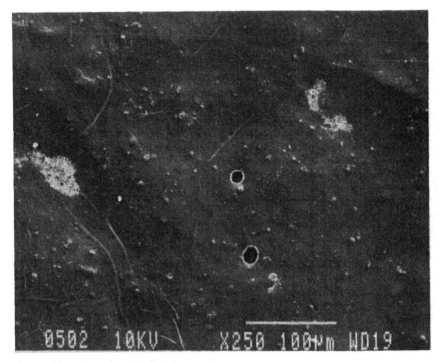

FIG. 6 Scanning electron micrograph of foam flaws in paper coating. White bar at bottom denotes 100 μm. The small white spots are intact dried bubbles in the coating. The dark circular areas are holes in the coating caused when the tops of larger dried bubbles are broken, probably in calendering. (Photograph courtesy of Glynis de Sylveira.)

modifications to avoid air entrainment often involve large capital investments.

The extensive foam problems in kraft brownstock washers, and the consequent costs of chemical antifoams required for foam control, have been reduced by changes in brownstock washing equipment. Most of the new equipment is added to the process line just upstream of the conventional rotary drum washers. Displacement washing in the bottoms of continuous digesters removes significant quantities of foaming materials with the black liquor, without air entrainment. Conventional blowing of the pulp slurry out of batch digesters is effectively a violent cascade; dissipating digester pressure more gradually [10], or ejecting the pulp so that it does not come in contact with air (for example, into the bottom of a vessel filled with stock) avoid this. Diffusion washers [11], in which wash liquor is pumped through the stock, further reduce air uptake, as do pressure knotters. These changes in process equipment decrease the amounts of air and surfactant in the stock,

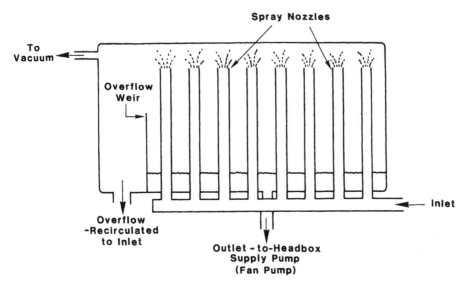

FIG. 7 Schematic of the Deculator vacuum deaerator (manufactured by Clark and Vicario). This apparatus is usually installed just upstream of the paper machine. Stock is sprayed into an evacuated chamber, entrained air boils off, and the deaerated stock passes through the fan pump to the headbox.

and thus reduce the amount of foam (and the need for chemical antifoams) in the washers. Further foam reduction has been reported when drum washers are replaced by horizontal pressure washers [12]. The new machinery designs have been aimed primarily at improving washer efficiency, with consequent improved product quality and decreased chemical losses and bleach chemical use; foam prevention was pursued as much for improved washing as for decreased cost of chemical antifoams.

Innovation has also reached the screen room, where pressure screens are reported to reduce foaming [13].

Commercial apparatus is available which breaks surface foam by beaters, bars, and sonic fields; entrained foam can also be removed by centrifugal force and vacuum deaeration. A sketch of a common vacuum deaerator (the Deculator) is shown in Fig. 7. Physical defoaming equipment is most useful when the foam is consistent, copious, and localized.

Water showers can be used to break surface foam. This does little to reduce foam downstream from the spraying site, and may significantly dilute the pulp. Nevertheless, it is sometimes the method of choice to control bubbles in the headbox of the paper machine.

In treating foam physically, one must consider the initial cost of equipment, lack of flexibility in treatment site, and lack of carrythrough of the defoaming effect downstream. If the technique, the situation, and the point of treatment are chosen carefully, mechanical foam control has the advantages that it is relatively insensitive to chemical variations in the stock, and it adds no new chemicals to an already complex system.

V. CHEMICAL CONTROL OF FOAM

A. Early Antifoams

The first chemical antifoams were aimed exclusively at breaking visible surface foam. Milk and cream were the forebears of modern emulsion-type antifoams. Kerosene, fuel oil, and other light petroleum oils enjoyed long-term popularity for their ability to break foam; they are still used occasionally, although odor, flammability, environmental pollution, and deterioration of rubber rolls and fittings are drawbacks. Pine oil and linseed oil found some use. Alcohols with C_7—C_{16} chains were effective but expensive antifoams, and were often mixed with kerosene for economy (and sometimes worked better for it).

In the 1950s, oil-in-water emulsion* antifoams were developed using fatty acids, fatty alcohols, esters, waxes, sulfated tallows, and saturated soaps [14]. Some were sold as preemulsified liquids and pastes, thus ensuring good control of emulsification but increasing the cost of shipping to the customer. Others were emulsifiable liquids and solids that were cheaper to ship but which had to be emulsified by the user, with less consistent results. Both types were dispersed in water before use.

B. Development of Brownstock Washer Antifoams

Also during the 1950s, silicone antifoams (polydimethyl siloxane oil, dispersed in water or light oil) were introduced to the industry. Their use on paper machines and in the latter stages in pulp mills was limited by the tendency of silicone in the paper to interfere with physical properties, printing, and coating [7]. The stability of these antifoams at high temperature and alkalinity, however, made these the first truly effective antifoams for kraft black liquor and brownstock washers (temperature up to 90°C, pH up to 12–13).

Dispersions of hydrophobic particles in light oil formed the next generation of brownstock washer antifoams. The first such antifoams were pat-

---

*No distinction is made in the pulp and paper antifoam literature between true emulsions and dispersions of a liquid crystal phase in a bulk liquid. Here, for simplicity, the term emulsion will refer to any colloidal dispersion of one liquid in another, regardless of the actual phase of the dispersed liquid.

ented in 1963 [15]; they consisted basically of silica particles, rendered hydrophobic by a high-temperature treatment with silicone, and dispersed in mineral oil. The early 1970s saw the introduction of oil-based brownstock washer antifoams using hydrophobic amide wax particles (e.g., [16]). Some antifoams contain both hydrophobic silica and amide wax particles, in light oil.

Just how these antifoams work is not fully understood (see Chapter 1 for discussion of mechanisms). Most workers are in agreement that the particulate components act by entering and disrupting the foam film [17], but the role of the oil is less well defined. Much of the patent literature refers to the oil as an "inert liquid" or "carrier oil," implying that it plays no active part in defoaming. However, the success of kerosene and other light oils as antifoams suggests that the oil in these brownstock washer antifoams may be a major contributor to the defoaming process.

The oil crisis of the early 1970s fueled the push to reduce the amount of oil in antifoams. "Water-extended" antifoams appeared, which were at first essentially oil-based antifoams with water emulsified into the oil phase, effectively diluting the concentration of solid defoaming particles. They were less expensive on a weight basis, but less efficient. Now, water-extended antifoams are similar to the oil-based compositions in solids content, but up to half of the oil is replaced by water (plus appropriate stabilizers and emulsifiers). When more than half of the oil is replaced with water, the antifoam becomes an oil-in-water emulsion; these formulations are termed "water based."

Table 5 gives a representative sampling of patents for brownstock washer antifoams [15,16,18–32], beginning with the first hydrophobic silica particle antifoams. Unless indicated otherwise, the preferred bulk liquid is a light petroleum oil (mineral oil). In many patents the preferred formulation contains no water, but water-extended and/or water-based formulations are also included, giving the option of eliminating part of the oil if desired.

It is difficult to relate patents to antifoams in the marketplace. Competition makes manufacturers reluctant to give detailed information about their formulations; when a product can be definitely associated with a specific patent, the range of formulations covered by the patent makes it impossible to determine exactly how the antifoam is made. Usually all that can be determined is the type of antifoam (e.g., silica, wax, or silica-wax) and the nature of the liquid base (e.g., oil-based, oil-based/water-extended, or water-based).

This lack of precise information on the compositions of commercial antifoams makes it difficult to compare antifoam efficiency rigorously. In addition, the variety in stock formulations and equipment means that the results of such a comparison in a mill can often be applied with certainty only to

TABLE 5 Selected Kraft Brownstock Washer/Black Liquor Antifoam Patents (Hydrophobic Particle Type)

Patent	Type of particles	Components/comments
F. J. Boylan (to Hercules Inc.), 1963 [15]	Hydrophobic silica	Silica treated with silicone Not exclusively for pulp and paper
W. R. Christian and R. Liebling (to Nopco Chemical Co.), 1965 [18]	Hydrophobic silica	Silica treated with amide, then with silicone Does not require additional spreading agents
E. Domba (to Nalco Chemical Co.), 1968 [19]	Hydrophobic silica	Silica treated with cyclic siloxane
H. Lieberman, C. A. Duharte-Francia, and J. W. Henderson (to Betz Laboratories Inc.), 1972 [20]	Aluminum oxide	May contain silicone as spreading agent May contain up to 10% water
T. F. MacDonnell (to Diamond Shamrock Corp.), 1972 [16]	Amide wax	Bisstearamide wax
J. H. Curtis and F. W. Woodward (to Diamond Shamrock Corp.), 1972 [21]	Amide wax Talc	Contains silicone Contains oil-soluble copolymer of vinyl acetate and fumaric acid
I. A. Lichtman and A. M. Rosengart (to Diamond Shamrock Corp.), 1972, 1973 [22]	Amide wax	Amide of tallow fatty acid Contains silicone Contains copolymer of vinyl acetate and fumaric acid esterified with tallow alcohol Contains triglycerides to make wax particles more amorphous Covers water-extended and water-based formulations
T. F. MacDonnell (to Diamond Shamrock Corp.), 1972 [23]	Hydrocarbon and/or amide wax	May contain silicone Contains copolymer of vinyl acetate and fumaric acid Covers water-extended and water-based formulations
H. J. S. Shane, J. E. Schill, and J. W. Lilley (to Hart Chemical Ltd.), 1973 [24]	Amide wax	Bisstearamide wax Contains silicone

TABLE 5 Continued

Patent	Type of particles	Components/comments
R. J. Michalski and C. C. Cochrane (to Nalco Chemical Co.), 1975 [25]	Amide wax	Beheramide wax Contains copolymer of vinyl acetate and ethylene
R. J. Michalski and R. W. Youngs (to Nalco Chemical Co.), 1975 [26]	Amide wax Hydrophobic silica	Methylene bisstearamide Contains silicone Silica added to reduce viscosity
H. Lieberman, A. J. Graffeo, and J. S. Kucsan (to Betz Laboratories Inc.), 1976 [27]	Amide wax	
J. V. Sinka and I. A. Lichtman (to Diamond Shamrock Corp.), 1977 [28]	Hydrophobic silica Amide wax	Amide wax of tallow fatty acid added to stabilize antifoam
R. K. Berg and D. S. Smalley (to Associated Chemical Inc.), 1977 [29]	Amide wax Hydrophobic silica	Fatty acid diamide Contains stearyl alcohol ethoxylate emulsifier Water-based (oil-in-water emulsion)
H. J. S. Shane and F. S. Schell (to Hart Chemical Ltd.), 1978 [30]	Amide wax	Ethylene bisstearamide Contains silicone Water-based (oil-in-water emulsion)
DIC Hercules Inc., 1983 [31]	Amide wax Hydrophobic silica	Ethylene bisstearamide Contains silicone Contains ethylene glycol for low temperature stability Water-extended
B. Danner (to Sandox GmbH), 1983 [32]	Amide wax Hydrophobic silica	Ethylene bisstearamide Contains vegetable or animal oil instead of mineral oil Contains no silicone Water-based (oil-in-water emulsion)

that particular mill, under the prevailing operating conditions, and are thus difficult to evaluate.

In some mill studies, water-extended brownstock washer antifoams are claimed to be superior to their oil-based analogs [33–35]. Viscosities are similar to those of oil-based antifoams, so that the same pumping and dispensing equipment can be used. The cost per unit weight is less, and benefits reported include

Lower antifoam use
Better washing
Reduced biological oxygen demand in effluent
Reduced deposition of antifoam on process equipment.

Costs savings of several hundred thousand dollars per year per mill have been claimed [33,34]. In the authors' experience, however, few mills realize such benefits; at present, oil-based brownstock antifoams are still usually considered the most cost effective.

At least one company [36,37] has a water-based brownstock washer antifoam which it claims is as cost-effective as oil-based antifoams, without many of their drawbacks. Water-based antifoams have, as yet, met with only limited acceptance.

Water-based/hydrocarbon oil-free antifoams similar to the original silicone emulsion antifoams that appeared in the 1950s are attempting a comeback. They circumvent the problems involved with mineral oil; however, the relatively expensive silicone renders them less cost-effective than the standard oil-based antifoams.

C. Other Pulp and Paper Antifoams

Other types of antifoams are used in pulp and paper production where conditions of temperature and pH are less extreme. The old paste-type antifoams (high-viscosity oil-in-water emulsions of fatty acids, esters, waxes, etc.) lost popularity largely because of handling problems, but, with the increase in oil prices in the 1970s and improvements in pumping and metering equipment, they have made something of a comeback. Blends of fatty acids, fatty alcohols, and/or polyethylene glycols, in mineral oil, are easier to handle, but have lost ground because of problems associated with the oil (see Sec. V. D.). Some antifoams are sold without carrier liquids, as solids or undiluted liquids.

Most modern antifoams are pumpable fluids. Antifoams that have no carrier liquid save on transportation costs and are relatively stable when stored at low temperatures ($<0°C$); when added to the stock undiluted, they require agitation to ensure thorough mixing. These carrier-free antifoams are usually fatty acids, fatty alcohols, and their derivatives, plus emulsifiers. Some contain flammable components, and must be handled appropriately.

Aqueous emulsion antifoams (typically of long-chain fatty acids, fatty alcohols, and esters) are gaining in popularity for several reasons. Water as a carrier is less expensive than oil. High concentrations of active ingredients (up to 50%, usually ~30%, versus ~10% for oil-based antifoams) can be tolerated before viscosity is a problem. Dispersion in stock is easier because the defoaming material is already emulsified. Problems associated with oil are reduced. In addition, on the paper machine they often appear to be more efficient at defoaming than oil-based antifoams [38].

Some aqueous dispersions of relatively hydrophilic fatty acids and alcohols incorporate water in the droplets; in recent years, these have been called "gel-particle" antifoams. Because the gel particles are hydrophilic, they function somewhat differently from the hydrophobic particles in a brownstock washer antifoam [39,40]. Apparently, the gel particle reacts with polyvalent cations in the stock (e.g., calcium, magnesium, aluminum) to form a hydrophobic shell around the particle, thus permitting it to enter the film of a foam bubble. When watched in a microscope the particle appears to "explode," disrupting the film and causing collapse or coalescence. Patents specifically for gel-particle antifoams (e.g., [41,42]) include aluminum salts in the formulation; the hydrophobic shells thus are formed around the gel particles before the antifoam is added to the process liquid.

Paper coating compositions resemble paints, and can use similar antifoams. Many pulp and paper antifoams can also be used in coatings, but very hydrophobic formulations should be avoided, as they can cause spots on the coated sheet.

D. Disadvantages of Chemical Defoaming

There are limitations and problems associated with the use of chemical antifoams; the problems are frequently aggravated by high dosages of the antifoams. With care, these difficulties can be minimized.

For high temperature applications, industry opinion favors oil-based or sometimes oil-based/water-extended antifoams. The tendency of these antifoams to deposit on process equipment [43], however, is a serious drawback to their use; the amount of deposition increases with dosage [44]. In addition, overuse of oil-containing antifoams has been associated with reduced strength, brightness, uniformity, and opacity of the finished paper, and with flaws in coating [7,45]. They should be used as far upstream in the pulp mill as possible, to maximize the washing of the antifoam from the pulp.

Hydrocarbon oils in some brownstock washer antifoams have been linked to trace quantities of polychlorinated dioxins and furans in products and effluents from bleached kraft mills [46]. This may be avoided by the use of oils free of the unchlorinated precursors of these toxins [47].

Kerosene and fuel oil share the deposition problem of oil-based anti-foams. In addition, they are flammable, work only on surface foam, cause deterioration of rubber parts, and give off odor in use and in the final product.

Oil-free antifoams avoid these problems. However, water-based emulsions are sensitive to temperature extremes and generally have a shorter life-time than oil-based or carrier-free antifoams. Emulsifiable, carrier-free antifoams are relatively stable, but require care when being dispersed to achieve consistent results.

VI. MEASUREMENT OF FOAM

Monitoring of surface foam levels is relatively straightforward. With open vats and screens, it can be done by eye; however, continuous monitoring with electronic or pneumatic sensors is more precise.

Various systems have been developed to measure entrained foam in pulp and paper stock; most are based on a measurement of the compressibility of the stock. For example, in one such device [48,49] a stock sample is isolated in an onstream sampling cylinder, then compressed with a test piston; the volume of compression and the pressure developed are fed into a microprocessor whose output governs antifoam feed. Because large, free-floating air bubbles contribute proportionately more to the measured entrained air volume, but often cause less trouble, than the smaller bubbles adhering to solid particles in the stock, the amount of entrained air that can be tolerated will depend on the size distribution of the bubbles. The maximum tolerance must be determined empirically for each mill, each site, and each stock/antifoam combination. Thus, while average tolerable entrained air limits are around 6%, actual limits vary from 2.5% to 10% [49].

A recent Finnish development [50] circumvents this by making use of the fact that ultrasonic transmissions through water are attenuated considerably more by small bubbles than by large bubbles. Initial trials look promising; the ultrasound device appears to react rapidly and precisely to troublesome changes in the amount of entrained air.

To determine the total amount of foam (surface and entrained) in a stock vessel, a system has been described [51,52] in which the measured level of the air-liquid or air-foam surface is compared with the theoretical level of unfoamed liquid calculated from measurements of the hydrostatic head at the bottom of the vessel. The difference gives a measure of the combined volumes of surface and entrained foam. If the depth of the surface foam is also measured, the amount of entrained air can be estimated.

VII. TESTING OF ANTIFOAMS

The suitability of an antifoam must ultimately be determined by mill trials, but there are several laboratory tests that can be used for preliminary screening. Most measure only surface foam, which may not reflect the effect of the antifoam on entrained air [49].

Typically, a quantity of stock, or stock plus antifoam, is put in a graduated cylinder. Air is introduced into the stock in the cylinder by sparging (bubbling controlled amounts of air into the stock through a frit in the bottom of the cylinder) [53–55], by cascading one aliquot of stock into another [56,57], by shaking the cylinder reproducibly [58], or by cascading the stock with a recirculating pump [14]. The amount of surface foam produced is measured and correlated with the presence of antifoam.

The same tests can also be used to screen the efficiencies of antifoams in paper coating formulations. In addition, a measure of the tendency of an antifoam to cause defects in the finished coating may be obtained by spreading a sample of the liquid coating plus antifoam on a clean glass plate, drying the mixture, and examining the resultant film for imperfections [59].

Antifoams may be screened in the laboratory for their tendency to contribute to pitch deposits by measuring deposition with a Vibromixer™ [60]. A new procedure now exists [44] which enables the antifoam to be tested in pulp stock rather than in water, which is an advantage as the presence of other components of the stock affects deposition rates [61]. If deposits have already formed in the mill, the amounts of antifoam components in the pitch can be analyzed [43].

VIII. HANDLING AND USE OF ANTIFOAMS

The manufacturer of an antifoam should be consulted for specific instructions about its storage and handling. In general, antifoams should not be exposed to temperature extremes. Water-containing antifoams are particularly sensitive to temperature, and should be kept between 5°C and 20°C. The storage life is a few months, but may be extended somewhat if the antifoam is stirred intermittently during storage; care should be taken to avoid entraining air during stirring.

Oil-based and carrier-free antifoams are more resistant to high and low temperature; however, extremes should still be avoided. Some carrier-free antifoams become viscous and hard to handle at low temperatures ($<0°C$). Antifoams containing volatile or flammable materials should be stored below 30°C. Carrier-free antifoams usually have storage lives of about a year.

Emulsion and dispersion antifoams, and most other pumpable liquid antifoams, may be added to the stock without dilution. Viscous or carrier-free antifoams require vigorous agitation to ensure complete dispersion when they are added to stock or diluted prior to use. Any diluted antifoam should be stirred frequently until used. During any agitation of the stock or antifoam, air entrainment should be avoided.

Traditionally, antifoam was added by an operator when surface foaming got out of hand. This is still done when necessary, but it is now customary to add antifoam continuously to the stock by metering pumps; positive displacement pumps with large diameter suction lines are recommended. Antifoams should be added to the stock at points where there is sufficient agitation or turbulence for complete mixing (e.g., into flowing stock lines or stirred vessels; before pumps). When the pulp forms a mat or web (e.g., on washers), diluted antifoam may be sprayed on. Each antifoam addition site should have its own pump and control.

Antifoam is metered into the stock at a constant rate with adjustments made as needed when the amount of foam changes. Some form of automatic foam sensing device is preferable to visual inspection (particularly if it gives a measure of entrained air); the sensor location may have to be determined by trial and error to optimize the correlation of sensor output with required antifoam dose.

The antifoam feed rate can be adjusted manually, in response to change in the amount of foam present. If frequent adjustments are necessary, however, it may be more economical to invest in an automated control system. Claims have been made [48,49,51,62] of antifoam savings of from 20% to 50% when manual control of antifoam feed to a sensor-monitored stock system was replaced with computerized control responding automatically to sensor output. This is not surprising; a busy operator may make only periodic inspections of sensor output, while the computer monitors sensor output continuously, responding immediately to higher foam levels with a minimum increase in antifoam addition rate before things get too bad, then decreasing antifoam feed as soon as possible.

With experience, the need for changes in antifoam dosages can sometimes be anticipated. Use of larger quantities of very resinous wood (e.g., pine, aspen) or recycled process liquids, or less aging of the chips (shorter time and/or freezing temperatures) will cause more foam, especially in the early stages of the pulp mill. Colder makeup water in the winter and increased amounts of dissolved or suspended material in the water during spring runoff may also have a predictable effect on foaming. Foam levels should be watched carefully when physical or chemical changes are made in the system, or when the type or brand of antifoam is changed.

IX. FUTURE RESEARCH AND DEVELOPMENT

The biggest developments in physical control of foaming in recent years have been in the areas of kraft brownstock washing and screening. Displacement washing, diffusion washers, horizontal pressure washers, pressure knotters and screens, and more gradual release of digester pressure have reduced greatly the amount of air entrained in the brownstock. The potential for savings from improved washing and reduced foam is so great that continued progress in these areas can be expected for some time to come.

Similarly, in the field of chemical defoaming, it is the brownstock antifoams that are attracting the most attention. Current oil-based antifoams are efficient at defoaming, but the oil causes numerous problems; the oil-free antifoams available presently are more expensive. Much research remains to be done in the development of oil-free brownstock antifoams that are both effective and inexpensive.

The industry has come a long way from the days of cream and kerosene. No doubt the challenge of foam control in pulp and paper mills will continue to be met creatively and effectively.

REFERENCES

1. Anonymous, *Pulp Paper J.* 42(1): 27 (1989).
2. Corpus Information Services, Market Study: Chemicals for Pulp and Paper in Canada, Southam Communications, Ltd., 1985.
3. E. Sjöström, *J. Appl. Polym. Sci.: Appl. Polym. Symp.* 37: 577 (1983).
4. G. A. Smook, *Handbook for Pulp and Paper Technologists*, p. 15. Joint Textbook Committee of the Pulp and Paper Industry (TAPPI/CPPA), 1982.
5. E. Sjöström, Int. Symp. Wood and Pulping Chem. (Japan) 3: 57 (1983).
6. L. H. Allen and M. C. Lapointe, *Nordic Pulp Paper Res. J.* 4: 94 (1989); 4: 229 (1989).
7. K. Minakuchi, *Japan Pulp Paper* 4(3): (1966).
8. H. Guo and B. A. W. Coller, *Tappi J.* 72(1): 50 (1989).
9. J. B. Ball and S. R. Forster, Proc. CPPA Western Branch Meeting, Jasper, 1984.
10. J. M. MacLeod and M. E. Cyr (assigned to Paprican and Irving Pulp and Paper Ltd.), US 4,814,041; March 21, 1989; filed March 18, 1987.
11. Anonymous, *Pulp Paper J.* 38(3): 22 (1985).
12. K. A. Coffey, *Paper Trade J.* 169(1): 67 (1985).
13. J. P. Campbell, *Pulp Paper* 56(7): 121 (1982).
14. I. Lichtman and T. Gammon, in Kirk-Othmer Encyclopedia of Chemical Technology, Vol. 7, 3rd ed., Wiley, New York, 1979, pp 430–448.

15. F. J. Boylan (assigned to Hercules Inc.), US 3,076,768; February 5, 1963; filed April 5, 1960; Also: Can 662,736; May 7, 1963; filed July 6, 1960.
16. T. F. MacDonnell (assigned to Diamond Shamrock Corp.), US 3,652,453; March 28, 1972; filed January 27, 1970.
17. R. H. Pelton, *Pulp Paper Canada* 90(2): T61 (1989).
18. W. R. Christian and R. Liebling (assigned to Nopco Chemical Co.), US 3,207,655; September 21, 1965; filed February 13, 1963.
19. E. Domba (assigned to Nalco Chemical Co.), US 3,388,073; June 11, 1968; filed December 12, 1963.
20. H. Lieberman, C. A. Duharte-Francia, and J. W. Henderson (assigned to Betz Laboratories Inc.), GB 1,267,479; March 22, 1972; filed May 22, 1970.
21. J. H. Curtis and F. W. Woodward (assigned to Diamond Shamrock Corp.), US 3,673,105; June 27, 1972; filed April 11, 1969.
22. I. A. Lichtman and A. M. Rosengart (assigned to Diamond Shamrock Corp.), US 3,677,963; July 18, 1972; filed September 30, 1970. Also: Can. 927,707; June 5, 1973; filed September 3, 1971.
23. T. F. MacDonnell (assigned to Diamond Shamrock Corp.), Can. 909,630; September 12, 1972; filed August 18, 1971.
24. H. J. S. Shane, J. E. Schill, and J. W. Lilley (assigned to Hart Chemical Ltd.), Can. 922,456; March 13, 1973; filed December 31, 1971.
25. R. J. Michalski and C. C. Cochrane (assigned to Nalco Chemical Co.), US 3,893,941; July 8, 1975; filed September 9, 1971.
26. R. J. Michalski and R. W. Youngs (assigned to Nalco Chemical Co.), Can. 971,851; July 29, 1975; filed June 5, 1972. Also: US 3,923,683; December 2, 1975; filed September 7, 1971.
27. H. Lieberman, A. J. Graffeo, and J. S. Kucsan (assigned to Betz Laboratories Inc.), US 3,935,121; January 27, 1976; filed June 29, 1973.
28. J. V. Sinka and I. A. Lichtman (assigned to Diamond Shamrock Corp.), US 4,021,365; May 3, 1977; filed January 24, 1973.
29. R. K. Berg and D. S. Smalley (assigned to Associated Chemists Inc.), US 4,032,473; June 28, 1977; filed June 4, 1975.
30. H. J. S. Shane and F. S. Schell (assigned to Hart Chemical Ltd.), US 4,088,601; May 9, 1978; filed November 11, 1974.
31. DIC Hercules Inc., Jap. 57 24,309 [83 24,309]; February 14, 1983; filed August 7, 1981.
32. B. Danner (assigned to Sandoz GmbH), DE 3,242,202; June 1, 1983; filed November 25, 1981.
33. R. T. Maher, Proc. Tappi Pulping Conf., p. 461, 1981.
34. F. Poltenson, *Pulp Paper* 60(2): 97 (1986).
35. L. F. Twoomey, Proc. Tappi Papermakers Conf. (Atlanta), p. 241, 1984.
36. L. F. Twoomey, *Southern Pulp Paper* 49(5): 33 (1986).
37. Nalco Chemical Co. Product Information.
38. R. H. Lorz, *Pulp Paper Canada* 88(10): 85 (1987).
39. R. C. Montani and F. J. Boylan, Proc. Tappi Annual Meeting, p. 293, 1980.
40. R. C. Montani and F. J. Boylan, Proc. Tappi Pulping Conf., p. 49, 1982.

41. F. J. Boylan (assigned to Drew Chemical Corp.), US 4,303,549; December 1, 1981; filed October 18, 1979.
42. F. J. Boylan (assigned to Drew Chemical Corp.), US 4,340,500; July 20, 1982; filed October 18, 1979.
43. G. M. Dorris, M. Douek, and L. H. Allen, *J. Pulp Paper Sci. 11*(5): 149 (1985).
44. N. Dunlop-Jones and L. H. Allen, *J. Pulp Paper Sci. 15*(6): 235 (1989).
45. A. M. Springer, J. P. Dullforce, and T. H. Wegner, *Tappi J. 69*(4): 106 (1986).
46. L. H. Allen, R. M. Berry, B. I. Fleming, C. E. Luthe, and R. H. Voss, *Chemosphere 19*: 741 (1989).
47. R. H. Voss, C. E. Luthe, B. I. Fleming, R. M. Berry, and L. H. Allen, *Pulp Paper Canada 89*(12): 151 (1988).
48. J. Morton, Trans. PIMA Meeting (Seattle), 1986.
49. S. J. Dougherty, *Tappi J. 72*(1): 50 (1989).
50. M. Karras, T. Pietikäinen, H. Kortelainen, and J. Tornberg, *Tappi J. 71*(1): 65 (1988).
51. G. L. Cooper, Proc. Tappi Annual Meeting (Atlanta), p. 71, 1988.
52. W. E. Sande, *Tappi J. 69*(3): 246 (1986).
53. ASTM Method D1881-86 (1986).
54. W. L. Carpenter and I. Gellman, *Tappi J. 50*(5): 83A (1967).
55. M. Penttilä, T. Laxén, and N.-E. Virkola, *Pulp Paper Canada 87*(1): 82 (1986).
56. ASTM Method D1173-80 (1986).
57. I. Yrjälä, R. Räsänen, and H. Bruun, *Paperi ja Puu 46*(4a): 257 (1964).
58. H. Hollaender, *Paper-Maker (London) 130*: 564 (1955).
59. Sandoz Chemicals Corp. Product Information.
60. Tappi Useful Method 223.
61. M. Douek and L. H. Allen, *J. Pulp Paper Sci., Trans. Tech. Sect. CPPA 9*(2): TR48 (1983).
62. M. Andrews, *Pulp Paper 56*(2): 153 (1982).

4

Application of Antifoams in Pharmaceuticals

ROLLAND BERGER Th. Goldschmidt AG, Chemische Fabriken, Essen, Germany

I.	Introduction	177
II.	Silicone Antifoam Materials and Their Toxicity	179
III.	Regulations	181
IV.	Formulations	184
	A. Tablets	184
	B. Suspensions	185
	C. Emulsions	185
V.	In Vitro and In Vivo Experiments	186
VI.	Conclusions	189
	References	189

I. INTRODUCTION

Pharmaceutical applications of polysiloxane (silicone) chemistry started in the early fifties, when polydimethylsiloxanes were first used as antifoams in the manufacture of drugs [1]. Since this introduction more than 35 years ago, an increasing number of fields of application have emerged [2–7]. One of the earliest applications of polysiloxanes, which is still of great interest, is their use in pharmaceuticals to combat intestinal gas.

The first therapeutic application of polysiloxanes against digestive upsets was done by Quin et al. [8], who examined the flatulence in ruminants. Hirschowitz et al. [9], Gasster et al. [10], and Rider et al. [11–13] later reported the successful application of polysiloxane-based antifoams in the prevention of intestinal foam during gastroscopy, which otherwise hindered the inspection of the gastrointensial tract [14].

Intestinal gas, entrapped in foam bubbles, may lead to any form of indigestion, discomfort, fullness, and pressure or pain in the gastrointenstinal tract accompanied by lack of appetite and vitality or general fatigue. These complaints are normally referred to as dyspepsia. The main indications of dyspepsia are hyperacidity and flatulence, which frequently occur together. The intestinal gas is mainly composed (99%) of nitrogen, oxygen, hydrogen, carbon dioxide, and methane, and the distribution of these gases may vary over a wide range [15,16].

Dyspepsia may result from different causes [17]. Air swallowing is recognized to be one of the most common sources of abdominal distension, especially by those people who eat and drink too fast, smoke, or are emotionally disturbed. Swallowed air can contribute up to 70% of the gastrointestinal gas [18].

Other causes of gastrointestinal gas are certain foods which enable the colonic bacteria to be particularly productive in the formation of gas. The substrate in the intensine and the predominant organism will influence the flatus, and the volume will depend on the quantity of substrate acted upon by the bacteria. Steggerda [19] has reported that the volume of flatus can be increased more than tenfold by ingestion of beans. The gas produced is mainly carbon dioxode.

Additionally a deficiency in digestive function may result in gastrointestinal malfunction. This can be caused either by reduced digestive enzymes or stress, an activity which is known to increase the production of mucus and to increase gastrointestinal motility, entrapping whatever gas is present in the bowel. A deficiency of pancreatic enzymes has been demonstrated to be one of the physiological changes of aging. So gastrointestinal complaints are more common in the elderly [20,21].

Stress is also known to lead to inflammation of the gastrointestinal tract, thus disturbing the absorption and the diffusion of intestinal gas into the blood. Gas therefore cannot be eliminated via the lungs.

As pointed out, flatulence is frequently accompanied by hyperacidity. A study by Graham et al. [22] reported that 50% of the American population have used an antacid at some time in their lives and that 27% take two or more doses a week. Seventy-five percent of the regular users take six or more doses every week. Dyspepsia therefore seems to be a very common complaint.

The described causes of dyspepsia have given rise to pharmaceuticals to combat the digestive upset. The pharmaceuticals normally contain antifoams, antacids, and/or enzymes.

II. SILICONE ANTIFOAM MATERIALS AND THEIR TOXICITY

Polysiloxanes are polymers with repeating units of organosiloxy groups with general formula $R_nSiO_{(4-n)/2}$ with a varying degree of substitution, and therefore varying functionality. The four basic structures of the repeating units and their generally used abbreviations are

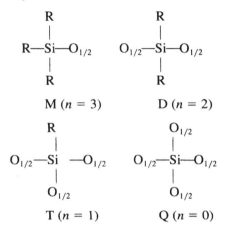

These repeating units in the polymers are derived from the corresponding halosilanes $R_nSiX_{(4-n)}$ by hydrolysis under the elimination of HCl. A mixture of the halosilanes is produced from elementary silicon and organic halides RX using the Rochow synthesis, followed by a distillation process to yield the pure monomers. Examples of the organic halides are especially CH_3Cl and in part C_6H_5Cl, leading to methylchlorosilanes or phenyl group containing chlorosilanes.

Depending on the type and on the ratio of the chlorosilanes used during the hydrolysis, cyclic, linear, branched or cross-linked oligomers and polymers are achieved [23]. Using water in excess, the polymers contain OH end groups, with the exception of the cyclic compounds. In a further reaction, the so-called equilibration reaction, the polymers can be tailored with respect to their molecular weight, for which the viscosity of the products is a measurement. The relationship between the molecular weight of the (linear) polymers and the viscosity can be described by the empirical equation

$$\log \eta \; (cSt/25°C) = 1.00 + 0.0123M^{0.5}$$

where η is the viscosity and M is the molecular weight. This equation was developed by Barry [24] and is valid for $M > 2500$. Other equations have been developed by Piccoli and Stark [25] and Weissler [26]. The equilibration reaction is an acid- or basically catalyzed random rearrangement of siloxy bonds providing a statistical distribution of molecular sizes. Products in the viscosity range of about 20 mPa s up to 10^6 mPa s are commercially available. As chain-limiting end groups, M units can be added. The equilibration process is characterized by the formation of a constant amount of cyclic oligomers with octamethylcyclotetrasiloxane (D_4) as the main compound of the cyclic. The low molecular weight polysiloxanes are then removed by distillation and the residual polymer mixture can be further purified when used for medical applications. It has been shown that higher contents of volatile, low molecular weight polysiloxanes lead to an increased appearance in the urine of humans, while other experiments have shown that polysiloxanes with molecular weights higher than 500 were poorly absorbed from the gastrointestinal tract [27].

For use in pharmaceuticals the linear polymers of the general formula

$R_3SiO[R_2SiO]_xSiR_3$ (MDxM)

with R = CH_3 (polydimethylsiloxane) are well established and referred to as dimethicone in the pharmaceutical literature. Dimethicone is an ideal raw material for antifoam formulations due to its low surface tension. The surface tension has values of about 20–22 mN m^{-1}, depending on the molecular weight (for the dimeric compound hexamethyldisiloxane (M—M) the surface tension decreases to about 16 mN m^{-1}). The low surface tension is an important property of antifoam raw materials which is essential for their mode of action [28–34]; see Chapter 1.

When used alone, dimethicones are ineffective antifoams. They gain their extraordinary antifoam properties when formulated with finely divided hydrophobic solids. These solids can in principle be organic [35–37] or inorganic. For application in pharmaceuticals to combat flatulence highly dispersed hydrophobed silica is commonly used. Silica is comprised of spherical particles with a mean particle size of about 12–15 nm when it is of the fumed variety. Other types of silica are precipitated from water glass but are not preferred for the formulation of pharmaceutical antifoams because they can contain excess alkali. Commercially available silica is surface-treated to yield hydrophobic types. The treatment can be done by the reaction of the OH groups at the surface with, for example, $(CH_3)_3SiCl$ or with polydimethylsiloxanes under the influence of heat and/or catalysts and mechanical shear forces. Treatment of the silica with organic substances is also known. Mixtures of dimethicone with silica are normally referred to as simethicone. Simethicone is either produced by mixing pretreated silica with dimethicone or by mixing untreated silica with dimethicone followed by an in situ treatment at elevated temperatures [38]. A mixture of untreated silica with dimethicone is characterized by a high viscosity which is shear sensitive. This means that the viscosity of such a mixture is strongly reduced under the influence of shear forces (thixotropoy). This effect is caused by interparticle action of the silica which is reduced under shear. The same mixture with an equal quantity of pretreated or in situ treated silica is less viscous and shows a smaller thixotropic effect. It has, however, been shown that simethicones containing pretreated or in situ treated silica are more effective as antifoams than those with untreated silica [39,40]. The role of the silica in simethicone antifoams is discussed elsewhere [31,41–43]. A detailed account is found in Chapter 1.

Besides the physicochemical properties which make dimethicone an ideal raw material for antifoams, their nontoxicity make dimethicones ideal for use in pharmaceuticals, especially for those which are designed for oral use.

Dimethicones are recognized as chemically inert and physiologically neutral. Early toxicological studies including dimethicone were conducted by Rowe et al. [44]. These studies indicated that commercially available dimethicone maintained in stock diets for a group of female rats at levels up to 1% for three months showed no significant abnormalities in the liver, spleen, kidneys, heart, lungs, pancreas, or stomach. The studies were extended to male and female rats by feeding 0.3% simethicone for two years showing no adverse effects on growth, mortality, or behavior [45]. During subacute studies with simethicone fed to dogs in doses from 0.3 g/kg up to 3 g/kg no symptoms of toxicity were noted [46].

Only mild effects of moist, loose stools were observed. The list of toxicological studies can be completed by the studies of Gloxhuber and Hecht [47], who carried out acute and chronic oral as well as intraperitonal, intravenous, and dermal sensitivity investigations on rats, cats, and rabbits, using dimethicone of different viscosities. In no case was any toxic action observed. McDonald et al. [48] conducted subacute oral toxicity tests using dimethicone of 50 up to 60,000 cSt at a level of 1% in the feedstock for rats over 90 days. Leucocyte counts and haemoglobin were normal. The effects of dimethicone on reproduction and fetal development in rats and rabbits and the mutagenic potential in mice were evaluated by Kennedy et al. [49]. No evidence of fetotoxicity and mutagenic effects were obtained. Cutler et al. [50] ran long-term studies with simethicone incorporated in the diet of mice at levels of 0.25% to 2.5% for 76 weeks. No significant histopathologic changes were observed. Single doses of silicones were given to rabbits, cats, dogs, and rats as high as 54 mL/kg. Except for diarrhea there were no indications of negative effects [51,52]. Silicone fluids were excreted unchanged and there was no absorption [53].

III. REGULATIONS

Because of their inertness and physiologically neutral behavior, simethicones are admitted as antiflatulents for "over the counter human use" by the

TABLE 1 Essential Specification for Dimethicone According to National Formulary [55]

Viscosity (at 25°C): 18–13750 cSt (USP method No.911) [57]
Specificity gravity: 0.946–0.975 g/ml (USP method No.841) [58]
Refractive index: 1.3980–1.4055 (USP method No.831) [59]
Loss on heating (15 g; 200°C; 4 h): <2% (except Dimethicone 20 cSt: <20%)
Heavy metals: <0.001% (10 ppm) (USP method No.231,II) [60]

Source: Ref. 55.

United States Food and Drug Administration (21 CFR Part 332). The maximum daily dose is limited to 500 mg. The antiflatulent may also contain "any generally recognized as safe and effective antiacid ingredient(s)" as sanctioned under 21 CFR Part 331. Moreover, polydimethylsiloxanes and silica are allowed as antifoam formulations in food and food processing (with concentration limitations!) [54].

The specification for dimethicone, silica, and simethicone can be found in the *United States National Formulary* [55] and the *U.S. Pharmacopeia* [56], respectively. The *National Formulary* is a sister compendium to the *U.S. Pharmacopeia*, describing pharmaceutical ingredients which can be used in the formulation of pharmaceuticals, while the *Pharmacopeia* itself is limited to drug substances and dosage forms.

Following the *National Formulary* specification, dimethicone is

a mixture of fully methylated linear siloxane polymers containing repeating units of the formula

$[—(CH_3)_2SiO—]_n$

stabilized with trimethylsiloxy end-blocking units

$[(CH_3)_3SiO—]$

wherein n has an average value such that the corresponding nominal viscosity is in a discrete range between 20 and 12,500 centistokes (cSt). It contains not less than 97.0 percent and not more than 103.0 percent of polydimethylsiloxane ($[—(CH_3)_2SiO—]_n$).

The essential specifications are described in Table 1. Silica is referred to as colloidal silicon dioxide in the *National Formulary* and defined as

submicroscopic *fumed* silica prepared by the vapor phase hydrolysis of a silicon compound. When ignited at 1000° for 2 hours it contains not less than 99.0 percent and not more than 100.5 percent of SiO_2.

The specifications are summerised in Table 2.

TABLE 2 Specification for Colloidal Silicon Dioxide According to National Formulary [55]

pH value: 3.5–4.4
Loss on drying: <2.5%
Loss on ignition: <2%
Content of arsenic: <8 ppm [61]

Source: Ref. 55.

According to the *U.S. Pharmacopeia* [56], simethicone is

a mixture of fully methylated linear siloxane polymers containing repeating units of the formula

$$[\text{—(CH}_3)_2\text{SiO—}]_n$$

stabilized with trimethylsiloxy end-blocking units of the formula

$$[(\text{CH}_3)_3\text{SiO—}]$$

and silicon dioxide. It contains not less than 90.5 percent and not more than 99.0 percent of polydimethylsiloxane ($[\text{—(CH}_3)_2\text{SiO—}]_n$), and not less than 4.0 percent and not more than 7.0 percent of silicon dioxide.

It should be noted that the SiO_2 content is restricted, while simethicone formulations for other than pharmaceutical applications may contain less or more SiO_2.

Simethicone is mainly specified [56] by the data given in Table 3. Similar descriptions for simethicone do not exist in other countries, so the *U.S. Pharmacopeia* definition [56] is normally accepted in many other states.

Specifications for dimethicone are also given in the United Kingdom (Pharmaceutical Codex), France, and Germany (DAB) [66], but the descriptions differ from those given in the *U.S. National Formulary* [55]. So, for example, dimethicone (dimeticon) is specified in the DAB 9 [66] as

TABLE 3 Essential Specification of Simethicone according to US Pharmacopeia [56]

Loss on heating (15 g; 200°C; 4 h): < 18%
[compared with the definition of NF XVII for dimethicone, simethicone can contain a significantly higher amount of volatile, low molecular weight polymers]
Heavy metals: <0.001% (10 ppm) (USP method No.231,II) [62]
Defoaming activity against octoxynol 9 [63] (1%)–solution using 20 ppm simethicone: foam collapse time <15 s

Source: Ref. 56.

TABLE 4 Specification for Dimethicone According to DAB 9 [66]

Consumption of NaOH (0.01 w) to estimate the content of acidic substances: <
  0.15 mL
Viscosity (25°C): 20 − 1000 cST
Content of mineral oils: Compared with 0.1 ppm quinine sulphate in 0.01 N sulphuric acid
Heavy metals: <5 ppm
Loss of heating (1 g; 150°C; 2 h: <0.3% (for dimethicone < 50 cSt)

Source: Ref. 66.

summarized in Table 4. Even if there is no listing for simethicone in countries like Germany there are numerous pharmaceutical preparations available. All these formulations undergo a separate registration by the health authorities. The so-called *Rote Liste* [67] summarizes pharmaceutical antiflatulent preparations and describes the composition of the active ingredients. Not all of these pharmaceuticals use basic dimethicone which corresponds to the DAB 9 description. Besides dimethicone with 500-cSt or 1000-cSt dimethicones with a viscosity of 2000 cSt or 3000 cSt are also included.

IV. FORMULATIONS

Pharmaceuticals combating flatulence are offered in different dosage forms [68–72]. Dosage forms like simethicone tablets, simethicone oral suspensions, and simethicone emulsions are defined by the *U.S. Pharmacopeia* [56]. Simethicone tablets seem to be the most common dosage form.

A. Tablets

Tablets are formulated as pressed mixtures of simethicone with powdered components like milk powder [73] or saccharides [74] where simethicone is the only active ingredient. Simethicone tablets available on the market may contain between 20 mg and 300 mg simethicone, using mainly 40 mg per tablet.

According to the *U.S. Pharmacopeia* [56] simethicone tablets "contain an amount of polydimethylsiloxane ($[—(CH_3)_2SiO—]_n$) that is not less than 85.0 percent and not more than 115.0 percent of the labeled amount of Simethicone." The defoaming activity, defined as foam collapse time of a foamed octoxynol 9 (1%) solution should not exceed 45 s using 200 ppm simethicone.

In most pharmaceuticals, however, simethicones are combined with other ingredients designed to overcome symptoms associated with dyspepsia. The main additional ingredients are antacids such as different aluminum salts (carbonates, hydroxides, phosphates, etc.), bismuth salts (aluminates, carbonates, subcarbonates, etc.), calcium salts (carbonates, phosphates, etc.), magnesium salts (oxides, hydroxides, carbonates, trisilicates, aluminumsilicates, etc.), sodium and potassium salts (bicarbonates, carbonates, tartrates, citrates, etc.) or combinations of these salts, because flatulence frequently is accompanied by hyperacidity [75–85].

In many cases simethicones are also combined with different enzymes especially lipases and/or proteases [62] to overcome gastrointensial malfunction caused by reduced digestive enzymes.

Although simethicones are found to have an increased antifoam activity when compared with dimethicone, there are numerous formulations available containing only dimethicone.

B. Suspensions

Simethicone oral suspension [56] is "a suspension of Simethicone in water. It contains an amount of polydimethylsiloxane ($[$—$(CH_3)_2SiO$—$]_n$) that is not less than 85.0 percent and not more than 115.0 percent of the labeled amount of Simethicone."

The defoaming activity, defined as foam collapse time of a foamed octoxynol 9 (1%) solution, should not exceed 45 s using 200 ppm simethicone. Again suspensions can be formulated containing additional ingredients like antacids and/or enzymes.

C. Emulsions

Simethicone emulsion according to the *U.S. Pharmacopeia* [56] is

a water dispersible form of Simethicone composed of Simethicone, suitable emulsifiers, preservatives and water. It may contain suitable viscosity-increasing agents. It contains an amount of polydimethylsiloxane ($—$$[(CH_3)_2SiO$—$]_n$) that is not less than 85.0 percent and not more than 110.0 percent of the labeled amount of Simethicone.

The defoaming activity, defined as foam collapse time of a foamed octoxynol 9 (1%) solution, should not exceed 15 s using 50 ppm of simethicone.

Suitable emulsifiers may for example be polysorbates with different amounts of ethylene oxide, monoglycerides, and sorbitan esters with different fatty acids, acetylated monoglycerides, or ethoxylated fatty alcohols and fatty acids as defined in *U.S. National Formulary* [55]. These emulsifying agents also comply with the U.S. Food and Drug Administration regulation [86] and are recognized as safe in food.

As viscosity-increasing agents agar, guar gum, starch xanthan gum, different cellulose derivatives as well as certain polyacrylates are listed in *U.S. National Formulary* [55].

Preservatives are used to protect the organic-based emulsifiers and thickening agents against biodegradation by microorganisms. Dimethicone and simethicone themselves are not biodegradable. According to the U.S. Pharmacopeia [56] the aerobic microbial counts should not exceed 100 per gram [87]. In addition, as for the simethicone itself, the heavy-metal content of simethicone emulsion should not be greater than 0.001% (10 ppm).

V. IN VITRO AND IN VIVO EXPERIMENTS

The different amounts of simethicone used as a quality control and the differences of maximum foam collapse time in the description of the *U.S. Pharmacopeia* [56] indicate, that the efficiency of the pharmaceutical antifoams will vary with their formulation. Unfortunately there is little information about the methods of production of these pharmaceuticals. The "quality" of the antifoams can therefore only be estimated by experimental trials. Apart from the detailed description in the *U.S. Pharmacopeia* [56] some other methods are described.

One of the most widely used in vitro test methods is that described by Rezak [88]. This method is a shaking test using a sodium lauryl sulfate solution (100 mL; conc. of surfactant: 2.5 gL^{-1}) in a graduated cylinder (volume, 250 mL). Using this method Cox and Nijland[89] have compared antiflatulant preparations available in the Netherlands. The tests were carried out by creating a certain foam volume (250 mL). Then the pharmaceuticals were added in an amount corresponding to 40 mg simethicone. After mixing, the foam volume and the foam collapse time were estimated. Cox and Nijland found that the formulations containing simethicone as the only drug are the most efficient. Preparations containing antiacids show reduced antifoam properties.

Carless et al. [90] have proposed another test method during their investigation of the effect of finely divided silica on the antifoam properties. They have introduced a dynamic foam test by aerating a stirred polyoxyethylene-stearyl-cetyl ether solution through a sintered glass disk. After creation of a certain foam volume the antifoam was added and the time measured until complete foam disappearance occurred. Using these tests, Carless et al. [90] showed that solvent extraction of dimethicone from simethicone tablets led to a considerably reduced efficiency of the tablets. Unfortunately Carless et al. [90] gave no test results with the extracted dimethicone, so that their conclusion that dimethicone is primarily responsible for the antifoam properties is in part contrary to test results, which have shown, that the pure dimethicone is only a poor antifoam and that simethicone shows a significantly enhanced efficiency [91]. Nevertheless the results show, that dimethicone, strongly absorbed on solid particles in the tablets as a thin film, is a weak antifoam. This may lead to the conclusion, that simethicone formulated with antacids to form a tablet may, due to the properties of dimethicone, spread over the particles thus loosing antifoam efficiency. These effects may be an explanation for the results found by Cox and Nijland [89].

Similar phenomenon are well known where simethicone antifoams are used in powdered detergents. If the antifoams are not prevented from spreading over the detergent powder (for example, by encapsulation) the efficiency is lost after very short storage times. For this application antifoam formulations have been developed where the polysiloxane is prevented from spreading [92–95].

While the antifoam effect may be reduced when simethicone is formulated with antacids the acid neutralization capacity of the antacids, on the other hand, is not significantly changed [84,96].

Aeration tests have also been carried out by Stead et al. [97]. Here the foaming medium constituted a solution containing hydroxyethylcellulose and sodium lauryl sulfate which was adjusted to a pH value of 1.2 with hydrochloric acid. Stead et al. [97] have compared three different antacid preparations (partly containing simethicone) and also assigned a ranking in respect to the antifoam properties. They have found that the combination of hydrotalcite with simethicone showed a greater antifoaming effect than aluminum hydroxide/dimethicone did. This ranking was also confirmed by in vivo tests with 12 patients.

A third method to test the efficiency of pharmaceutical simethicone preparations is described by Loebel [98]. Using sodium lauryl sulfate as the surfactant, a 1% solution is circulated with a flow inducer. The circulated solution falls back into its reservoir, thus creating foam. Adjusting the flowrate a constant foam volume is produced. Powdered simethicone tablets are then added, and the foam volume is estimated after certain times of circulation. The tests are repeated with standard gastric and gut juice corresponding to DAB 7 [66]. The results found, indicate, that for the different foaming systems the ranking of the antifoams with respect to their efficiency did change. Nevertheless, the most efficient preparation against sodium lauryl sulfate was the best in the other media too. The differences found should have been caused by the different formulations and the production technology respectively.

The tests described above can be done in laboratories. Initial tests have always been conducted using synthetic foam stabilizers, where sodium lauryl sulfate is preferred. Unfortunately, extensive test series which are possible in vitro using these tests cannot be done in vivo. So for in vivo tests usually only one antiflatulence preparation is tested in a "double blind" study. Although the efficiency of simethicone was established about 35 years ago, there are numerous recent publications demonstrating the efficiency of new formulations. Efficiency of the pharmaceuticals is normally determined by the reduction of discomfort and pain and/or by sonografic, radiological, or endoscopic examinations.

Gladisch et al. [99], for example, described the efficiency of two simethicone preparations of different formulations, used for premedication during an abdominal ultrasound investigation with 40 patients. They have found that both pharmaceuticals improved the quality of imaging of the abdominal vessels and the epigastric structures. Similar results were found by Brockmann [100] during his sonographic examination and by Fixa et al. [101] during their radiological examinations.

Konstantinidis [102] demonstrated improvements in endoscopy by using a dimethicone preparation which permitted good assessment of the mucosa of the stomach and the duodenum in about 90% of all cases. The investigation has been done with 578 patients.

Significant improvement of the symptoms of dyspepsia has been found in the treatment of 43 elderly female patients with an antacid/simethicone/enzyme combination by Gotzes [103](reduction of symptoms, 87%), by Vivat [104] using an aluminum hydroxide/magnesium hydroxide/cholestramine/simethicone formulation (reduction of ulcer size; 80%; 28 patients), and by Frühsorger and Fuchs [105] applying aluminum sodium carbonate dihydroxide/simethicone (improvement, 79%; 60 patients (male and female)).

Comparable results were obtained by Eveld [106] using an enzyme/simethicone combination (improvement, 80%; 39 patients; placebo, 38%; 40 patients), by Auld [107] treating 22 male and female patients with an enzyme/simethicone agent (patients' response to therapy, 100%) or by Schroeder [108] applying an enzyme/simethicone preparation (improvement, 100%; placebo improvement, 21%).

Besides the reduction of symptoms caused by dyspepsia some authors reported additional effects of simethicone during their examinations. Birtley et al. [75] have found a protection of the mucosa against aspirin-induced gastric irritation in rats, while Neumeyer and Rogozinski [109] observed improvement from pain caused by duodenic ulcer (12 patients).

Simethicone is not only useful for combatting foam formation in the gastrointestinal tract, but it also improves the properties of double-contrast barium enema [110].

There have been a few attempts to use materials other than simethicone as antifoams in pharmaceutical preparations. Davies et al. [111] have tested lecithin as an antifoam agent. Compared with simethicone a 20-fold greater amount has to be used to obtain comparable results.

VI. CONCLUSIONS

Silicone antifoams (simethicones) have been used in pharmaceuticals to combat digestive upsets or to enhance the inspection of the gastrointestinal tract for more than 35 years. They still play an important role today. This is demonstrated by numerous publications over the years which have again and again demonstrated the effectiveness of simethicone as an antifoam against abdominal foams. Unfortunately, there is no information about the surface active species in gastric juices which stabilize foam in the gastrointestinal tract. Simethicones have gained their position due to their chemical inertness and their neutral physiological behavior. It has to be noted that there are regulations, which define the materials to be used in pharmaceutical applications. The *U.S. Pharmacopeia* in connection with its sister compendium the *National Formulary* seems to be the most important, which is accepted nearly everywhere.

REFERENCES

1. L. E. Casida, Industrial Microbiology, Wiley, New York, 1968.
2. A. Rembaum and M. Shen, Biomedical Polymers, Marcel Dekker, New York, 1971.
3. N. Kossovsky and J. P. Heggers, *CRC Crit. Rev. Biocompat. 3*:53 (1987).
4. H. W. Weissenburger and M. G. von der Hoeven, US 3499909, (1970).
5. A. F. Asker and C. W. Whitwirth, *J. Pharm. Sci 63*: 1630 (1974).
6. F. S. Rankin, *Chimicaoggi*, 37 (1986).
7. C. E. Creamer, *Pharma. Technol.*, 79 (1982).
8. A. H. Quin, J. A. Austin, and K. Ratcliff, *J. Am. Vet. Med. Assoc. 114*: 313 (1949).
9. M. D. Hirschowitz, R. J. Bolt, and H. M. Pollardt, *Gastroenterology 27*: 649 (1954).
10. M. Gasster, J. O. Westwater, and W. E. Molle, *Gastroenterology 27*: 652 (1954).
11. M. E. Dailey and J. A. Rider, *J. Am. Med. Assoc. 155*: 859 (1954).
12. J. A. Rider and H. C. Moeller, *J. Am. Med. Assoc. 174*: 2052 (1960).
13. J. A. Rider, *Ann. N. Y. Acad. Sci. 150*: 170 (1968).
14. E. Brunk, *Diagnostika 9*: 637 (1976).
15. I. A. D. Bouchier, *Practitioner 224*: 373 (1980).
16. M. M. van Nees and E. L. Cattau, *Am. Fam. Physician 31*: 198 (1985).
17. G. Trieb and E. Nusser *Med. Welt 24*: 1045 (1973).
18. J. Weiss, *Curr. Therapeutic Res. 16*: 909 (1980).
19. F. R. Steggerda and J. F. Dimmick, *Am. J. Clin. Nutrit. 19*: 120 (1966).
20. J. Weiss, S. Weiss, and B. Weiss, *Am. J. Gastroenterology 40*: 528 (1963).
21. M. L. Riccitelli and J. Weiss, *J. Am. Ger. Soc. 8*: 469 (1965).
22. D. Y. Graham, J. L. Smith, and D. J. Patterson, *Am. J. Gastroenterology 78*: 257 (1983).

23. W. Noll, Chemie und Technologie der Silicone, Verlag Chemie, Weinheim, 1968.
24. A. J. Barry, *J. Appl. Phys. 17*: 1020 (1946).
25. W. A. Piccoli and F. O. Stark, *J. Am. Chem. Soc. 77*: 5017 (1955).
26. A. Weissler, *J. Am. Chem. Soc. 71*: 93 (1949).
27. R. R. LeVier, *L'actualité chimique*, 89 (1986).
28. W. D. Harkins, *J. Chem. Phys. 9*: 552 (1941).
29. J. V. Robinson and W. W. Woods, *J. Soc. Chem. Ind. 67*: 361 (1948).
30. S. Ross, *Rensselaer Polytech. Inst. Eng. Sci. Ser. 63*: 1 (1950).
31. G. Koerner, "Goldschmidt informiert . . ." *3/70*, 24 (1970).
32. R. D. Kulkarni, E. D. Goddard, and B. Kanner, *J. Colloid Interface Sci. 59*: 468 (1977).
33. S. Ross and G. Nishioka, *J. Colloid Interface Sci. 65*: 216 (1978).
34. This book, Chapter 1.
35. K. Haubennestel, DEOS 3245482 (1982).
36. E. Pirson and J. Schmidlkofer, DEOS 2720512 (1978).
37. I. A. Lichtmann and F. E. Woodward, US 3793223 (1974).
38. S. Nitzsche and E. Pirson, DEOS 1545185 (1971).
39. K. Klein and J. Maluzi, BP 1204383 (1967).
40. S. Ross and G. Nishioka, Emulsions, Latices, Dispersions, in (P. Becher and M. N. Yudenfreund, eds), Marcel Dekker, New York 1978, p. 237.
41. P. R. Garrett, *J. Colloid Interface Sci. 69*: 107 (1979).
42. A. Dippenaar, *Int. J. Miner. Process 9*: 1 (1981).
43. G. C. Frye and J. C. Berg, *J. Colloid Interface Sci. 127*: 222 (1989).
44. V. K. Rowe, H. S. Spencer, and S. L. Bass, *J. Ind. Hyg. Toxicol. 30*: 332 (1948).
45. V. K. Rowe, H. S. Spencer, and S. L. Bass, *Arch. Ind. Hyg. 1*: 539 (1950).
46. G. D. Child, H. O. Panguin, and W. B. Deichmann, *Arch. Ind. Hyg. 3*: 479 (1951).
47. Ch. Gloxhuber and G. Hecht, *Arzneimittel Forschng 5*: 10 (1955).
48. W. E. McDonald, G. E. Lainer, and W. B. Deichmann, *AMA Arch. Ind. Health 21*: 514 (1960).
49. G. L. Kennedy, M. L. Keplinger, and J. C. Calandra, *J. Toxicol. Environ. Health 1*: 909 (1960).
50. M. G. Cutler, A. J. Collings, I. S. Kitt, and M. S. Sharratt, *Food Cosm. Toxicol. 12*: 443 (1974).
51. E. Largent, M. Blackstone, and J. Roth, *U.S. Air Force Med. Service*, 1 (1950).
52. J. C. Calandra, M. L. Keplinger, E. J. Hobbs, and J. L. Taylor, *Polymer Preprints 17*: 1 (1976).
53. R. Dailey, FDA Report No. FDA/BF-79/29 (1978).
54. FDA 21 CFR 174.340 of the United States Food and Drug Administration.
55. NF XVII; Official Monograph publ. by the United States Pharmacopeial Convention Inc. (1990).
56. USP XXII; Official Monograph publ. by the United States Pharmacopeial Convention Inc. (1990).

57. General Test and Assays, U.S. Pharmacopeia XXII, p. 1619, (1990).
58. General Test and Assays, U.S. Pharmacopeia XXII, p. 1609, (1990).
59. General Test and Assays, U.S. Pharmacopeia XXII, p. 1609, (1990).
60. General Test and Assays, U.S. Pharmacopeia XXII, p. 1523, (1990).
61. General Test and Assays, U.S. Pharmacopeia XXII, p. 1520, (1990).
62. General Test and Assays, U.S. Pharmacopeia XXII, p. 1524, (1990).
63. octylphenol ethoxylated.
64. A specification for the registration of Simethicone in Germany is in preparation.
65. J. E. F. Reynold, Martindale—The Extra Pharmacopoeia, 28th ed. Pharmaceutical Press, London, 1982.
66. DAB = Deutsches Arzneimittel Buch.
67. "Rote Liste 1989", editor: Bundesverband der Pharmaceutischen Industrie e.V., Frankfurt, 1989.
68. Smith Kline & French Lab GB 1129260, (1968).
69. O. Honecker, DEOS 2040425, (1972).
70. W. Bilhuber and P. Rados, DEOS 2517585, (1975).
71. J. A. Rider, USP 4198390, (1980).
72. J. A. Rousseau, EPA 0068566, (1983).
73. M. Wischniewski, H. Pfanz, and S. Funke, GB 1154256, (1969).
74. V. B. Surpuriya and C. A. Colleen, DEOS 2940905, (1980).
75. FDA 21 CFR 331 of the United States Foods and Drugs Administration.
76. J. Schnekenburger, *Arzneimittel Forsch.* 24: 142 (1974).
77. G. D. Kerr, *J. Int. Med. Res.* 2 (Suppl (2)): 1 (1974).
78. C. van Dop, G. M. Overliet, and H. M. Smits, *Pharma. Weekblad 111*: 1093 (1976).
79. R. A. Yokel, *Am. J. Hosp. Pharm.* 34: 200 (1977).
80. A. Singh and H. C. Mital, *Pharm. Acta Helv.* 52: 319 (1977).
81. P. Hagemann and K. Gamper, *Schweiz. Rundschau Med.* 69: 847 (1980).
82. D. Drake and D. Hollander, *Ann. Intern. Med.* 94: 215 (1981).
83. M. Stowasser, Z. *Allgem. Med.* 57: 1262 (1981).
84. K. Thoma and H. Lieb, *Pharm. Zeitung 129*: 121 (1984).
85. M. E. MacCara, F. J. Nugent, and J. B. Garner, *Can. Med. Assoc. J. 132*: 523 (1985).
86. FDA 21 CFR. 178.3400.
87. General Test and Assays, U.S. Pharmacopeia XXII, p. 1479, (1990).
88. A. M. Rezak, *J. Pharm. Sci.* 55: 538 (1966).
89. H. L. M. Cox and C. J. Nijland *Pharma. Weekblad 111*: 973 (1976).
90. J. E. Carless, J. B. Stenlake, and W. B. Williams, *J. Pharm. Pharmac.* 25: 849 (1973).
91. R. N. D. Birtley et al. *J. Pharm. Pharmac.* 25: 859 (1973).
92. H. F. Fink et al., EPA 097867, (1982).
93. P. A. Morgan, EPA 0091802, (1983).
94. H. Reuter, W. Seiter, DEOS 3436194, (1986).
95. P. W. Appel et al., EPA 0256833, (1988).
96. F. Mittelstaedt and E. Grabener, *Akt. Gastrologie 6*: 563 (1977).

97. J. A. Stead, R. A. Wilkins, and J. J. Ashford, *J. Pharm. Pharmac. 30*: 350 (1978).
98. K. D. Loebel, *Ärtzliche Praxis 28*: 1194 (1976).
99. R. Gladisch, R. Elfner, B. Massmer, and H. Ulrich, *Ultraschall 6*: 114 (1985).
100. P. Brockmann, *Therapiewoche 31*: 5882 (1981).
101. B. Fixa, O. Komarkova, and R. Krizek *Ther. Gegenw. 111*: 217 (1972).
102. Th. Konstantinidis, *Akt. Gastrologie 5*: 447 (1976).
103. H. Gotzes, *Therapiewoche 26*: 6177 (1976).
104. J. Vivat, *Acta Therapeutica 8*: 79 (1982).
105. A. Frühsorger and E. Fuchs, *Therapiewoche 36*: 3737 (1986).
106. H. Eveld, *Med. Mschr. 31*: 424 (1977).
107. J. M. Auld, *Curr. Therap. Res. 26*: 55 (1979).
108. W. Schroeder, *Therapiewoche 30*: 4945 (1980).
109. G. Neumeyer and H. J. Rogozinski, *Z. Allg. Med. 55*: 539 (1979).
110. R. Virkki, P. Mäkelä, and M. Koramo, *Europ. J. Radiol. 1*: 134 (1981).
111. P. J. Davis, K. A. Khan, and S. M. Sallis, *Pharm. Acta Helv. 51*: 378 (1976).

5

High-Performance Antifoams for the Textile Dyeing Industry

GEORGE C. SAWICKI Dow Corning Europe, Brussels, Belgium

I.	Introduction	193
II.	Causes of Foam in Textile Dyeing	194
	A. Textile dyeing auxiliaries	194
	B. The dyeing process	198
III.	Foam Control in Textile Dyeing	202
	A. Foam stabilization	202
	B. Antifoams for conventional dyeing processes	203
	C. Antifoams for jet dyeing processes	211
IV.	Conclusion	218
	Acknowledgment	219
	References	219

I. INTRODUCTION

Fortunately for the human race, simple incorporation of air in pure water or water containing inorganic salts does not create a durable foam. It is only through the introduction of organic surface-active agents or surfactants that a durable foam will be generated. The stability or persistency of the foam and the volume of the foam will depend on the nature and concentration of the surfactants present and the degree of air entrainment within the process. Therefore, any process that involves the introduction of air into a surfactant-containing aqueous system is liable to have a foam problem. Many such processes exist in the textile industry and preventing the formation of unwanted foam is often critical to their efficiency.

Since foam can only occur if air is incorporated into an aqueous, surfactant-containing bath, the first step in tackling a foam problem should be a complete review of the process to minimize turbulence and air entrainment, for example, extending feed pipes to below the liquor level in tanks and kettles, checking that agitators are well immersed in the liquid, checking pumps to ensure that air is not being fed into the flow of liquor, and removing air from the textile material prior to processing. The second step should be to review all the surfactant-containing ingredients used in the process, for example, reducing the concentration of each of these ingredients to an absolute minimum, identifying alternative products that contain surfactants which produce less foam or a less stable foam, and minimizing the carryover of ingredients from one process step to the next. By taking the necessary corrective actions the foam problem should have been minimized, but if the foam level is still unacceptable then the addition of the appropriate antifoam will be necessary.

II. CAUSES OF FOAM IN TEXTILE DYEING

A.   Textile Dyeing Auxiliaries

The largest use of antifoams in the textile industry is during the dyeing of fibers and fabric. The medium used is predominantly water, because of its suitability, availability, and low cost, and the process itself can be described as the transfer of water-soluble or water-dispersible dyes from the aqueous medium onto and into a given fiber. To obtain uniform or even dyeing there must be unimpeded transfer of the dye between the medium and the fiber. This can only be achieved if foam is eliminated from the dye bath, thus ensuring that the surface of the fiber is in direct contact with the dyeing medium throughout the dyeing process. In many cases this requirement can only be met through the addition of the appropriate antifoam.

The choice of dye will very much depend on the required color, the nature of the fiber being dyed, and the dyeing process selected. One feature that all the dyes have in common is that they are all water soluble at some stage during the dyeing process. While the dyes are present in the dye bath in their water-soluble state, they can be classified as surface-active materials since they consist of a hydrophobic color component and hydrophilic, water-solubilizing groups. As surface-active species they are able to promote and stabilize foam. However, in general, the foam that can be generated by the dye molecules themselves is neither copious nor stable.

The main categories of dyes are acid dyes, basic dyes, disperse dyes, reactive dyes, direct dyes, vat dyes, and sulfur dyes. Normally, the nature of the fiber dictates the category of dye that is chosen. Acid dyes are used primarily to dye fibers with cationic groups such as nylon, wool, silk, and modified acrylics. Basic dyes are used to dye fibers with anionic groups such as acrylics and modified polyesters. Disperse dyes are used to dye hydrophobic fibers such as polyesters and nylon. Reactive dyes are used to dye fibers such as cotton, nylon, wool, and silk that contain chemical groups with which they can react, and direct dyes, vat dyes, and sulfur dyes are used to dye cotton. However, simply selecting the appropriate dye for a given fiber will not ensure level or even dyeing. In order to achieve this aim, the dye bath must also contain the appropriate level of various other ingredients, commonly referred to as textile-dyeing auxiliaries.

The first example is the use of leveling agents with acid dyes. Acid dyes are sodium salts of dye molecules which contain one or more sulfonic acid groups, the presence of the sulfonic acid groups providing the necessary solubility in aqueous media. They are used extensively in dyeing fibers such as nylon which contain potential cationic groups. In the case of nylon, these dyes are applied under acidic conditions, so that the amine dye adsorption sites present on the fiber are protonated, thereby making bonding possible through complex formation between the cationic dye adsorption sites and the anionic dye molecules [1]. In practice, mixtures of dyes are more commonly used than single dyes to produce the required shade. This is an added complication, because the dyes with the higher affinity for the limited number of amine groups on nylon will be adsorbed preferentially, and because they are more firmly bonded tend not to level by migration effectively. This also leads to highlighting of irregularities in the yarn introduced by irregular tension during weaving or nonuniform heat setting [2]. Anionic leveling agents, such as colorless sulfonic acids [2], are usually added to the dye bath to improve the leveling and migration properties of the dyes. These materials preferentially adsorb on the dye receptor sites and temporarily block them, slow down the rate of exhaustion of the dye and by so doing help achieve more level dyeing [1]. An alternative method is to complex the dye with a cationic surfactant and thereby reduce the rate of adsorption of the acid dye on the fiber. When using this approach, nonionic surfactants must also be added to maintain the dispersion of the complex in the dye bath [2]. The presence of these surface-active, leveling agents will have a significant impact on the foam propensity of the dye bath.

Basic dyes are water-soluble dyes that contain a cationic dye molecule. They are used in dyeing acrylic fibers which have structural anionic groups present [1]. Under neutral or basic conditions, bonding occurs through complex formation between the anionic dye adsorption sites of the acrylic fibers and the cationic dye molecules. Diffusion of these dyes to these sites is negligible below the glass transition temperature (about 65°C in water), but is rapid at the normal dyeing temperature of 96–100°C. Typical dyeing time is 1–2 hr but can be shortened by dyeing under pressure. Leveling of basic dyes can be a significant problem that can often be traced back to the different rates of exhaustion of the individual components of the particular dye shade. To overcome the poor leveling, colorless, cationic surfactants are usually added to the dye bath. They preferentially adsorb on the dye receptor sites, slow down the rate of adsorption and this in turn leads to improved uniformity of color [1]. An alternative method is to add anionic surfactants which can form complexes with the dye. This reduces the amount of cationic dye available for adsorption and hence reduces the rate of adsorption of the dye on the fabric. To achieve adequate dispersion of the anionic surfactant-dye complex in the dye bath, nonionic surfactants must also be added [2]. As with acid dye leveling agents, these additional surface-active materials will have a significant effect on the foam propensity of the dye bath.

Unless chemically modified, hydrophobic fibers such as polyester lack reactive dye sites and are dyed almost exclusively with disperse dyes. These dyes have a very low solubility in water and have to be added to the dye bath as very fine, uniform, stable dispersion. To achieve the correct level of dispersion, the dyes are wet-milled in the presence of high levels of surface-active, dispersing agents such as lignin sulfonates to give a dye in the liquid or paste form, or subsequently spray-dried to give the powder form [3]. This implies that every time a disperse dye is added to the dye bath, a significant quantity of surface-active, dye dispersants will also be added. In some cases, additional dye dispersant may need to be added to the dye bath to ensure that the disperse dye remains properly dispersed. This is most likely when dyeing pale-to-medium shades, when the quantity of dye dispersant added via the dye is insufficient to give the required concentration in the dye bath [4]. Dispersing agents based on naphthalene sulfonic acid–formaldehyde condensation products and especially cresol sulfonic acid–formaldehyde condensation products have been shown to be particularly effective, and even at high concentrations show little or no retarding [5].

The levelness of polyester dyeing is not only dependent on the stability of the disperse dyes in solution, but also on the affinity of the fiber for the dyes and the ability of the dyes to migrate in the fiber at the dyeing temperature. By adding leveling agents such as ethoxylated fatty acids and ethoxylated nonyl or tributyl phenols [5], it is possible to increase the solubility of the disperse dyes which in turn leads to improved dye migration and increased dye penetration. In general, anionic surfactants are used to improve the dispersion stability of the dye and nonionic surfactants are used as leveling agents. Although the latter improve leveling, they are not effective enough to cover highlighting of irregularities in the yarn introduced by irregular tension during weaving or nonuniform heat setting.

At temperatures below 100°C the rate of dyeing of polyester with disperse dyes is very slow. To accelerate the process, dye carriers or accelerants are added to the dye bath. These include materials such as o-hydroxydiphenyl, methyl salicylate, tripropyl phosphate, butyl benzoate, and trichlorobenzene [6]. Their addition can increase the dyeing rate by as much as 50 times that achievable in their absence. These, essentially water-insoluble, organic compounds, swell the fiber and allow the molecularly dispersed dye to diffuse into the fiber more rapidly [6]. As well as increasing the rate of diffusion into the fiber, they also strip excess dye from the fiber. The dyeing process is complete when the rate of uptake of dye equals its rate of removal and wash fast, level dyeing is achieved. To overcome the slow dyeing rate, high-temperature and pressure processes have been developed. In these processes, much lower levels of dye carrier are required since the rate of diffusion of the dye both into and out of the fiber occurs far more rapidly at these elevated temperatures. Legislative pressure is hastening the trend toward higher temperature and pressure since very few of the carriers can be regarded as environmentally desirable. Dye carriers are normally supplied as emulsions to achieve the correct level of dispersion in the dye bath and prevent the essentially insoluble, active material from separating out and causing spotting or uneven dyeing. In the case of solid, carrier-active compounds, solvents are used to solubilize the materials and enhance the activity of the finished product. This means that whenever a dye carrier is introduced into a dye bath, surface-active emulsifiers such as sodium dodecylsulfate [2] used in formulating the emulsion will also be added.

Clearly the composition of the dye bath is more than just a mixture of fabric, dye, and water, and will necessarily include other additives such as dyeing auxiliaries. This implies that the dye bath will contain a number of different surface-active materials, and it is these materials that promote and stabilize foam in various dyeing processes. The surface-active materials will include the actual dyes themselves, when they are present in their water-soluble state, but more importantly the surface-active components of the dyeing auxiliaries. Some examples already given include anionic, nonionic, and cationic leveling agents, dye dispersants in the formulation of disperse dyes and emulsifiers used in formulating dye carriers. In addition there will also be other auxiliaries such as wetting agents to ensure that the fiber or fabric comes into direct contact with the dyeing medium, and fiber lubricants to reduce chafe marks.

A description of the composition of a dye bath would not be complete without at least some indication as to the concentrations of the various dyeing auxiliaries that are likely to be present. Taking one of the major dyeing processes, namely the dyeing of polyester with a disperse dye, a typical dye bath could contain [5]

0.5–5% by weight of substrate of a disperse dye
0.3–1 g/L leveling agent
0.5–2 g/L dispersing agent
0.0–1 g/L dye carrier

This means that significant quantities of surface-active materials are intentionally added to achieve level or even dyeing.

Finally, there is one further source that is often neglected and that is accidental carryover of detergent surfactants from the previous scouring step, usually resulting from inadequate rinsing prior to the dyeing step. This means that virtually every dye bath is the possible source of a significant foam problem.

B. The Dyeing Process

Within the dyeing machine, there must be sufficient movement of the liquor to ensure uniform penetration of the fiber assemblies, uniform temperature throughout the liquor, and rapid dilution of the concentrated dyestuff and auxiliaries before they come into contact with the fiber assemblies. There are three main ways of achieving these goals [7].

1. Circulating the Dye Liquor through Stationary Substrate

In package dyeing machines, yarn is wound around a stainless steel "former" to give a "cheese." These cheeses are loaded onto perforated steel tubes mounted in the dyeing vessel, and dye liquor is then pumped through the perforated steel tubes, the formers, and the fabric before flowing back to the pump for the process to continue. In beam dyeing, cloth is wound onto a horizontal, hollow, perforated cylinder and dye liquor is pumped radially through the beam. To ensure level dyeing, package and beam dyeing machines have an automatic reverse switch which changes the flow direction of the liquor on a preset schedule. In loose stock dyeing machines, the fibers are packed into a vessel between two perforated plates. The liquor is then circulated through the bottom plate, through the mass of fibers and out through the top plate, before returning to the pump. Since level dyeing is not so critical with loose stock dyeing, machines tend only to pump the dye liquor in one direction.

2. Movement of Substrate without Mechanical Movement of Liquor

This technique is used in chain warp dyeing, jig dyeing, beck dyeing, and continuous dyeing. In chain warp dyeing, several warp ropes are pulled through a series of vats of dye liquor until the correct shade has been obtained. In jig dyeing several hundreds of metres of cloth are wound on a beam roller, and then transferred onto another roller via a dye bath. Once the cloth has been wound onto the opposite beam roller, the direction is reversed and the cloth is rewound onto the original roller. This is continued until the cloth is dyed to the correct shade. In beck or winch dyeing, an elliptical roller lifts the fabric from the vertically faced side of the dye filled beck and transfers it to the gradually sloping side of the machine. The fabric slides down along the slope before once again being lifted out by the roller. The elliptical roller arranges the fabric in folds before it falls back into the dye liquor. The fabric is circulated through the dye bath in this manner until the correct shade is obtained. The modern beck is constructed of stainless steel to be operated at atmospheric pressure or if constructed as a heavy-walled, sealable unit it may be used at high temperatures and pressures. In continuous dyeing, the dye is pad applied to the fabric. The dye is applied by passing the fabric through the dye bath, excess dye is then removed by passing it between a pair of squeeze rollers before going on to be processed to achieve adequate fixation of the dye. This technique is widely used in dyeing long lengths of nonstretch, woven goods.

3. Movement of Substrate and Circulation of Liquor

This procedure is used in jet dyeing. In this batch process the circulation of the liquor is actually used to move the fabric [8,9]. The fabric, which is in rope form, is transported by the movement of liquor through a venturi jet (Fig. 1). The cloth travels vertically through the center of the jet, and at the throat of the jet is met by a circumferential, converging jet of dye liquor which emerges from a ring-shaped slot completely surrounding the fabric (Fig. 2). The fluid simultaneously penetrates and propels the fabric and produces tremendous interchange between the fluid and the fabric. The fabric emerges from the cloth guide tube where it is met by a doffing jet which changes the direction of motion of the material and helps pile the material in loose folds in the cloth storage chamber, so no crushing or rope marking can occur. Here the fabric is completely submerged, but the high-volume flow through the chamber ensures the mass of fabric is kept moving. As the cloth emerges from the chamber it passes through metering rolls which both control the cloth speed and reduce the tension on the fabric, as now it only has its own weight to bear to the rolls and not the full height to the jet. A centrifugal pump recirculates the entire dye bath through the jet every 20–30s. Operating temperatures up to 160°C are combined with the high fluid energy and very short dye times result. Modern jet dyeing machines are fairly gentle to the fabric, while providing the required intimate contact between the dye and the fabric necessary for uniform dyeing [9]. The big advantage that this technique has over beck dyeing is the energy saving that

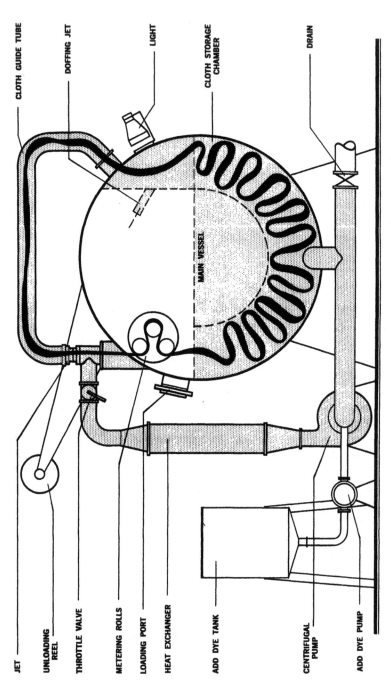

FIG. 1 Schematic cross section of a Gaston County Jet Dyeing Machine. (From Ref. 8. Reproduced with kind permission of Gaston County Dyeing Machine Co., North Carolina.)

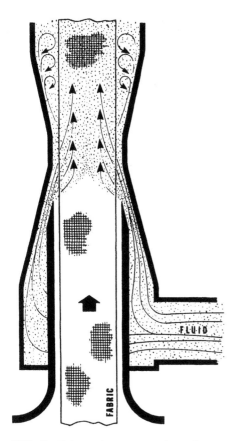

FIG. 2 Schematic cross section of the main jet in a Gaston County Jet Dyeing Machine. (From Ref. 8. Reproduced with kind permission of Gaston County Dyeing Machine Co., North Carolina.)

results from the increased cloth-to-liquor ratio, especially when using high-temperature and -pressure dyeing conditions. Typical cloth-to-liquor ratios would be of the order of 1:10 for a full load. Under certain conditions, cloth-to-liquor ratios as high as 1:5 can be used. The high circulation rate ensures the homogeneity of the dye bath from a chemical and temperature standpoint and leads to more level dyeing than with other dyeing techniques such as jig or winch dyeing [9].

The Gaston County jet dyeing machine described above is one of many different designs available. What is common to all jet dyeing machines is the use of high-velocity circulating liquor to transport the fabric rope through the machine. The high velocity is achieved either via a large circulating pump and venturi, as in the Gaston County machine, or via an overflow system. In the latter, the venturi is replaced by a system in which liquor is pumped from the lower half of the machine to a reservoir in the upper half. Liquor then flows down a narrow tubular sluice under gravity carrying the fabric with it. Again a driven winch is used to assist and control fabric circulation [9].

The jet dyeing machines described above are known as partially flooded machines since the dye liquor only fills part of the total volume. Due to the high level of liquor turbulence, caused by the rapid rate of fabric and liquor circulation in the jet dyeing machine, such machines represent a major foam control problem. If foam is allowed to form, the pump may well start cavitating. This in turn will lead to loss of pump efficiency, and tangling and stoppage of the fabric rope. In order to continue the dyeing process, the dyeing machine must be cooled to below the boil, the tangle located and dealt with. The liquor must then be reheated to the dyeing temperature. This is both costly and time consuming. In order to eliminate the foam problem fully flooded machines were introduced. In these machines the total free volume of the machine is filled with the dye liquor. As a result no air can be entrained into the circulating liquor and hence no foam will be formed. Such a development also allowed the use of dyes and dyeing auxiliaries that could not be used in partially flooded machines because of their high tendency to promote foam.

Unfortunately, because the fabric is in constant motion, tangling is a frequent occurrence [9]. This coupled with a need to reduce energy consumption has led to a move back to partially flooded machines in which lower liquor volumes are required.

In every one of the partially flooded processes described, there is sufficient turbulence to entrain air in the dyeing medium. Since the dyeing medium already contains a significant number of surface-active materials, foam formation during the dyeing process is inevitable. The volume of foam and its persistence will depend on the nature and the concentration of the surfactants present, and the degree of turbulence within the process. Persistent foam will cause uneven dye adsorption, fabric flotation, pump cavitation, fabric tangling, wasted volume in process equipment, and general housekeeping difficulties. It is therefore not surprising that the largest use of antifoams in the textile industry is in the textile dyeing area.

III. FOAM CONTROL IN TEXTILE DYEING

A. Foam Stabilization

A foam consists of gas entrapped within thin films or lamellae of a liquid. Foaming is always accompanied by an increase in the interfacial surface area of the system and hence its total free energy. Since systems always seek to reduce free energy, foams must be thermodynamically unstable. Foam formation requires a net input of energy and it is the mechanical agitation present in the process that provides the energy needed to entrain the gas initially. The ease of foam formation in an aqueous medium is determined in part by the requirement that stabilizing surface tension gradients exist. The surface-active materials present in the continuous phase can adsorb at newly created surfaces to create such gradients thereby stabilizing the gas that has been entrained and hence the foam that has been produced.

Depending on the conditions prevailing, the stability of the foam generated can vary from milliseconds to almost unlimited duration. The presence of materials in the dye bath that can adsorb at the freshly created interface implies that foam formation is inevitable when air is entrained in the dye liquor. However, the stability of the foam generated is dependent on a number of other factors. These include

Surface elasticity
Surface viscosity
Bulk viscosity
Electrical double layer and entropic repulsion
Gas diffusion
Evaporation

The relative importance of each of these will depend on the materials that are present in the dye bath. Unfortunately, it is only possible to guess at the order since these factors have not been investigated for dye bath compositions. A more comprehensive coverage of the theory of foam stabilization is given in Chapter 1.

B. Antifoams for Conventional Dyeing Processes

In many practical applications the foam lasts long enough to interfere with the physical and chemical processes being carried out, and steps must be taken to prevent it occurring. It is this need that has led to the development of a vast number of antifoaming agents whose primary role is to destabilize the foam as it is being generated through air entrainment.

Though antifoams are primarily judged by the persistence of their antifoaming action in a given foaming medium, this is by no means the only criteria for choice. Other equally important attributes include

Must eliminate existing foam as well as prevent further foam from forming
Must be storage stable
Must be chemically stable under dyeing conditions
Must be easy to disperse in the dye bath

Lack odor
Lack color
Must not react or interfere with dyes or auxiliaries present in the bath
Must not deposit on fabric and machinery causing spotting or staining of
   fabric
Must be safe to both humans and the environment

Such conditions can often be met by proper formulation of the product. However, no one product is ever likely to give the optimum performance in all the possible dye bath conditions that could be met. Ideally, to avoid any subsequent problems, one should test the compatibility of the antifoam with all the other dye bath components under the conditions that will be encountered within the process prior to use in production. Such preliminary testing will avoid the risk of instability in the dye bath leading to imperfect dyeing.

1. Typical Antifoams Used in Textile Dyeing

In textile processing a number of different categories of antifoaming agents are used. These include formulated products based on fatty acids, fatty alcohols [10], and their ethoxylates, polyoxyethylene and polyoxypropylene block copolymers [11], dispersions of mineral oils or silicone fluids and hydrophobic silica [12], silicone glycols [13], and phosphoric acid esters [14–16] among others. The choice is often governed by the specific surface-active agents present in the bath and the conditions that will be encountered in the dye bath.

Unfortunately, each class of foam control agent has inherent deficiencies [17]. Alkali earth metal or aluminum salts of fatty acids can leave undesirable deposits on the processing equipment which in extreme cases can cause pipes and jets to clog. Aliphatic alcohols have relatively poor foam control performance and in addition have an unpleasant odor that can cause nausea to workers, even at low concentrations in the atmosphere surrounding the textile bath. Polyoxyethylene and polyoxypropylene block copolymers have relatively poor foam control performance which significantly limits their use. Emulsions of hydrophobic oils and silica are particularly effective as foam control agents; however, they are susceptible to destabilization when used in processes where the dye liquor is subjected to high temperatures, above 100°C, and high shear. If the emulsions break in use, the hydrophobic oil deposits on the textile resulting in undesirable spotting or staining of the goods being dyed, and simultaneously the desired foam control action is lost. This problem can be overcome by using silicone glycols, rather than hydrophobic oils and silica, but such materials have much poorer foam control ability. Phosphoric acid esters also have relatively poor foam control performance which means that their use is limited to selected dyeing processes.

2. Improving Antifoams through Formulation

The challenge to the textile auxiliary formulator is to provide the textile dyer with a foam control agent that can be used in a wide variety of dyeing processes without any negative impact on either the dyeing process or the goods being dyed. As yet there is no universal solution, however, significant progress has been made over the last two decades through better formulations.

The first example shows that through proper formulation it is possible to overcome some of the drawbacks associated with using the aluminum salt of a fatty acid as the foam control agent [18], by combining the fatty acid salt with surfactants, aliphatic alcohols, and hydrophobic mineral oils. A typical formulation would be

4% Aluminum distearate
3% Phosphoric acid esterified with polyoxyethylated *p*-nonyl phenol
6% 2-ethyl-*n*-hexanol
87% Paraffin oil

by weight. Unlike previous formulations containing aluminum stearates such formulations are very stable and will control the foam when dyeing wool with 1:1 and 1:2 metallic, acid, or reactive dyes, dyeing nylon with acid or disperse dyes, dyeing polyester with disperse dyes, dyeing cotton with reactive and direct dyes, and acrylics with cationic dyes. An alternative to using aliphatic alcohols is to use a propoxylated, branched, aliphatic acid, such as 1,2-propylene glycol mononeodecanoate [17]. Though the foam control performance is similar to that of isooctanol, it has a much lower odor. Typical addition levels that would be required are from 0.2% to 0.5% by weight based on the weight of the textile-dyeing liquor. Although it can be used in combination with virtually all dyes, it is particularly effective when used in combination with acid dyes in either batch or continuous dyeing processes.

To overcome the deficiencies of polyoxyethylene–polyoxypropylene block copolymers, it has been proposed to combine them with a silicone antifoam compound and use a silicone glycol copolymer as the dispersing agent [19]. By using this composition it is possible to obtain a stable, highly effective foam control agent. A typical composition would be

52% Polypropylene glycol with a molecular weight of 4100
40% Silicone antifoam compound consisting of polydimethylsiloxane fluid, silicone resin, and silica
8% Silicone glycol copolymer as dispersing aid by weight

The formulation of the silicone glycol copolymer being

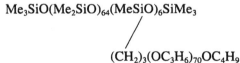

$$Me_3SiO(Me_2SiO)_{64}(MeSiO)_6SiMe_3$$
$$|$$
$$(CH_2)_3(OC_3H_6)_{70}OC_4H_9$$

The dispersing aid maintains the dispersion of the silicone antifoam compound in the polypropylene glycol.

3. Formulating Antifoams into Textile Dyeing Auxiliaries

Rather than adding the foam control agent separately to the dye bath, it is becoming more common to formulate the antifoam into the textile-dyeing auxiliary to give a low-foaming auxiliary. The formulation of such an auxiliary could be [1,20]

15–30% Nonionic, anionic, or cationic textile-dyeing auxiliary
10–50% Fatty acid esterified with a polyoxyethylene–polyoxypropylene block copolymer
55–70% Water or water-soluble solvent

by weight. The typical formulation of the foam control agent [1] would be the esterification product of 2 moles of tallow acid and a polyoxyethylene–polyoxypropylene block polymer having a molecular weight of 2600. Preferably, the fatty acid should contain at least 30% by weight of oleic acid in order to improve the solubility of the ester in the formulation to give a clear, stable, homogeneous mixture. The textile-dyeing auxiliary is a leveling or wetting agent normally added to ensure even and level dyeing. Typical nonionic dyeing auxiliaries that could be used in the above formulation are polyoxyethylated and/or polyoxypropylated saturated fatty acids, fatty alcohols, fatty amines, fatty amides, alkyl phenols, and glycols. Typical anionic auxiliaries are the sulfated or sulfonated versions of the nonionic surfactants already described, alkylbenzene sulfonates and alkyl sulfonates. Typical cationic auxiliaries are salts of the alkyl pyridinium ion, methylalkyl piperidinium ion, and tetraalkyl ammonium ion. The resultant mixture when added to the dye bath gives level dyeing, and the dye bath exhibits practically no foaming. However, these formulations can only be used in dyeing processes where the pH remains between 2.5 and 8.0 during dyeing. Most importantly, they cannot be used in alkaline dye baths because the fatty acid ester foam control agent becomes saponified and hence inactivated. This presents a limitation as regards the dyeing processes in which they can be used.

When dyeing synthetic fibers with insoluble or sparingly soluble dyes, uneven dyeing with poor fastness to rubbing can often occur due to inadequate dispersion of the dyes in the dye liquor, for example, in high-temperature dyeing of polyester fibers with disperse dyes. In order to overcome this problem, dispersing agents are generally added in conjunction with leveling agents. However, these auxiliaries normally have only either dispersing or leveling properties, and it is necessary to use mixtures of both to achieve the best results. It has been found that alkali metal or ammonium salts of the addition products of polypropylene glycol and polyhydric aliphatic alcohols, such as glycerol, reacted with sulfamic acid not only have the ability to stabilize the dye dispersion, so that neither aggregation nor precipitation of the dyestuff occurs during the dyeing process, but also have a low-foaming tendency [21]. This leads to level dyeing which is fast to rubbing.

However, in order to achieve even lower foaming, it is recommended to incorporate adipic acid di-2-ethylhexyl ester and polydimethylsiloxane as antifoaming agents. The amount of antifoaming agent used being 15% to 60% by weight based on the weight of the auxiliary. Water-miscible alcohols are also added to improve the dispersibility of the formulation in the dye bath.

Using a similar approach, it is possible to obtain a low-foaming, wetting agent by formulating together a wetting agent and a foam control agent [22–24]. A preferred formulation [22] is

10–20% Anionic surfactant
15–35% Insoluble or sparingly soluble propoxylate
0.1–2% Silicone antifoam compound
40–75% Water

by weight. The anionic surfactant could be for example

$$C_9H_{19} \quad \longrightarrow\!\!\!\bigcirc\!\!\!\longrightarrow \quad O(CH_2CH_2O)_{1-3}SO_3X$$

where X is hydrogen, ammonium, or an alkali metal ion. Such surfactants are well established as wetting and leveling agents in textile-dyeing processes. However, when they are used by themselves, they produce copious quantities of undesirable foam. The insoluble or sparingly soluble propoxylate referred to above is either the reaction product of 1 mole of a polyhydric alcohol (such as ethylene glycol, polypropylene glycol, glycerol, pentaerythritol, sorbitol, trimethylolethane, or trimethylolpropane) with 30 to 120 moles of 1,2-propylene oxide or the reaction product of a monoalkylamine, monoalkylolamine, or polyalkylenepolyamine (such as monoisopropanolamine or ethylene diamine) and 1,2-propylene oxide. When used in conjunction with the anionic surfactant, the foaming tendency is lower than when the surfactant is used on its own. The silicone antifoam compound provides an additional means of reducing the foam generated by the anionic surfactant and this helps to enhance the wetting of the textile goods by the wetting agent. Optionally, a water-insoluble, nonionic surfactant may also be included, such as the reaction product of 2-ethyl-n-hexanol and ethylene oxide or p-nonyl phenol and 3 moles of ethylene oxide among others [22].

Such compositions can be used in both alkaline and acid baths in the pH range from 1 to 12 and over a wide temperature range from 20°C to 120°C. They can be successfully used in a wide variety of processes, such as dyeing

wool with 1:1 or 1:2 metallic, acid, or reactive dyes, exhaustion and continuous dyeing processes for nylon fibers with acid or disperse dyes, dyeing
polyester fibers with disperse dyes, dyeing cellulosic fibers with reactive
and direct exhausting dyes, and dyeing acrylic fibers with cationic dyes.

4. Antifoams Based on Water-Insoluble Oils and Silica

Of the different types of foam control agents described earlier, the most
effective are those based on dispersions of hydrophobic silica in water-insoluble oils. With relatively mild surfactants and low-shear conditions, the
most cost-effective water-insoluble oils are vegetable oils, or more commonly, mineral oils such as paraffin oils, naphthenic oils, kerosene, and
similar petroleum fractions [12]. A spreading agent is often incorporated to
improve dispersion throughout the foam system. When appropriately formulated, they are very effective at knocking down existing foam [25].

With the low surface tension and high agitation encountered in many textile dye baths, more effective foam control is achieved by using polydimethylsiloxane fluids as the water-insoluble oil. They are most often used
where prevention of foam for the duration of the process is required and
either high levels of agitation and/or high levels of low-surface-tension surfactants are present. They consist of proprietary combinations of polydimethylsiloxane polymers or silicone fluids:

$$Me_3SiO(Me_2SiO)_nSiMe_3$$

and finely divided, hydrophobic silica, and are more commonly known as
silicone antifoam compounds. The fluids are colorless, essentially odorless,
chemically stable, unreactive materials. They are insoluble in water, difficult
to emulsify, have a very low surface tension, typically 21 mN m^{-1} at 25°C,
and spread spontaneously over the surface of most surfactant-containing solutions. In addition, they have been shown to be safe to both humans and
the environment. However, on their own they show little or no foam-inhibiting ability in aqueous foaming media. It is only when these fluids are combined with finely divided, hydrophobic silica particles that an effective foam
inhibitor is produced [12]. An example of a formulated silicone antifoam
compound is given below [26]:

88% Polydimethylsiloxane fluid having a viscosity of 350 cSt at 25°C
9% Siloxane resin composed of $(CH_3)_3SiO_{0.5}$ and SiO_2 units in which the
    ratio of $(CH_3)_3SiO_{0.5}$ units to SiO_2 units is within the range of 0.6:1 and
    1.2:1
3% Silica aerogel

by weight. The method described to prepared the product is to mix the polydimethylsiloxane fluid and silicone resin with heat at 70°C to 120°C for

TABLE 1 Typical Formulation of a 10% Active Silicone Antifoam Emulsion

Percent by weight	Ingredient
10.0	Silicone antifoam compound
4.0	Emulsifiers:
	polyoxyethylene monostearate and glyceryl monostearate
2.0	Stabilizers: water-soluble thickeners
84.0	Water
As required	Preservatives

1 to 3 hr. Then after adding the silica heat the resultant mixture to 130°C to 175°C for a period of approximately 2 hr.

Even though there has been a significant amount of speculation concerning the individual roles of the fluid and the silica [27–29], it is well established that a combination of these two materials provides an antifoam that will perform effectively in virtually all the foam control applications met in the textile industry. The art of formulating an effective silicone antifoaming agent is to combine the silicone fluid and silica in such a way that the most efficient foam control agent is produced.

However, the products are rarely used as such in textile processes because of the difficulty of dispersing such high-viscosity materials effectively in the textile bath at very low addition levels. To achieve the required level of dispersibility, these products are introduced into the process as emulsions or dispersions. This also overcomes the difficulty in handling the addition of the very small quantities of antifoam compound required to control the foam, and the potential risk of spotting and staining problems if overdosed. Dispersibility of the antifoaming agent in the foaming medium is important because it affects the ability of the antifoam to perform its foam-breaking function. If the antifoam is not adequately dispersed, it will merely sit on the surface of the bath as an insoluble mass of liquid and no appreciable antifoaming action will be observed. To achieve the correct level of dispersibility in the bath, the combination of silicone fluid and silica is supplied as a 10% to 30% active emulsion. A formulation for a typical 10% emulsion is given in Table 1 [30]. Textile mills in general prefer to use low-solids emulsions as they can be used as received and thereby minimize the potential for overuse and losses due to waste in handling. These emulsions are used in a variety of textile-dyeing processes, as well as many textile preparatory and finishing operations, and effluent treatment processes. In textile dye baths they are commonly used where temperatures encountered are below 100°C. They are not recommended for processes that require high-temperature dispersibility and/or shear-stable antifoam products [30].

The intrinsic foam control ability of an antifoam emulsion is directly related to the performance of the active material. However, the efficiency of the resultant emulsion will also depend on formulating variables such as the particle size distribution of the active material, the level of active material, the type of emulsifiers and thickeners used in manufacturing the emulsion, and their concentrations. It has been found that if the particle size of the active ingredient in the emulsion is less than 2 microns, the product shows poor antifoam efficiency [31]. On the other hand, if the particle size is of the order of 50 microns or more, the emulsion has poor shelf stability unless it contains very high levels of thickeners to retard separation. In addition, large particle size emulsions will "oil out" on dilution in the foaming medium resulting in oil spotting on the textile goods and machinery. In practice, the particle size is carefully controlled somewhere between these extremes through the careful selection of the emulsifiers, thickeners, and the emulsification equipment used in their manufacture [32].

Another important consideration is the suitability of the emulsion formulation for the end application. A good example is the choice of thickener system used to retard creaming in these emulsions [32]. Some of the earliest emulsions used methyl cellulose as the thickening agent. This is a very effective thickening agent, but unfortunately it is insoluble in water above about 80°C. When emulsions of this type were used in textile-dyeing processes at or near 100°C, the methyl cellulose would come out of solution taking the antifoam active ingredient with it. This resulted in dye stains caused by adsorption of the dye on the insoluble cellulose and subsequent deposition on the fabric and loss of foam control efficiency due to depletion of the active material from the foaming medium. To make matters worse more antifoam emulsion would be added to compensate for the loss of foam control efficiency. Modern, silicone antifoam emulsions use nongelling, cellulose ethers such as hydroxy ethyl cellulose, alginate derivatives and synthetic carboxy vinyl polymers as thickeners [32]. With improved emulsion formulations, they have shed their poor image and are used successfully in preparatory, finishing and dyeing processes. The levels of addition will very much depend on the concentration and nature of the surface active ingredients present in the process. Typically it would be 50–300 ppm active material in open beck dyeing and 1–10 ppm in wastewater treatment where the level of agitation and surfactant concentration is relatively low.

It is preferable that the antifoaming agent is added at the start of the dyeing process, because many of the foam problems are caused by air entrainment in the system when circulation begins. If possible it should be added even before the dyes and the auxiliaries [33]. However, this must be done carefully to avoid overuse with subsequent risk of oil spotting. Alternatively, the antifoaming agent may be added, as a very dilute dispersion, continuously or semicontinuously during the course of a particular process. This is normally the safest and most cost-effective method [32].

It must be stressed that other industries have different needs, and so not all silicone antifoam emulsions will meet the stringent requirements encountered in controlling foam in textile processes.

C. Antifoams for Jet Dyeing Processes

The introduction of jet dyeing machines for rapid dyeing of polyester presented a major foam control challenge to the antifoam formulator, especially when dyeing darker shades where higher levels of disperse dyes are required. Excessive foaming was a serious problem in production because of the high-shear and high-temperature conditions encountered, especially in partially flooded machines. Even by carefully selecting low-foaming, disperse dyes and textile-dyeing auxiliaries the level of foam was often too high and would interfere with the dyeing process. For example, the presence of foam would cause the centrifugal pump to cavitate, which in turn impaired the action of the jet and reduced the speed of fabric circulation through the machine. In addition this could also lead to fabric flotation, tangling, and fiber breakage as well as uneven dyeing [13]. Jet dyeing machines also exposed weaknesses in the available antifoam emulsion systems. Under the very severe conditions of high shear and high temperature encountered in these machines, the emulsions would often break and the oil phase would accumulate on the inside of the jet causing oil spotting on the fabric. This gave silicone antifoam emulsions a poor image in the industry until new formulations were developed.

1. Antifoams Based on Silicone Glycol Copolymers

The first approach was to develop more stable emulsions with smaller particle size that could better withstand the high-shear conditions present in the jet dyeing process. Though these overcame the spotting problem, the antifoam efficiency was greatly reduced. The more successful approach was to utilize the unique properties of a class of materials commonly known as silicone glycol copolymers or silicone polyethers [13]. There are essentially four structures that represent those most commonly found in the patent literature:

Structure A

$Me_3SiO(MeGSiO)_m(Me_2SiO)_nSiMe_3$

Structure B

$GMe_2SiO(MeGSiO)_m(Me_2SiO)_nSiMe_2G$

Structure C

$R_aSi[(OSiMe_2)_n(OSiMeG)_mOSiMe_3]_{4-a}$

Structure D

$R_aSi[(OSiMe_2)_n(OSiMeG)_mOSiMe_2G]_{4-a}$

where R is typically a methyl group, a is typically 1, and G is typically represented by the following formulae:

$$-(CH_2)_3(OC_2H_4)_x(OC_3H_6)_yOR'$$

or

$$-(OC_2H_4)_x(OC_3H_6)_yOR'$$

where R' is an end-capping group.

By varying the relative proportion of dimethylsiloxane, polyoxyethylene, and polyoxypropylene in the molecule, it is possible to dramatically alter its solubility characteristics. For example, it is possible to manufacture a polymer that is soluble in cold water but not in hot water. The temperature at which this occurs is known as the cloud point. Below its cloud point the molecule acts as a conventional surfactant, whereas above its cloud point it acts as a foam control agent. Furthermore, the higher the temperature of the foaming medium the greater is the antifoaming efficiency of the silicone glycol copolymer. This class of material became the basis of an acceptable antifoam for the jet dyeing process. During the high-temperature dyeing stage, the material is insoluble and acts as an effective antifoam. Whereas after the dye liquor has been cooled, the material becomes soluble once more and can be completely removed by subsequent rinsing. This means that by formulating products based on this class of polymer it is possible to achieve foam control without the risk of oil spotting. It is important to point out that the temperature of the dye bath must be below the cloud point of the polymer when it is added to achieve rapid and even dispersion of the antifoam in the bath through solubilization. Otherwise, the optimum performance benefits will not be achieved.

In general, the silicone glycol copolymers with the higher water solubility and higher cloud points have lower foam control efficiency than those with lower cloud points. However, as the solubility is decreased to achieve better foam control performance, deposition of the antifoam on the dyed fabric becomes more pronounced with the resultant risk of staining or spotting on the fabric [34]. In addition, the antifoaming action of these cold-water-soluble silicone glycol copolymers is inferior to combinations of polydimethylsiloxanes and silica. In order to overcome these problems, several defoaming compositions have been proposed. For example, the antifoam action can be increased significantly by incorporating a suitable silica as in the case of hydrophobic oils [35]. A relative comparison has been established by determining the volume of a 10% by weight aqueous dispersion of antifoam that is necessary to control the foam of 2.5 L of 0.01% aqueous solution of sodium alkyl sulfate that is being circulated through an open vessel at a rate of 10 L/min at a temperature of 90°C. The results obtained were as follows:

Conventional silicone antifoam compound 0.05 mL
Silicone glycol copolymer (no silica) 1.00 mL
Silicone glycol copolymer (with silica) 0.15 mL

As can be seen the silicone glycol copolymer is significantly less efficient than the conventional silicone antifoam compound. However, through the addition of 2% fumed silica, the performance can be dramatically improved. The formula of the polymer used in this set of experiments was

$$Me_3Si(OSiMe_2)_{14}(OSiMeG)_8OSiMe_3$$

where G has an average molecular weight of about 1500 and can be represented by the formula $—(OC_nH_{2n})_yOR$ containing equal weights of oxyethylene and oxypropylene units and R is a n-butyl group.

The method of preparation was to add the fumed silica to the silicone glycol copolymer with stirring and subsequently heating the mixture at 100°C to 120°C for 1 hr. The composition was also shown to control the foam in a jet dyeing machine throughout the whole process, when the temperature of the dye liquor was raised to 125°C, when dyeing was being carried out for 8 hr, and when the pressure was built up and released. Careful examination of the dyed goods showed no color unevenness nor any traces of silicone spotting.

Alternative structure for the silicone glycol copolymer have been proposed [36]:

$$GMe_2Si(OSiMe_2)_a(OSiMeG')_b(OSiMe_2)_aOSiMe_2G$$

where

$$G' = —(OSiMe_2)_aOSiMe_2G$$
$$G = —(OC_3H_6)_n(OC_2H_4)_mOR$$

The foam control composition as described consists of 92.5% by weight of the glycol, where $a = 6$, $b = 2$, $n = 21$, $m = 6$, and R = C_4H_9, and 7.5% by weight of a hydrophobic fumed silica with a surface area of 200 m^2/g. To this composition may also be added one or more nonionic emulsifiers with HLB 8–14 to improve the dispersion in the aqueous liquor. Typical emulsifiers being polyoxyethylated fatty alcohols, fatty acids, fatty acid triglycerides, and glyceryl monofatty acids.

Another structure for the silicone glycol copolymer that has been described [34] is

$$GMe_2Si(OSiMe_2)_{30}(OSiMeG)_4OSiMe_2G$$

where G = $—(CH_2)_3(OC_2H_4)_{10}OMe$. However, rather than formulating this polymer with silica, it is formulated with a mixture of a polyoxyalkylene glycol derivative such as C_8H_{17}—$(OC_2H_4)_{10}(OC_3H_6)_3OMe$ and a nonionic surfactant such as polyoxyethylene sorbitan monolaurate having 10 oxyethylene units. The relative proportions being 50% by weight of the silicone glycol copolymer, 30% by weight of the polyoxyalkylene derivative, and 20% by weight of polyoxyethylene sorbitan monolaurate. When the silicone glycol copolymer was added to a high-temperature dye bath on its own it only managed to control foam for 20 min with some staining on the polyester fibers occurring, whereas when the same weight of the above composition was used, foam was controlled for 53 min with no staining observed. As a result such formulations are claimed to be particularly suitable for high-temperature jet dyeing of polyester and other fibers. Usually 10–500 ppm of the antifoaming composition is sufficient to control foam in a typical dyebath. It should be noted that in the above silicone glycol copolymer structure the polyoxyalkylene groups are bonded to the silicon atom through a Si—$(CH_2)_3$—O—C linkage. This is because the copolymers are more hydrolytically stable than those in which the polyoxyalkylene groups are bonded directly to the silicon atom through a Si—O—C linkage. The poorer hydrolytic stability is particularly noticeable in alkaline media where it manifests itself as less durable antifoaming activity.

It is also possible to formulate the silicone glycol copolymers with the reaction product of a silicone resin and a polyoxyalkylene [37]. A typical formulation is 87% by weight of a silicone glycol copolymer with the structure

$$Me_3SiO(Me_2SiO)_{108}(MeGSiO)_{10}SiMe_3$$

where G = —$(CH_2)_3O(C_2H_4O)_{12}(C_3H_6O)_{12}OCCH_3$, 13% by weight of a 50% xylene solution of the reaction product of a resin, consisting of $(CH_3)_3SiO_{0.5}$ units and SiO_2 units in a ratio of 0.67 : 1, and polypropylene glycol with a molecular weight of 4100. The formulation was tested by adding 200 ppm to a 0.1% polyoxyethylene sorbitan monooleate solution at 86°C under high shear. After running the recirculation pump for 6 min, the foam height was 5.1 cm when the silicone glycol copolymer was used on its own, whereas with the reaction product present the foam only reached 2.5 cm. The time for the foam to collapse at the end of the test after the recirculation pump has been switched off was 12 and 5 s, respectively. Clearly the reaction product enhances the antifoaming behavior of the silicone glycol copolymer. More significantly though, when the reaction product was present, there was no sign of the antifoam composition plating out and depositing on the surface of the apparatus, whereas plating out did occur with the polymer on its own. Plating out is the formation of an insoluble layer covering part of the surface of the water that becomes visible when the recirculation pump has been turned off. This is important when using these types of products as jet dyeing foam control agents because plating out of the antifoam leads to staining and spotting of the fabric during dyeing. It has also been shown that it is possible to incorporate these types of products into the dye carrier formulations to give a low-foaming textile dyeing auxiliary.

In order to achieve greater antifoam efficiency, these polymers have been formulated with silicone antifoam compounds [32]. By formulating products in this way it is possible to increase low-temperature antifoam efficiency and achieve better foam control at the depressurization stage, while retaining high-temperature performance and nonstaining properties. Such materials have one major advantage over conventional silicone antifoam compounds described earlier, and that is they are water dispersible. This means that the textile dyer is able to simply disperse small quantities in cold water and add them directly to the dye bath. The silicone glycol copolymers also ensure complete dispersibility of the antifoam throughout the dyeing process. Further, they retain their antifoaming ability even at the very end of the dyeing cycle when the dye bath is cooled and depressurized. This stage is the most critical with respect to foam formation, because foam is rapidly generated by the sudden escape of dissolved air and the antifoam must not only limit the amount of foam generated but also rapidly knock it down. These types of formulated products are generally not quite as effective as conventional silicone antifoam emulsions; however, their efficiency is sufficient to allow the textile dyer to employ disperse dyes, in partially flooded jet dyeing machines, that otherwise would have to be excluded in view of their foaming behavior. Furthermore, they can be used without the risk associated with emulsion breakage. The use of such materials is now not just confined to jet dyeing since progressively more applications are being developed that need antifoam products that can withstand high-shear, high-temperature conditions.

An example of such a formulated product is given below [38]. It consists of four components:

A. Silicone glycol copolymer

$$Me_3Si(OSiMe_2)_{75}(OSiMeG)_7OSiMe_3$$

where

$$G = -(CH_2)_3(OC_2H_4)_{21}(OC_3H_6)_{21}OOCCH_3$$

B. Silicone resin copolymer which is the reaction product of a copolymer consisting of SiO_2 units and $(CH_3)_3SiO_{0.5}$ and a hydroxylated polyoxypropylene polymer.

C. Silicone antifoam compound prepared from polydimethylsiloxane, precipitated silica, and siloxane resin.

TABLE 2 Antifoam Performance of Products Based on Silicone Glycol Copolymer

Test number	Test composition				Foam profile
	A	B	C	D	
1	0	0	0	0	15 s to reach 7.6 cm
2	100	0	0	0	5 min to reach 3.8 cm
3	50	50	0	0	5 min to reach 3.8 cm
4	94	5	1	0	3 min to reach 6.3 cm
5	90	5	5	0	5 min to reach 3.8 cm
6	90	5	3	2	30 s for the foam to reach 2.5 cm and more than 5 min without the foam reaching top of test vessel.

Source: Ref. 38.

D. Silicone glycol copolymer

$Me_3Si\ (OSiMe_2)_{10}(OSiMeG)_3OSiMe_3$

where

$G = -(CH_2)_3O(C_2H_4O)_{12}H$

The optimum proportions are 90 parts by weight of A, 5 parts of B, 3 parts of C, and 2 parts of D.

The antifoam composition was tested in an 0.1% aqueous solution of polyoxyethylene sorbitan monooleate. The foaming solution was continuously circulated at 86°C and the foam height and the time for the foam to reach that height measured.

The bath was examined for signs of separation and plating out of the components from the test composition. Ideally, there should be no separation or plating out whatsoever on the sides of the test vessel. This particular test method exposes the antifoam composition to high amounts of agitation, emulsification, and shear to simulate the conditions likely to be encountered in a jet dyeing machine. The results are given in Table 2.

Though composition 6 did not have such good foam control performance as compositions 2–5, the performance with respect to the intended end use was rated excellent because the foam was controlled without any plating being observed. The presence of the silicone glycol copolymer (ingredient D) increases stability of the composition at the expense of antifoam performance. This is yet another example where the foam control performance is in part sacrificed by the formulator in order to minimize or eliminate a performance defect.

2. Nonsilicone-Based Antifoams

There are also foam control formulations that have been proposed that contain neither polydimethylsiloxane fluid nor silicone glycol copolymers. They are claimed to be effective under jet dyeing conditions and consist of combinations of ethylenebis (stearic amide), paraffin wax, hydrophobic silica, nonsilicone hydrophobic oils, and a nonionic surfactant [39,40]. For example [39], a dispersion of ethylenebis (stearic amide) in rapeseed oil is added to a dispersion of polyethylene wax in mineral oil and stirred until homogeneous. To this is added an emulsifier, which consists of an equimolar mixture of $C_{17}H_{33}CO(OC_2H_4)_{6.5}OCOC_{17}H_{33}$ and $C_{17}H_{33}CO(OC_2H_4)_{6.5}OH$, hydrophobic silica, and finally a further addition of mineral oil. The resultant product is a stable, fluid dispersion that can be diluted with water containing 3.3% by weight of ethylenebis(stearic amide), 2.3% polyethylene wax, 3.0% hydrophobic silica, 57.1% mineral oil, 26.1% rapeseed oil, and 8.2% emulsifier. To test the efficacy of this antifoam composition, prewashed polyester fabric was dyed in a Gaston County Mini-Jet machine with a capacity of 650 L and 30 kg of fabric using the following dye liquor ingredients:

0.95% C.I. Disperse Red (based on the weight of the substrate)
1 g/L Leveling agent, C_{16-18} fatty alcohol condensed with 30 moles of ethylene oxide
1 g/L Dispersing agent, Turkey red oil
2 g/L Ammonium sulfate
200 mL Formic acid to pH 5
0.65 g/L Antifoam composition, as described above
450 L Demineralized water
17 kg Polyester, prewashed and prefixed

The goods-to-liquor ratio was 1:26 and the liquor circulation rate was 75 L/min. The liquor was heated from 60 to 126°C over a period of 2 hr. On reaching 126°C dyeing was continued for 30 min, then the liquor was cooled to 60°C over 30 mins. No foam formation was observed over the entire process. Inspection of the dyed goods showed no spots or uneven areas due to deposition of the antifoaming agent. Ethylenebis(stearic amide), paraffin wax, hydrophobic silica, and hydrophobic oils are each known to have antifoaming properties in their own right. However, this particular combination gives unexpectedly good antifoaming properties under high-temperature conditions. The compositions are claimed to be particularly suitable for use in textile treatment liquors containing disperse dyes at high temperatures, in the range 105–150°C, and under pressure.

An alternative foam control system which is also nonsilicone is a dispersion based on alkali earth metal salts of a fatty acid, dialkyl ester of an unsaturated dicarboxylic acid, mineral oil, anionic and nonionic emulsifiers, and ethylenebis(stearic amide) [41]. For example, 2.0% by weight of magnesium stearate, 37% maleic acid esterified with 2-ethylhexanol, 35.5% mineral oil, 24% nonionic and anionic surfactants, and 1.5% ethylenebis(stearic amide). Instead of the alkali earth metal salt of a fatty acid it is also possible to use either the reaction product of a polyol, an anhydride of an aliphatic dicarboxylic acid, and a polyalkylene glycol fatty acid adduct or the reaction product of a polyol, an anhydride of an aliphatic dicarboxylic acid, and a polyalkylene glycol fatty alcohol adduct [42]. Such compositions are claimed to be effective under the high-shear conditions encountered in jet dyeing and so eliminate the necessity to continuously meter in the antifoam to avoid spotting of fabric as in the case of antifoams based on silicone emulsions.

As a result of the improvements made in formulation, it is now normal to add these self-dispersible antifoaming agents to the dye tank along with the other dyeing auxiliaries. For example, a typical dyebath composition for dyeing polyester fabric in a jet dyeing machine would contain a disperse dye, leveling agent, dye dispersant, ammonium sulfate, formic acid to pH 5, and antifoaming agent [39]. Ideally, little or no foam is observed over the entire dyeing cycle and inspection of the dyed goods shows no spots or uneven coloration.

IV. CONCLUSION

The textile industry in common with other manufacturing industries is striving to increase productivity and reduce energy consumption. At the same time it is making every effort to meet increasing, environmental pressures. One response to these pressures will be the development of wet dyeing processes that involve the use of lower liquor ratios. This means that it will be possible to dye with greater energy efficiency, because less energy is necessary to heat the dye bath and to circulate the dye liquor when using lower liquor ratios. Another advantage is that less effluent is produced that requires processing. This implies that, in future, even higher concentrations of surfactants are likely to be present in dye baths, and consequently there will be an even greater tendency for these baths to foam. Therefore, there will be a need for even more effective, formulated, foam control agents that can deliver antifoaming without affecting the quality of the dyeing process.

Foams are also being used advantageously in textile processing, because where a stable foam is created with an adequate distribution of textile treatment systems within it, application of such systems to textile fabric can be very cost effective as drying costs are dramatically reduced. For example, in the foam finishing of fabric, products such as reactant resin formulations for cotton and polyester/cotton "wash and wear" finishes can now be effectively applied to fabric from an aqueous foam [43]. The low water content of the fabric after foam application compared with conventional dip/nip padding operations gives higher productivity and lower energy usage during drying and curing of the resin. The same principle has been used in the dyeing of small articles such as garments and socks in rotating drum machines using dyestuffs dispersed in foams. This process, known as the Sancowad process and developed by Sandoz, had some success a few years ago. If such a process could be improved then new and different foam control agents would be required by the industry, because now foam needs to be controlled after the dye has been applied rather than preventing it occurring at all.

ACKNOWLEDGMENT

I thank Charlie Smith for his support in compiling the references at the outset.

REFERENCES

1. M. Daeuble, K. Oppenlaender, and R. Fikentscher (assigned to Badische Anilin—& Soda-Fabrik Aktiengesellschaft), US 3,830,627; August 20, 1974; filed September 8, 1971.
2. R. H. Peters, Textile Chemistry, Vol. 3, Elsevier Scientific, Amsterdam, 1975.
3. G. Prazak, *Am. Dyestuff Reporter 59*: 44–46 (1970).
4. I. Slack, *Canadian Textile J. 79*: 46–51 (1979).
5. G. Weckler, Conference Proceedings, AATCC 1978 National Technical Conference, pp. 140–145.
6. V. S. Salvin et al., *Am. Dyestuff Reporter* Nov., 23–24 (1959).
7. Encyclopedia of Chemical Technology, 3rd ed., Vol. 8, Wiley, New York, 1979, pp. 280–350.
8. W. T. Carpenter, *Textile Chemist and Colourist 1*(12):22–23 (1969).
9. A. Hodgson, Shirley Institute Publication S.35, 1979.
10. H. M. Tobin, *Am. Dyestuff Reporter 68*(9): 26–28 (1979).
11. I. R. Schmolka, *J Am. Oil Chem. Soc. 59*: 322–327 (1982).
12. R. E. Patterson, *Textile Chem. Color 17*(9): 181–184 (1985).
13. J. A. Colquhoun, *Textile Industries (Atlanta) 137*(8): 100–105 (1973).
14. Intl. Textile Bulletin, *Dyeing/Printing/Finishing, 2nd Qtr., 30*: 16 (1984).
15. H. Distler and D. Stoeckigt, *Tenside Detergents 12*(5): 263–265 (1975).
16. *Intl. Dyer & Textile Printer 162*: 351 (1979).
17. R. H. Via and S.C. Taylors (assigned to Milliken Research Corporation), US 4,428,751; January 31, 1984; filed March 29, 1982.
18. F. Redies, B. Redies, D. Tuerk, and C. Gille (assigned to Ciba-Geigy AG), DE 2,943,754; May 14, 1980; filed October 30, 1979.

19. J. W. Keil (assigned to Dow Corning Corporation), US 3,984,347; October 5, 1976; filed December 19, 1974.
20. M. Daeuble, K. Oppenlaender, and R. Fikentscher (assigned to BASF AG), DT 2,102,899; January 13, 1977, filed January 22, 1971.
21. A. Berger and H.U. Berendt (assigned to Ciba-Geigy Corporation), US 4,132,525; January 2, 1979; filed April 8, 1977.
22. Ciba-Geigy AG, GB 1,522,121; August 23, 1978; filed June 11, 1976.
23. H. Abel and A. Berger (assigned to Ciba-Geigy AG), DE 2,625,706; July 30, 1981; filed June 9, 1976.
24. H. Abel and A. Berger (assigned to Ciba-Geigy AG), DE 2,625,707; April 22, 1982; filed June 9, 1976.
25. H. F. Fink, H. Fritsch, G. Koerner, G. Rossmy, and G. Schmidt (assigned to Th. Goldschmidt Aktiengesellschaft), GB 1,453,383; October 20, 1976; filed September 5, 1974.
26. L. A. Rauner (assigned to Dow Corning Corporation) US 3,455,839; July 15, 1969; filed February 16, 1966.
27. R. D. Kulkarni, E. D. Goddard, and D. Kanner, *Ind. Eng. Chem. Fundam.* *16*(4): 472–474 (1977).
28. R. D. Kulkarni and E. D. Goddard, *Croatica Chem. Acta 50*: 163–179 (1977).
29. R. D. Kulkarni, E. D. Goddard, and M. R. Rosen, *J. Soc. Cosmet. Chem.* *30*(2): 105–125 (1979).
30. K. W. Farminer and P. B. Herter, Proceedings AATC Natl. Tech. Conf., 1974 Conf. Publication, Am. Assoc. Text. Color Chemists, pp. 337–342.
31. S. Ross, Rensselaer Polytechnic Institute Bulletin, Engineering and Science series No. 63, Troy, New York, 1950.
32. D. N. Willing, *Am. Dyestuff Reporter 69*(6): 42–51 (1980).
33. A. N. Derbyshire and A. T. Leaver, *J.Soc. Dyers Colour. 91*: 253–258 (1975).
34. A. Abe (assigned to Shin-Etsu Chemical Co., Ltd.), US 4,042,528; August 16, 1977, filed March 25, 1975.
35. W. W. Cuthbertson (assigned to Imperial Chemical Industries Ltd.), US 3,700,400; October 24, 1972; filed May 3, 1971.
36. H. F. Fink, G. Koerner, G. Rossmy, and G. Schmidt (assigned to Th. Goldschmidt AG), DT 2,443,853; April 1, 1976; filed September 13, 1974.
37. J. A. Colquhoun (assigned to Dow Corning Corporation), US 3,912,652; October 14, 1975; filed December 17, 1973.
38. J. W. Churchfield (assigned to Dow Corning Corporation), US 3,746,653; July 17, 1973; filed May 15, 1972.
39. B. Danner (assigned to Sandoz Ltd.), GB 2,112,767; July 27, 1983; filed November 22, 1982.
40. B. Danner (assigned to Sandoz Patent—GmbH), DE 3,505,742; September 12, 1985; filed February 20, 1985.
41. H. Abel, C. Guth, and H.U. Berendt (assigned to Ciba-Geigy AG), EP 263,069; April 6, 1988; filed September 23, 1987.
42. H. Abel and R. Toepfl (assigned to Ciba-Geigy AG), EP 207,002; December 30, 1986; filed June 23, 1986.
43. J. J. Whiting and G.A. Richardson, Chemspec Europe 89 BACS Symposium 1989, pp. 11–13.

6
Foam Control in Detergent Products

HORST FERCH* AND WOLFGANG LEONHARDT   Applied Technologies
Silicas and Silicates, Degussa AG, Hanau, Germany

I.	Introduction	221
II.	Foam Generation in Washing Machines	222
III.	Surfactant Types Used in Laundry Detergent Formulations	226
IV.	Types of Detergent Antifoams	227
	A. Introduction	227
	B. Soaps as antifoams	229
	C. Nitrogenous antifoams	232
	D. Phosphoric acid esters	235
	E. Hydrophobed silica/ hydrophobic oil mixtures	236
V.	Conclusion	263
	References	263

I. INTRODUCTION

The development of foam with a product or during a process is sometimes desired and sometimes not wanted at all. Shaving creams or the flotation of ore are examples where the formation of foam is of some significance. Beer which is not frothy is not very popular for aesthetic reasons.

On the other hand, foams can cause serious problems during many technical processes, for instance, in the purification of waste water, during certain applications in the coatings industry, or in the production of mineral

---

*Retired.

oil. It is therefore important to develop special additives or technical devices which prevent excessive generation of foam.

Suppression or control of foam plays an important role during fabrics washing processes. The development of the first detergents at the beginning of the present century paralleled the development of washing machines. A considerable amount of foam was generated by the detergent in those days and was understood to be a sign of good cleansing power. It is still the case nowadays under certain washing conditions.

Significant changes have taken place in the detergent sector over the past few decades. The washing process has been continuously optimized in order to satisfy the customers's wishes and to meet legislative demands, particularly during the last 20 years. The greatest changes have been made both in the builder system and in the surfactant system. Here the builder is a material which complexes divalent metal ions to prevent preciptation of insoluble salts of the anionic surfactant which is usually present in the surfactant system.

The tub-type machines formerly used in the European washing-machine sector are gradually being replaced by drum-type washing machines. Due to the nature of their design, the drum-type washing machines are very susceptible to the generation of foam with the result that much significance has to be attached to this factor. Control of foam has therefore become an integral part of the work of those involved in developing new detergent formulations.

The main aim of this article is to review developments in the formulation and application of foam control compositions for heavy-duty laundry detergents.

The practical part, however, does not intend to detail the performance and composition of every suds-suppressing system for each powdered laundry formulation. Emphasis is placed on their use in heavy-duty detergent powders for European drum-type washing machines.

It is hoped that this contribution may open up profitable areas of research and development for those who are not familiar with this special subject.

II. FOAM GENERATION IN WASHING MACHINE

A closer look at the factors determining the washing process will allow a better understanding of the problems related to the development of foam.

The "laundry," or articles to be washed, the method of washing, and the detergent are all integral parts of the washing process. These partners are well coordinated in order to ensure that optimum results are attained during the cleansing of the laundry.

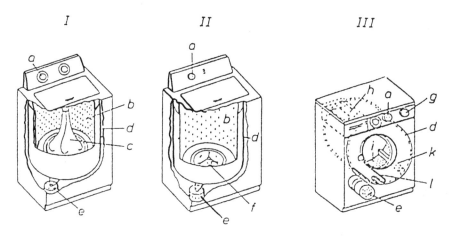

FIG. 1 Schematic drawing of different automatic washing machines
I = agitator-type (U.S.A); II = pulsator-type (Japan); III = drum-type (Europe)
a = timer; b = basket; c = agitator; d = outer tub; e = motor; f = pulsator;
g = thermostat; h = detergent dispenser; i = drum; k = carrier baffle;
l = heating coil. (From Ref. 1.)

Washing habits vary considerably around the world, and this is manifested in the widely differing methods of washing. In underdeveloped countries, for example, washing of laundry is usually done by hand. Here copious amounts of foam are associated by consumers with successful cleaning performance by the detergent. Detergent formulations for such applications should therefore be high foaming.

Washing machines are used almost without exception to clean the laundry in the industrial nations of Europe, North America, and in Japan. Typical designs of such washing machines are shown in Fig. 1 [1].

There are two basic designs of washing machine: the tub-type washer used in North America and Japan and the European drum-type washing machine [2]. Wash conditions used with these different types of machine in the three markets are summarized in Table 1.

North American and Japanese washing machines generally tend to be top-loaders, because they are invariably filled from above. The outer tub is made out of enamel-coated tin or plastic (Japan) and contains the perforated washtub which turns around a vertical axle. In the American washing machines the mechanical action is conveyed by means of a rotating agitator which is situated vertically in the middle of the washtub. In contrast, Japanese machines are equipped with a flat rotating disklike impeller fitted to the base

TABLE 1 Washing Habits and Conditions in Different Parts of the World

Washing conditions	USA/Canada	Japan	Western Europe
Washing machine	Agitator type	Impeller type	Drum type
Heating coils	No	No	Yes
Amount of wash liquor (1)	Extra small: approx. 35	Low: 30	Low:16–20
	Medium: approx. 50 Large: approx. 65 Extra large: approx. 80	High: 45	High: 25–30
Total water consumption (1) (regular heavy cycle)	140	150	120
Wash liquor ratio	1:15–1:30	1:20–1:30	1:5–1:25
Washing time (min)	10–15	5–15	60–70 (90°C) 20–30 (30°C)
Washing, rinsing and	20–25	15–35	100–120 (90°C)
Spinning time (min)			40–50 (30°C)
Washing temperature (°C)	Hot: 50	10–40	90
	Warm: 27–43		60
	Cold: 10–27		40
			30
Wash cycle/month	22	60	18
Water hardness (ppm CaCO$_3$)	Relatively low 100	Very low 50	Relatively high 250
Method of detergent addition	Mostly by hand into the basket	Mostly by hand into the basket	Dispenser[b]
detergent dosage[a] (g/L)	1.5;	1.3;	7–10;
(g/kg fabric)	35–50	30–40	60–80

[a]In USA and Japan without bleaching components
[b]Pre-wash and main-wash cycle and fabric softener (and chlorine bleach in Southern Europe)
Source: Ref. 2.

of the washtub; this ensures that the wash liquor is kept in motion and leads to a through soaking of the laundry.

Neither of these two types of washing machine is equipped with any kind of heating element. In Japan, the washing is done at the temperature of the drinking-water supply as it comes from the tap. American washing machines are directly connected to the hot-water supply in the house and a cold-water connection allows the temperature of the water to be regulated according to the wash program chosen. Detergent is measured by hand directly into the washtub. Here, the following order is recommended: detergent, laundry, and then the water.

Foam-related problems are not to be expected with these Japanese and American tub-type washing machines because of limited mechanical agitation leading to low rates of air entrainment, low washing temperatures, and low detergent concentrations (see Table 1) [3].

Two different designs of European drum-type washing machines have established themselves. These are top-loaders and front-loaders. The former are filled from above through a hatch in the drum, and the latter are filled directly into the open drum through a door at the front of the washing machine.

The washing drum is perforated and revolves around a horizontal axle in varying cycles, reversing at intervals; it is situated lengthwise within a receptacle made out of stainless steel. This longitudinal outer tub is just slightly larger than the actual washing drum and is only about one third full of washing liquor during the washing process. Inside the drum are carrier baffles which lift the laundry lengthwise as the drum revolves, causing the laundry to be dropped back down into the water from the highest point. This results in an intensive mechanical agitation and ensures the close contact of the laundry with the washing liquor.

Depending on the wash program chosen the wash liquor, which is cold at the outset, is heated to and maintained at the desired temperature by means of built-in heating coils. The maximum temperature lies at approximately 95°C. The machine is programmed to automatically flush detergent out of the dispenser into the washing machine. This detergent dispenser contains three to four compartments into which the measured dose of the detergent for the prewash and main-wash cycles is filled together with any aftertreatment aids, such as fabric softeners and chlorine bleach. The latter is only used in some parts of southern Europe.

The design of European drum-type washing machines conspires with high washing temperatures and high detergent concentrations to produce relatively severe foam control problems. An important design factor is the nature of the mechanical agitation. This is responsible for entrainment of copious amounts of air and the development of undesirable foam. Air entrainment occurs in a number of different ways:

When the laundry is lifted out of the washing liquor by the carrier baffles situated inside the drum
When the laundry falls back into the washing liquor
When the washing liquor flows through the perforations in the washing drum

The amount of foam generated must be regulated in order to attain good washing results. Too much foam results in insufficient contact between the

TABLE 2   Formulations of Detergents (Heavy-Duty Powders)

Ingredient	USA/Canada/ Australia	Japan	Western Europe
Surfactants	8–22	19–25	8–13
Builders	20–35	18–30	20–45
Bleaching agents	a	a	15–35
Auxiliary substances			
Foam boosters	0–2	—	—
Foam depressants	—	—	0.3–5
Bleach activators	—	—	0–3
Bleach stabilizers	—	—	0.2–0.5
Antiredeposition agents	0–0.9	0–2	0.5–1.5
Anticorrosion agents	5–15	5–15	2–8
Enzymes	b	0–0.8	0–0.5
Optical brighteners	0.1–0.8	0.15–0.8	0.1–0.8
Fabric softeners	0–5	0–5	—
Perfumes	c	c	c
Fillers and water	Balance	Balance	Balance

[a]Bleaching is carried out separately with a liquid chlorine bleach during the rinsing cycle.
[b]Only very few detergents in the US contain enzymes up to 0.75%.
[c]Perfumes are present.
Source: Ref. 4.

laundry and the washing liquor. The entire mechanical washing process is impeded because the foam prevents the laundry from being able to fall freely into the washing liquor. In extreme cases foam and washing liquor may even flow out of the top of the machine.

The need to regulate the amount of foam generated during the washing process is, therefore, solely a problem incurred with European drum-type washing machines and is absolutely essential if optimum washing results are to be achieved.

III. SURFACTANT TYPES USED IN LAUNDRY DETERGENT FORMULATIONS

The washing conditions described in Sec. II have resulted from decades of development and reflect differences in textiles, wearing habits, washing machines, water hardness, washing temperatures, and duration of wash cycles. The "chemical" part of the washing process, i.e., the detergent, takes all these factors into consideration. It leads to different detergent compositions as shown in Table 2 [4].

Surfactants are the most important ingredients in the detergent and have gained further significance over the past 40 years. The choice of surfactants for detergent applications depends on a series of factors; the main ones are

Good detergency (cleansing) properties
Good handling
The manufacturing process for detergents
Economy
Toxicological safety
Ecological safety

Two main classes of surfactants are used in detergent applications: anionic and nonionic surfactants. The former group of surfactants is used most due to a favorable cost/benefit ratio. Good solubility, good soil removal and good dispersion effects render these anionic surfactants a valuable component of modern detergents [5,6]. Following a reduction in washing temperatures, the development of new textiles and the replacement of phosphates in detergents, nonionic surfactants have gained considerable importance in the past 10–15 years. Even at lower concentrations they are comparable to anionic surfactants with respect to cleansing power, especially the dispersion of soil.

Surfactant mixtures consisting of anionic and nonionic surfactants are generally employed in detergents. Amphoteric surfactants are also used for special applications and the special properties of cationic surfactants, which are often applied in combination with layer silicates, produce the fabric softening effect.

Some of these mixtures consist of several surfactants and tend to display foaming behavior which is not satisfactory for practical purposes. Despite the fact that the structural parameters of the surfactants are known, it is an extremely complex matter to relate these to the foaming properties of a surfactant. In general, this cannot therefore be done, and the foam behavior of surfactants in a detergent cleaning context in usually assessed empirically.

IV. TYPES OF DETERGENT ANTIFOAMS

A. Introduction

Early development of antifoams for laundry detergents was hampered by ignorance of the physicochemical processes involved. Consequently an empirical approach had to be adopted. Even today, it is difficult to establish clear structure-effect relationships concerning the foam-inhibiting properties of a given antifoam system. The desirable properties of an antifoam can be

TABLE 3 Classification of Antifoam Substances

Type	Substance
Carboxylates	Fatty acids
	Fatty acid esters
	Soaps
Phosphates	Monoalkylphosphoric acid esters
	Dialkylphosphoric acid esters
	Fluoroalkylphosphoric acid esters
Nitrogene-containing	Melaminresins
compounds	Amides
	Amines
	Ureas
Hydrocarbons	Mineral oils, -waxes
	Paraffins
	Synthetic oils, -waxes
	Edible oils, -waxes
	Natural waxes
	Microcrystalline waxes
Organic silicon compounds	Polydialkylsiloxanes
	Silicone resins
Fluoro compounds	Partially/totally fluorinated alcohols and carbonic acids

listed as follows:

High efficiency at a low concentration

Applicability within a wide temperature range and with different washing times

Insensitivity to the water hardness

Good price/performance ratio

High stability in the detergent

Toxicological and ecological safety

A wide range of compounds has been tested, some of which have proved to be suitable for practical use. These compounds are classified according to chemical type in Table 3. This list is not claimed to be complete.

The composition of antifoams has become more complex in the past 15 years. In part this has been due to the use of particulate substances, such as hydrophobed silica in admixture with hydrocarbons or polydimethylsiloxanes. In many such systems the addition of hydrophobed silica improves the antifoam efficiency markedly (see Chapter 1).

Today's antifoams can, therefore, be classified as silica bearing and silica free. This fact will be considered in the following overview, and the most

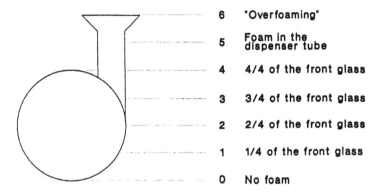

6	"Overfoaming"
5	Foam in the dispenser tube
4	4/4 of the front glass
3	3/4 of the front glass
2	2/4 of the front glass
1	1/4 of the front glass
0	No foam

FIG. 2   Description of the foam rate system introduced by Schmadel [11].

important developments discussed on the basis of the patent literature referred to. Practical examples (foam curves) will be used to illustrate certain correlations. The measurement is carried out in a front-loading washing machine (W433, Miele/Germany) which has an inbuilt electronic mechanism to measure via conductivity the amount of foam generated during the washing process. The foam curves obtained are evaluated on the basis of the foam-grading system proposed by Schmadel [7], which is illustrated in Fig. 2.

B. Soaps as Antifoams

Organic carboxyl compounds, such as carboxylic acids and their alkali salts (soaps) were the first compounds to be used as antifoams in low-foaming detergent formulations. Early work on the antifoam behavior of such materials was due to Harris [8] and Peper [9]. The effect of soap on the foam of sodium dodecylbenzenesulfonate and cetyl sulfate was examined as a function of the pH and water hardness (calcium ion concentration). The experiments showed that soap or fatty acids have an antifoaming effect only when the conditions favor the formation of slightly soluble calcium soaps and when there is no formation of mixed surface films consisting of calcium soaps and other surfactants. Islands of insoluble and solid calcium soaps are supposed to form, which, due to their inflexible nature, destabilize foam films [10].

Both water hardness and builder systems have a decisive influence on the effectiveness of soaps. As can be seen from Fig. 3, there must be an adequate calcium ion concentration available to enable sufficient calcium soaps to form [11]. This is also the cause of problems with excessive foam formation in so-called soft-water areas when soap is used as an antifoam. More-

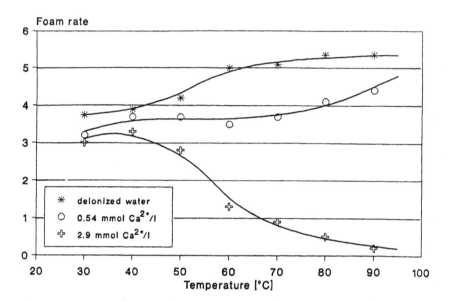

FIG. 3 Antifoam behaviour of sodium behenate as a function of water hardness. A: deionized water; B: 0.54 mmol Ca^{2+}/L; C: 2.9 mmol Ca^{2+}/L. Detergent: heavy-duty detergent with 3.1% linear alkylbenzene sulfonate (LAS); dosage: 100 g pre-wash and 100 g mainwash; washing machine: Miele W 433 (drum-type, front-loader); wash load: 3.5 kg clean cloth. (From Ref. 15.)

over, if soap is the only antifoam, then only builder substances can be used which have relatively low stability constants. In the presence of the builder, the calcium ion concentration must be high enough to allow formation of the calcium soap. This is the case with sodium tripolyphosphate. However problems may be encountered with, for example, nitrilotriacetic acid (NTA) or new builder combinations in today's phosphate-free detergents. An example showing the effect of NTA is given in Fig. 4 [12].

Soaps with chain lengths of 12 to 22 C atoms [13,14] are usually employed. Long-chain soaps (C_{20-22}) have however been found to be particularly effective for wash temperature above 60°C. Low foam profiles are produced with these soaps when sodium tripolyphosphate is used as builder. A typical soap blend and/or the fatty acid blend from which it is derived, has the following composition: 5–15% myristic acid (C_{14}), 25–35% palmitic acid (C_{16}), 15–25% stearic acid (C_{18}), 20–30% arachidic acid (C_{20}) and 10–20% behenic acid (C_{22}).

Precipitation of finely divided calcium soap is facilitated if the sodium soap dissolves readily at the relevant temperature. That is the wash temperature should exceed the Krafft temperature of the sodium soap. Therefore

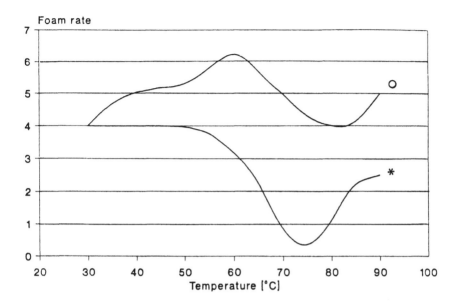

FIG. 4 Foaming behavior of soap-regulated detergents based on 40% NTA (O) or 40% tripolyphosphate (*); dosage 7.5 g/L. (From Ref. 12.)

at low wash temperatures only short-chain-length soaps are effective. However, the solubility of the corresponding calcium soap precipitate in the presence of STP increases with increasing temperature rendering it ineffective. Foam control at higher temperatures then requires the presence of long-chain-length soaps the calcium soaps of which are still insoluble.

Figure 5 shows a comparison between tallow soap and behenic soap [15]. There is a distinct reduction in the efficiency of the tallow soap just before the boil wash temperature is attained, and the machine overfoams. The foam remains at a low level when behenic soap is used. Foam formation at the beginning of the washing process is attributable to the low solubility of both sodium soaps so that the rate of formation of finely divided calcium soap antifoam is inhibited.

The concentration of soap in detergent ranges from 3% to 6%. Processing is effected together with the other detergent constituents in the detergent slurry which is homogenized in a crutcher and dried in a spray tower. A loss in activity when the detergent is stored, such as occurs in the case of the antifoam systems still to be discussed, is not observed.

The main disadvantages for soap are the relative weakness of the antifoam ability (as revealed by the high levels required) and its dependency on water hardness. There is also the danger that calcium soaps are deposited on the

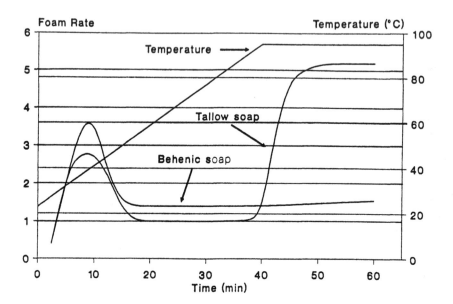

FIG. 5 Comparison between Tallow (C_{16-18}) and Behenic (C_{20-22}) soap in a STPP-reduced detergent (8% LAS, 4% nonionic surfactants, 16% zeolite and 22% STPP) with a soap content of 3.5%. Washing machine: Miele W433 (drum-type, front-loader); dosage: 8.5 g detergent/L; water hardness: 2.85 mmol Ca^{2+}/L.

fabric contributing to organic incrustations and a high degree of yellowing.

Some improvements have been claimed to have been achieved by selective modification of the fatty acid molecules from which the soaps are derived. These include fatty acids with side chains [16], isomeric fatty acids [17] and di- or trimerized oleic acids [18].

The amount of soap required for foam control can be reduced if nonionic surfactants, characterized by a high degree of ethoxylation, are used as co-surfactant in the laundry detergent [19,20]. As a result of this a larger quantity of anionic surfactants can be used to yield a noticeable improvement in the cleaning efficiency.

C. Nitrogenous Antifoams

One class of effective antifoams comprises nitrogenous compounds such as amines, urea derivatives, amides, and trialkylmelamines. Only the two latter compounds have gained a certain significance and should, therefore, be discussed. Both mono and diamides, (1) and (2), are used as antifoams.

Monoamides [21,22]

$$R—CONHR_1 \qquad\qquad\qquad\qquad (1)$$

$$R = C_{13-21}; \qquad R_1 = H, C_{1-22}$$

Diamides [23–26]

$$R_1\!-\!\underset{\underset{O}{\|}}{C}\!-\!\underset{\underset{R_2}{|}}{N}\!-\!R\!-\!CH_2\!-\!\underset{\underset{R_4}{|}}{N}\!-\!\underset{\underset{O}{\|}}{C}\!-\!R_3$$

$$R_1\!-\!CH_2\!-\!\underset{\underset{R_2}{|}}{N}\!-\!\underset{\underset{O}{\|}}{C}\!-\!\underset{\underset{O}{\|}}{C}\!-\!\underset{\underset{R_4}{|}}{N}\!-\!CH_2\!-\!R_3 \qquad (\underline{2})$$

$$R_1\!-\!CH_2\!-\!\underset{\underset{R_2}{|}}{N}\!-\!\underset{\underset{O}{\|}}{C}\!-\!R\!-\!\underset{\underset{O}{\|}}{C}\!-\!\underset{\underset{R_4}{|}}{N}\!-\!CH_2\!-\!R_3$$

$$R = C_{1-9}; \quad R_{1,3} = C_{11-23}; \quad R_{2,4} = H, C_{1-24}$$

It has been claimed that monoamides ($\underline{1}$) are highly efficient antifoams. In a typical, highly foaming detergent (17% alkylbenzene sulfonate, 3% of a nonionic surfactant, 45% tripolyphosphate, 8% sodium silicate, 1% CMC, 22% sodium sulfate, and 4% water) good results can apparently be achieved with as little as 1% octadecanamide. These materials may be introduced through the slurry. Thus after neutralization of the alkylbenzene sulfonic acid with caustic soda a well-homogenized premixture of nonionic surfactant and the amide are added and homogenized. Other powder constituents are then added. The slurry prepared in this way is then spray-dried.

Antifoam diamides ($\underline{2}$) are substances which can have high melting points (mp), for example, distearyldiamide: mp 123–137°C. They are exclusively used in combination with additives such as polysiloxanes, oils, waxes, other hydrocarbons, or even soaps. They are only effective when they are finely dispersed in the detergent solution so that they are quickly transported to air-water surfaces of the foam. The desired foam profile depends on the chain length [23]. Thus if R_1 and R_3 have chain lengths of C_{12-14}, then the resulting compound has an optimum antifoam range at 20–60°C. Longer chains (C_{18-22}) have an optimum range between 60 and 100°C. With dosage levels of the antifoam compound in the range of 0.1% to 3%, this antifoam system is highly efficient.

The advantage of diamides is that their effect is not dependent on the type of surfactant or the water hardness. Moreover, relatively low feed concentrations require different processing methods. Processing in a detergent slurry together with other constituents is, in fact, possible, but leads to activity losses, since the required disperse distribution is not achieved. A better and more effective method is the production of a homogeneous premixture by means of dispersion processes with a high shearing action, and subsequent further processing of this dispersion by spraying onto the washing powder [23–25,27] or by deposition on a carrier substance [25,27]. An example of an effective antifoam system consists of diamide and a polysiloxane which, after deposition on a sucrose or gelatinized starch carrier, is protected by coating with paraffin wax and thus has a long-term efficiency [24].

Combinations of 5–20% distearyldiamide with 20–50% soft/hard paraffin (mp 20–55°C) and 25–60% microparaffin (mp 60–80°C) have been found to be very effective, 0.1–0.3% of this combination being sufficient for effective antifoam action in a detergent formulation [27]. Approximately 10% of this antifoam are deposited [27] on a carrier (light sulfate). Subsequent coating is not necessary due to setting of the wax matrix. Particular attention is drawn to the good long-term storage stability in the washing powder.

Another group of nitrogenous antifoams (3) is formed by the trialkylmelamines [28–36]:

$$
\begin{array}{c}
\text{R}-\text{HN} \diagdown \underset{\text{C}}{} \diagup \text{N} \diagdown\diagdown \underset{\text{C}}{} \diagdown \text{NH}-\text{R} \\
\| \qquad | \\
\text{N} \diagdown \underset{\text{C}}{} \diagup \text{N} \\
| \\
\text{NH}-\text{R} \qquad \text{R} = \text{C}_{7\text{-}22}
\end{array}
\tag{3}
$$

Melamines of cyanuric chloride and fatty amines are available [37–39]. These are characterized by their insolubility in water and, depending on the substitution, by melting points of between 30°C and 90°C. As with the diamides, these are not processed through slurry. The wash active substances combined with the high temperatures would emulsify the melamine too much and cause it to become inactive.

A two-stage process has been found to be practicable. First, a liquid to slightly pasty emulsion is produced from melamine resin, emulsifier, and other additives. This emulsion is sprayed directly onto the washing powder in a second stage, or onto a carrier, which is then incorporated into the washing powder [33,34]. The advantage of the latter variant is that the carrier "dilutes" the active antifoam, as a result of which the antifoam is distributed uniformly and more homogeneously in the washing powder. At the same time, this measure improves storage stability.

Examples of the abovementioned emulsions are listed as follows [33]:

Example 1   Example 2
18% melamine   40% melamine
3% C_{16-18} fatty alcohol with 10 mol EO   20% fatty alcohol (see left)
9% C_{10} fatty acid (capric acid)   40% paraffin oil
70% water

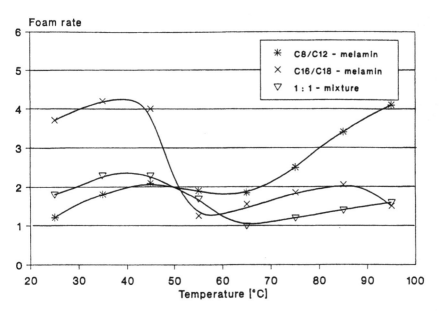

FIG. 6 Antifoam behavior of melamines with short-chain and long-chain alkyl groups in comparison to their 1:1 mixture. Detergent: heavy-duty STPP-based detergent; detergent dosage: 100 g main wash; melamin dosage: 2.6%; wash load: 3.5 kg clean cloth; water hardness: 2.8 mmol Ca^{2+}/L.

When the residual alkyls are properly selected, a desired, temperature-independent foam profile can be obtained (Fig. 6). Melamines with short-chain residual alkyls are effective up to about 70°C, while melamines with long-chain alkyls only begin to have an antifoam effect above about 50°C.

D. Phosphoric Acid Esters

A class of compounds used for antifoam purposes is formed by phosphoric acid esters (4). These are comparable to the fatty acid/soaps [40] in forming slightly soluble or insoluble salts with multivalent cations. The mode of action of the phosphoric esters is therefore probably similar to that of fatty acids in that insoluble precipitates of calcium salts are usually involved.

$$
\begin{array}{c}
O \\
\parallel \\
HO\!-\!P\!-\!OR_1 \\
\mid \\
OR_2
\end{array}
\qquad\qquad (\underline{4})
$$

$R_1 = C_{9-20};\qquad R_2 = H, C_{9-20}$

The disadvantage of higher manufacturing costs relative to the soaps is compensated to a far-reaching degree by greater effectiveness, Thus phosphoric acid esters are effective in dosages of as little of 0.5–3% [41,42].

Efficiency can be further increased by combining phosphoric acid esters with other additives or by modifying and optimizing the method of processing. Above all, the disadvantage already encountered with the fatty acids of efficiency being dependent on the water hardness, should be taken into consideration. Nonionic surfactants can be used as additives [43–45]. Thus for example detergent mixtures which contain 0.1–0.5% C_{16-24} phosphoric acid esters and 1–3.5% tallow alcohol with 11 mol ethyleneoxide are described [46]. The phosphoric esters and the nonionic surfactants are melted together, homogenized and then sprayed onto the washing powder. Combinations with soap [47] and hydrocarbons, such as mineral oils, or waxes of synthetic, mineral, vegetable or animal origin [48–52], are also described.

$$
\begin{array}{c}
\text{O} \\
\parallel \\
\text{HO---P---(EO)}_n\text{---OR}_1 \\
\mid \\
\text{X}
\end{array}
\qquad\qquad (\underline{5})
$$

$$X = OH, \quad R_2O(EO)_n; \qquad R_{1,2} = C_{12-24}; \qquad m,n = O,1-6$$

The use of modified phosphoric esters ($\underline{5}$) which contain a polyoxyethylene chain between the residual alkyls and the phosphoric acid ester group has been described [53,54]. The direct employment of the slightly soluble calcium salts of these alkyl phosphoric esters can be particularly advantageous. These salts may be dispersed in hydrocarbon wax. To obtain optimum efficiency the particle size of the calcium salts should be between 0.1 and 0.5 μm [49]. The melting point of the waxes employed has a considerable influence on the antifoam efficiency and should be below the maximum washing temperature. The antifoam is either hot-sprayed, in the form of the dispersion described above, onto the finished washing powder or processed together with other powder components to form antifoam agglomerates. This can be done using known processes, such as extrusion or pan agglomeration.

Use of phosphates in detergents is increasingly controversial, especially in Europe. Less and less interest is therefore being shown in phosphoric acid esters, and these are disappearing from detergent formulations.

E. Hydrophobed Silica/Hydrophobic Oil Mixtures

Silicone- and mineral oil–based antifoams activated by the addition of silicas represent one of the most important class of antifoams employed today. Here neither the silica nor the oil is effective alone. The mixtures, however, exhibit synergy provided the silica is rendered hydrophobic. This type of be-

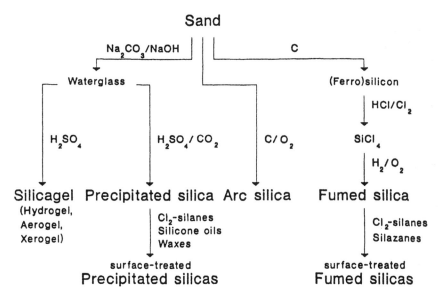

FIG. 7   Manufacturing of different silicas from quartz sand.

havior is also shown when, for example, intrinsically hydrophobic particles such as calcium alkyl phosphate or ethylenebis (stearamide) are mixed with mineral oils. The mechanistic aspects of this behavior are reviewed in detail in Chapter 1.

1. Silica Manufacture, Properties, and Application

The synthetic silicas frequently used today are tailormade γ-ray amorphous products. Depending on the manufacturing process, these can be divided into the following groups [55–57]:

Pyrogenically or thermally produced silicas (fumed silicas)
Wet-chemically produced silicas (precipitated silicas)
Aftertreated silicas

Silica sols occupy a special position, since they are not present in a particulate state. They can, however, be considered as the precursor to silica gels.

A raw material common to all synthetically manufactured silicas is quartz sand of as high purity as possible. For manufacturing wet-chemically produced silicas, water glass is prepared, either thermally or hydrothermally, from quartz sand. By contrast the basis for the pyrogenic products is silicon tetrachloride, produced from silicon or ferrosilicon (Fig. 7).

Only the fumed and precipitated silicas have acquired significance for use as antifoams. Fumed silicas are highly disperse silicas which form at high

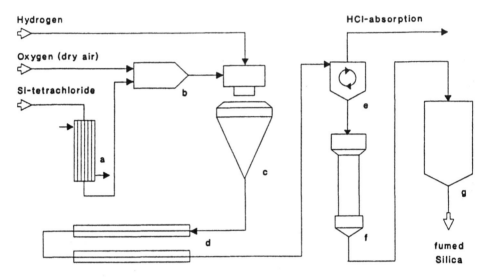

FIG. 8 Scheme of the production process of Aerosil (fumed silicas): a = evaporator; b = mixing chamber; c = burner; d = cooler; e = cyclone; f = deacidizing; g = silo.

temperatures due to coagulation from the gas phase. This process is described as flame hydrolysis (Aerosil process [58–60]). A homogeneous mixture of silicon tetrachloride, steam, hydrogen, and dry air is burnt. The energy and water necessary for hydrolysis is supplied by the oxyhydrogen flame. The silica is separated in cyclones and neutralized with water vapor and air in a fluidized-bed reactor. The product which has a bulk density of 20 g/L is pneumatically conveyed to packaging machines, where it is simultaneously compressed to 60–100 g/L (Fig. 8).

Precipitated silicas are formed by the reaction of alkali silicate (water glass) with a mineral acid (1).

$$Na_2(SiO_2)_{3.3} + H_2SO_4 \rightarrow 3.3\ SiO_2 + Na_2SO_4 \qquad (1)$$

By varying the chemical and physical conditions during precipitation, as well as the method of filtration, drying, and grinding, a number of different products can be manufactured, whose characteristics fulfil the requirements for certain fields of application. The type of precipitation process, in which further silica is deposited on the primary particle formed, thus leading to a growth in the particle size, is essentially responsible for the fact that only discontinuous manufacturing processes have achieved industrial-scale significance.

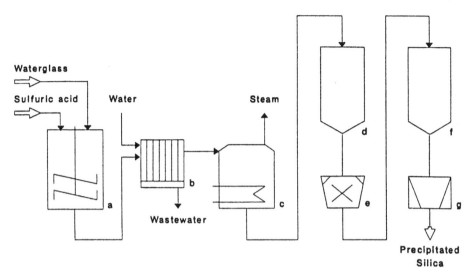

FIG. 9   Scheme of the production process of precipitated silicas: a = precipitation reactor; b = filtering apparatus; c = drying; d = intermediate product silo; e = grinder; f = silo; g = densifier.

A water-glass solution is mixed with acid, usually sulfuric acid, under defined temperature and stirring conditions in large mixing vats (Fig. 9) [61,62]. It is also possible to use carbonic acid and hydrochloric acid [63,64]. The suspension which results is filtered off (usingly chamber presses, plate and frame presses, automatic filter presses, or band and rotary filters) and washed free of salt. The filter cake (solids content 15–25%) is dehydrated on band, turbine, revolving, or rotary driers, or also by spray-drying after prior liquefaction of the filter cake [65,66].

The properties of the amorphous silicas manufactured according to the processes described above, are determined by the particle size, particle structure, and the chemistry of their surface, which is characterized by the presence of silanol groups Si—OH. These groups are responsible for the hydrophilic nature of silica surfaces. These surface silanol groups can be chemically reacted with various reagents to render the silica surface hydrophobic.

In the case of fumed silicas this is usually effected with chlorosilanes in a fluidized bed reactor at 400°C [67]. Precipitated silicas can be hydrophobed with alkylchlorosilanes in the precipitated suspension. In this case filtration, washing and drying are followed by tempering to 300–400°C [68,69]. The hydrophobing mechanism is illustrated in Fig. 10. In addition to chlorosilanes, silazanes, and, particularly in the case of precipitated sil-

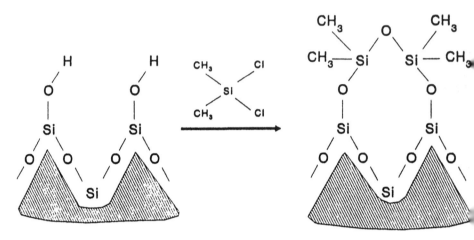

FIG. 10   Reaction between dimethylchlorosilane and the silanol groups of silica surfaces.

icas, polydimethylsiloxanes (silicone oils) are used [70]. Treatment with silicone oils, which is also known as "in situ hydrophobing" has achieved particular significance with respect to the use of silicas in antifoams, since the silicone oil is not only a hydrophobing agent but also an important component of the antifoam formulation.

A less important method of surface modification is the so-called coating process. Here the hydrophobic agent is only physically adsorbed on a silica surface so that a chemical reaction with the silanol groups is not involved. Fused-on waxes are used [71] for this purpose, or, in the case of precipitated silicas, wax dispersions [72].

The essential characteristics of the fumed and precipitated silicas are shown in Table 4. The most pronounced differences are the aggregate and agglomerate size and the lack of porosity of the fumed silicas, which only have an external surface. With aftertreated silicas, a further characterizing parameter is the degree of hydrophobicity.

In recent years, hydrophobic, precipitated silicas have proved to be extremely effective in silicone- and mineral oil–based antifoams, special antifoam silicas having been developed [56]. The essential characteristics are the degree of hydrophobicity, agglomerate size and specific surface (Table 5). A favorable influence is possibly exerted on the antifoam efficiency by the surface roughness. However, the degree of hydrophobicity has the most decisive influence on the efficiency. This can be determined either by measuring the contact angle at the air/water surface or by measuring the "methanol wettability."

TABLE 4 Comparison of Important Properties of Synthetic Amorphous Silicas

		Silica type	
Parameter		Fumed	Precipitated
BET-surface	$[m^2/g]$	50–600	30–800
Primary particle size	[nm]	5–50	5–100
Agglomerate size	[μm]	a	1–40
Density	$[g/cm^3]$	2.2	1.9–2.1
Tapped density (DIN 53 194)	[g/l]	50–100	50–500
Drying loss (DIN 53198)	[%]	<2.5	3–7
Ignition loss (DIN 55921)	[%]	1–3	3–7
pH value (DIN 53200)		3.6–4.3	5–9
DBP adsorption $(cm^3/100 \ g)$		250–350	150–320
Percentage inner surface		Nil	Low
Structure of aggregates/ agglomerates		Chainlike agglomerates	Slightly aggregated, spherical particles
Thickening effect		Very strong	Existing
Average pore size (nm)		Nonporous	>30

aFumed silicas do not have a defined agglomerate size

The contact angle is measured on pressed silica tablets (Fig. 11). A drop of water is placed on the surface of the pressed silica and the contact angle determined optically by means of a goniometer. Hydrophobic surfaces are characterized by an angle of >90°.

Methanol wettability is a measurement which relies upon the reduction of the surface tension of water by addition of methanol. If the proportion of methanol is high enough, the critical wetting tension of a hydrophobed

TABLE 5 Properties of Antifoam Silicas

Parameter	Requirement
Hydrophobicity	Contact angle >90° Methanol wettability >50%
Particle size	0.2–5 μm
Specific surface BET	>50 m^2/g
pH	>8

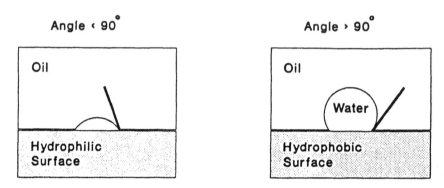

FIG. 11 Definition of the contact angle of water droplets placed on the surface of hydrophilic or hydrophobic silicas (pressed tablets immersed in oil).

silica can be achieved in that the material is wetted by the solvent. Methanol wettability therefore simply involves measurement of the proportion of hydrophobed silica dispersed into a water/methanol mixture as a function of the composition of the mixture. Typical plots are shown in Fig. 12.

2. Hydrophobed Silica/Hydrocarbon Antifoams

The range of hydrocarbons employed in antifoams includes solid and liquid products of mineral (petrochemical), natural or synthetic origin. The term *hydrocarbon* is not quite precise in this context and only serves to simplify and facilitate the following explanations.

Hydrocarbon waxes are characterized by melting points above 20°C and low saponification values of ≤60. Use in detergents requires that the material melt and disperse in the wash. This means that the melting point must be below wash temperatures which are always ≤95°C and often <50°C. The optimum melting point range is therefore 30–70°C. An important criterion is the type of "crystallinity." Microcrystalline waxes as described by Warth [73] are very effective. These waxes have smaller crystals than paraffin waxes, which is attributable to the higher portion of branched-chain hydrocarbons and to a higher molecular weight.

Vaseline (petroleum jelly) with a melting range of 35–50°C has found wide application. Synthetic waxes of the Fischer-Tropsch type (named after the process), earth or peat wax (such as ozokerite, ceresin, montan wax) or natural waxes such as beeswax, carnauba, or candelilla, can also be used. So-called Japan waxes, ester waxes or high-molecular weight polyethylene glycols are seldom used in hydrocarbon antifoams. Liquid hydrocarbons are also used, even if only to a limited degree. Apart from mineral oils of naphthenic or paraffinic origin, vegetable or animal oils can be used. Common to all these compounds is a low melting point of 40–50°C, with a minimum

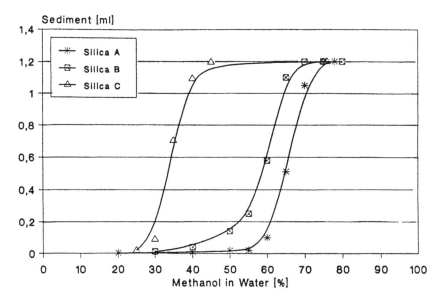

FIG. 12 Methanol wettability of different types of hydrophobic silicas. With rising content of methanol in water an increasing amount of silica is wetted which will form a bigger sediment. The antifoam behavior of the silicas is shown in Fig. 13.

boiling point of 110°C. The boiling point is important in that, at the maximum washing temperature of 95°C (boil wash), the antifoam components must not be volatilized. Adverse effect due to odor and color mean that only mineral oils such as spindle or paraffin oils have been able to achieve significance in practice.

Very few hydrophobed silica/hydrocarbon antifoams are described in the detergent patent literature (Table 6). Compared to this, waxes are widely claimed in connection with silicone oil antifoams, either as a component of the active system or as organic carrier. This will be discussed in detail in connection with the silicone oil defoamers.

A common feature of the claimed antifoams listed in Table 6 is mixture of a surface-active organic component with the hydrocarbon. For instance, glycerol monostearate [74,75,79] in portions of 50–70%, or so-called Guerbet alcohols [76,78], which are present in antifoams to an amount of 60–80% are claimed. The function of these substances (of HLB ≪10) is to assist dispersion of the antifoam.

The antifoam system usually comprises a mixture of a wide variety of hydrocarbons which, at most, has only few liquid constituents at ambient temperature and which has components with melting points in some cases just above 70°C. These mixtures are manufactured by mixing the individual

TABLE 6 Review of Patent Literature Dealing with Synthetic Silicas and Hydrocarbons (1978–1985)

Silica[a]	Hydrocarbon[b]	Other organic ingredients	Incorporation[c] Method	Claimed advantages	Reference
PS	LH, SH, MW	Fatty acid-ester	1, 3, 5	Robustness	74
PS	SH, MW	Nonionic surfactant	2, 3, 4	Effective with complex surfactant systems	75
FS, PS	SH	Guerbet alcohol	4, 5	Similar to silicon anti-foams, broad application range	76
PS	LH, SH	—	4, 5	Low amount of LH	77
FS, PS	SH, MW	Guerbet alcohol	2, 4, 5	Effective with high amounts of anionic surfactant	78
PS	LH, SH	Fatty acid-ester	4, 5	Storage stability	79

[a]PS = precipitated silica; FS = fumed silica
[b]LH = liquid hydrocarbon (mp < 20°C)
SH = solid hydrocarbon (mp > 30°C)
MW = microcrystalline wax
[c]1 = spray drying; 2 = spray cooling; 3 = spraying on detergent powder; 4 = spraying on carrier; 5 = antifoam granules

244

components as well as the emulsifier in the melt and subsequently incorporating the hydrophobic silica. In a first step, the silica is intermingled using a blade or beam agitator, which is followed by the actual dispersion. Dispersion is effected with high-shear mixers such as, for instance, Ultra-Turrax or the Gaulin Homogenisator.

A variety of methods are used to incorporate the antifoam in detergent powder. Here it should be noted that the method of incorporation has a decisive influence on the extent to which antifoam activity is maintained and the length of time the antifoam remains stable in the washing powder after storage. In contrast with soaps, incorporation in the detergent slurry with subsequent spray drying to produce washing powder is not recommended. The hydrophobed silica/hydrocarbon antifoam would be inactivated during preparation of the detergent slurry. The reasons for this are not well understood.

The following methods have been shown to provide more successful incorporation:

Spraying onto the washing powder
Spray cooling
Spraying onto a carrier
Production of antifoam agglomerates

Spraying onto the powder is the simplest method. However, the dosing accuracy must be high because relatively small quantities (approx. 0.1–1%) must be uniformly sprayed onto the powder [74,75].

Direct spray-drying of the antifoam dispersion has also been suggested. This is known as "spray cooling" because it is effected in spray towers with cooled air [75,78]. It can only be employed with dispersions whose melting point is 40°C and higher. As with direct spray-on, problems are encountered with the homogeneous incorporation into the detergent powder of small quantities of antifoam powder, particularly since the latter is difficult to handle. It also has a high tendency to agglutinate on storage, so that antifoam powders produced according to this method should be further processed immediately.

From an applications engineering point of view, the manufacture of antifoam agglomerates and the spraying of the antifoam dispersion onto carriers is advantageous. Preference is also given to these two methods in practically all the relevant patents. Antifoam agglomerates can be manufactured by mixing the hot, finished antifoam dispersion in a kneader together with salts which are preferably contained in the detergent. The simplest way is to cool during mixing, the mass then being ground [76,77] or extruded [81]. Agglomeration of the powdery constituents with the dispersion is also possible, e.g., in a pan granulator.

The method of spraying onto a carrier requires separate preparation of the carrier usually by spray-drying. This leads to highly absorptive granules which are impregnated with the antifoam dispersion in a subsequent step. Tripolyphosphate, carbonates, silicates, and sulfates can be used as carriers. A carrier specially designed for this purpose consists of 5–15% water glass, 30–35% tripolyphosphate, 25–45% sodium sulfate, and 7–18% water [78]. Carriers of quite another type are gelatinized starch and perborate monohydrate [77,79]. These are both coarse-particle substances with a particle size of 400–600 μm, and are extremely absorptive.

The advantage of the antifoam aggregates produced by these two methods is that the antifoam is fixed in a surfactant-free, water-soluble or at least easily dispersible matrix which means improved storage stability. Furthermore, the matrix serves to "dilute" the antifoam, so that instead of 0.1–1% antifoam up to 5% antifoam agglomerate is now to be added, which is a distinct advantage as far as homogeneous and uniform incorporation in the washing powder is concerned.

3. Hydrophobed Silica/Polydimethylsiloxane Antifoams

a. Properties. The most effective and most significant antifoams are the so-called silicone antifoams. These are characterized by the fact that they are not dependent on the water hardness and have a broad spectrum of efficiency as far as the most diverse surfactant systems and the most varied application conditions are concerned. What distinguishes them from all other antifoams, including hydrocarbon-based antifoams, is their high efficiency at low dosages. Depending on the type of detergent, quantities of below 0.1% can be sufficient to achieve a satisfactory foam profile in the washing process. The majority of silicone oils used are polydimethylsiloxane polymers which are also known as "silicone fluids." These linear polysiloxanes are manufactured by selected hydrolysis of dimethyldichlorosilane and trimethylchlorosilane, the linear siloxane chains being formed during condensation.

$$n\text{Me}_2\text{SiCl}_2 + 2\text{Me}_3\text{SiCl} + (n + 1)\,\text{H}_2\text{O} \longrightarrow$$

$$\text{Me}_3\text{SiO} \left[\begin{array}{c} \text{Me} \\ | \\ -\text{Si} - \text{O} \\ | \\ \text{Me} \end{array} \right]_n - \text{SiMe}_3 + 2(n + 1)\text{HCl} \tag{2}$$

The parameter n, given in the formula (2) for the degree of polymerization, varies from about 20 to 2000, at times, to even higher values. This is also combined with a wide range of viscosities of from 50 cSt at $n = 20$ up to 200,000 cSt at $n = 2000$ [80–82]. The siloxane skeleton makes it possible to densely pack the methyl groups which dominate intermolecular forces and, consequently, determine the surface tension. The latter is relatively insensitive to n for $n > 20$ and then has values of approximately 20–22 mN m^{-1} (at 20°C) [83]. This lies within the range of surface tensions of crystalline hydrocarbons, the crystal lattice of which forces dense packing of the methyl groups [84]. Liquid paraffin hydrocarbons have surface tensions higher by about 10 mN m^{-1} than silicone oils.

Other physical-chemical characteristics of the silicone oils are [83]

Clear, colorless liquids
Spontaneous dewetting and spreading onto the surface of the majority of
    aqueous surfactant solutions
Insoluble in the aqueous medium
Incompatible with surfactants
Difficult to emulsify
Chemically inert

These characteristics make it easy to understand why silicone oils are ideal antifoam liquids. Due to their extremely low surface tension, silicone oils dewet and spread spontaneously over the surface of most surfactant solutions. Thus the surface tensions of these solutions usually vary between 25 and 40 mN m^{-1} at 20°C which clearly exceeds the air-oil surface tension of a silicone oil [85]. The high efficiency of silicone antifoams can be derived from this since the dewetting of antifoam into a foam film surface is a necessary property (see Chapter 1).

In addition to the chain-form polydimethylsiloxane polymers, so-called siloxane resins have also achieved certain significance. These are three-dimensional, cross-linked polymers formed by $Me_3SiO_{1/2}$ and $SiO_{4/2}$ units. The resins are manufactured from trimethylchlorosilane and silicone tetrachloride by means of cohydrolysis or condensation. Another variant is the hydrolysis of alkyltrichlorosilanes. The resin usually has a portion of 0.5–5% free silanol groups.

Silicone antifoams usually consist of mixtures of polydimethylsiloxane and hydrophobed silica. The antifoam may be prepared by mixing hydrophilic silica with polydimethylsiloxane and subjecting the mixture to heat treatment to ensure surface reaction between the silica and the polydimethylsiloxane to produce hydrophobed silica in situ. Alternatively silica hydrophobed by other methods may be directly mixed with the polydimethylsiloxane.

As with hydrocarbon-based antifoams manufacturing is divided in two steps: preparation of antifoam dispersion and incorporation in the washing powder. Much attention is paid to the latter in the patent literature. This is illustrated by Table 7 where a list of silicone antifoam patents for detergent powder application is presented. Practical experience has shown that the method of incorporation of silicone antifoams has a decisive influence on antifoam efficiency, particularly with reference to maintaining activity when the washing powder is stored for a long period.

The first silicone antifoams with hydrophobic silica as promoter were described at the beginning of the sixties [112,113]. Here the silica was hydrophobed in situ by heating a silica/siloxane dispersion for several hours to temperatures of 150–220°C, with a possible addition of Lewis acids such as $AlCl_3$, $FeCl_3$, $SnCl_4$, $TiCl_4$, BF_3, $ZnCl_2$ as catalysts [112]. As we now know, this heat treatment modifies the surface of the silica. The activated silanol groups of the silica are present in a "silanate" form and can be reacted with the siloxane polymer in such a way that the basic silanate group is deposited on the silicon atom of the siloxane polymer [114]. A penta coordinated state of transition is passed through [115] and a new siloxane unit forms during heating, the chain-form siloxane polymer being decomposed. As a result of this reaction between the silica and the siloxane polymer, a silica with a hydrophobic surface is obtained.

$$\begin{array}{c} \text{Silicone oil} \\ + \\ \text{hydrophilic silica} \end{array} \xrightarrow[0.5\ \text{hr}]{\text{RT}} \alpha\text{-dispersion} \qquad (3)$$

$$\alpha\text{-dispersion} \xrightarrow[1-10\ \text{hr}]{40-300°C} \beta\text{-dispersion} \qquad (4)$$

In a first stage, the silica is added to the silicone oil and a so-called "α-dispersion" is produced under high-shear stirring conditions (3). Mixers which allow chargewise production are available, for example the Ultra-Turrax. The Dispax-Reaktor or Gaulin-Homogenisator are widely used for continuous processes. The silica employed is either "activated" by treatment with alkali hydroxides or the alkali or other catalysts functioning as Lewis acids are added during dispersion [80,81,116–118]. Silicone oils with viscosities below 5000 cSt are generally used at a ratio of from 85% to 97% to 3–15% in the mixture with the silica. Siloxane resins are also sometimes used which can react with the silica via the still freely available silanol groups (up to 5%) [81]. The dispersion is heated under inert gas for a specific period of time to temperatures usually in the range of 150–200°C. Temperature and time are dependent on the catalyst added. In the majority of the cases reaction is completed after only 1–2 hr at temperatures below 200°C. The resultant dispersion of a hydrophobic silica in silicone oil is also known as a "β-dispersion" (4).

An example is given in Fig. 13. Two precipitated and one fumed silica were hydrophobed in situ with a 1000-cSt silicone fluid (DC 200, Dow Corning). The silica portion of the antifoam was 5%. The silica was activated to the highest degree by soda lye to give a high pH and was consequently converted with the silicone oil. A measurement for this is the methanol wettability. The curves of the methanol wettability of the three silicas are given in Fig. 12 indicating that silica A is the most "hydrophobed" one. A foam test in a washing machine performed to determine antifoam efficiency showed silica A to be the most effective (Fig. 13).

In situ hydrophobing has the advantage of high flexibility, i.e., adaptation to the individual requirements. However, it is a relatively expensive process. It is simpler and usually more cost favorable to manufacture the antifoam directly from a hydrophobic silica and the silicone oil. Special antifoam silicas which are described above are available for this purpose. If different silicas are examined for their initial (immediately after mixing with polydimethylsiloxane) antifoam efficiency it becomes apparent that degree of hydrophobicity primarily determines antifoam efficiency. This is illustrated in Fig. 14 where it is clear that antifoam efficiency correlates with hydrophobicity as indicated by methanol wettability. As we have indicated the method of incorporation of silicone antifoams in detergent powders is of critical importance. This is a consequence of the tendency of silicone antifoams to lose their effectiveness when mixed with detergent powders. Deactivation processes are particularly noticeable with very low levels of antifoam. Usually, the antifoam is added to the detergent powder in the form of a 10–50% antifoam adjunct. The antifoam can diffuse from these concentrates into the powder matrix. This process has been suggested by Sawicki [83] as a possible mechanism to explain deactivation.

Here the loss of effectiveness is explained by a change in the droplet size of the antifoam as a result of such diffusion. Thus diffusion into the powder matrix supposedly results in release of "smaller" antifoam droplets when the powder is dispersed in the wash liquor. Such droplets are supposed by Sawicki to be less effective than the "larger" droplets released before diffusion has taken place. No direct evidence for this hypothesis is however given where measurement of the relevant droplet size has been made.

On the whole then there would appear to be no clear explanation for the deactivation of silicone antifoams when in contact with detergent powder.

b. Incorporation Methods. In essence, three methods have been developed to prevent deactivation, which can be subdivided into a number of variants. A compilation is given in Table 8. These methods are

Embedding the defoamer in an organic matrix, preferably of substances with a melting point above 35°C

Table 7 Review of Patent Literature Dealing with Synthetic Silicas and Silicone Oils (1971–1988)

Silicas[a]	Polysiloxane[b]	Hydrocarbon[c]	Other ingredient[d]	Method of[e] processing	Claimed advantages	Reference
FS^α, PS^α	LVS	—	—	C1	Granulated antifoam	82
FS^α	LVS/SR	—	PEG 4,000, TA25EO	A2, A3	Storage stability, insensitivity toward water hardness	86
FS^α	LVS/SR	MW	PEG 4,000, TA25EO	A2, A3	Low dosage, synergism: silicone carbonwax	87
FS^α, PS^α	LVS	MW	Nonionic, CMC,	PVA B3	Good powder dispersability, homogeneously distributed antifoam	88
FS^α, PS^α	LVS	—	Nonionic, CMC	B2	Storage stability, no fabric graying	89
FS^α, PS^α	LVS/SR	—	—	C2	Good powder dispersability, no activity loss during processing	90
FS^α, PS^α	LVS/SR	—	—	C2	Good powder dispersability, no graying of fabrics	91
FS^β	LVS	SH/MW	Sucrose, TiO₂, CMC	C1	Storage stability	92
FS^α, PS^α	LVS	—	CMC, MC, PVA	B2	Storage stability, homogeneously distributed antifoam	93
FS^α	LVS/HVS,SR	—	Nonionic, siloxane-oxyalkylene polymer	A1, A3, A4	Storage stability, good dispensing properties	94

FS$^\alpha$, PS$^\alpha$	LVS	SH	Gelatinized starch	C1	Improved storage stability	95
FS$^\beta$	LVS	MW	Nonionic, PEG	A3	Mineral oil/silicone oil antifoam effective at high surfactant levels	96
FS$^\alpha$	LVS/HVS	SH/MW	Diamide, alkylphosphate, gelatinized starch	C1	No organic solvent for processing of the hydrocarbons	97
FS$^\alpha$	HVS	—	Nonionic, CMC	B1	Storage stability	98
FS$^\alpha$	LVS/HVS	SH/MW	CMC/MC	B3	Storage stability, no surfactants in antifoam	99
FS$^\alpha$	LVS/HVS	—	Siloxane-oxyalkylene-copolymer	A3	High robustness in combination with 0.5–1.5% soap	100
PS$^\alpha$	LVS	—	Siloxane-oxyalkylene-copolymer, TA11EO	A4	Softening effect remains uneffected	101
FS$^\alpha$	LVS/SR		PEG, soap, TA25EO	A3	Storage stability, good antifoam properties	102
PS$^\beta$	LVS	SH/MW	Gelatinized starch, perborate, zeolite, salts	C1	Storage stability, low- and high-temperature antifoam	103
PS$^\beta$	LVS	LH	Gelatinized starch, diamide, alkylphosphate	C1	Storage stability	104
FS$^\beta$, PS$^\beta$	LVS/HVS	SH	CMC, MC, starch	A3	Storage stability, simple process	105
FS$^\beta$, PS$^\beta$	HVS	—	Fatty acid alcohol	A2, A3	Storage stability, simple antifoam	106

251

Table 7 Review of Patent Literature Dealing with Synthetic Silicas and Silicone Oils (1971–1988)

Silicas[a]	Polysiloxane[b]	Hydrocarbon[c]	Other ingredient[d]	Method of[e] processing	Claimed advantages	Reference
FS$^\beta$, PS$^\beta$	HVS	—	Glycerol-monoester	A2, A3	Storage stability	107
PS$^\beta$	LVS/HVS	LH/SH/MW	Fatty acid, soap, diamide, alkylphosphate	A2, A3	Storage stability because of soap matrix	108
PS$^\beta$	LVS/HVS	LH/SH/MW	Special soda ash	C2	Porous carrier, storage stability	109
FS$^\beta$	LVS	—	PEG 5,000–10,000, TA25EO	A3	Storage stability, protective envelope against surfactants	110
FS$^\alpha$, PS$^\alpha$	HVS	—	Nonionic, CMC, PA, PVA, PVP	B3	Good foam pattern because of combination silicone-anionics	111

[a] FS = fumed silica, PS = precipitated silica; α = hydrophobic and hydrophilic, β = only hydrophobic.
[b] LVS = low-viscosity siloxane <5000 cSt; HVS = high-viscosity siloxane > 5000 cSt; SR = silicone resin.
[c] LH = liquid hydrocarbon (mp < 20°C), SH = solid hydrocarbon (mp > 30°C); MW = microcrystalline wax.
[d] PEG = polyethyleneglycol; TA = tallow alcohol; CMC = carboxymethylcellulose; MC = methylcellulose; PA = polyacrylate; PVA = polyvinylalcohol, PVP = polyvinylpyrrolidone.
[e] Differentiation according to the compilation given in Table 8.

252

SILICA	A	B	C
Type	preci- pitated	preci- pitated	fumed
BET-surface [m²/g]	170	190	200
Mean agglome- rate size [µm]	5	7	-
pH-value	8,3	6,3	4,0

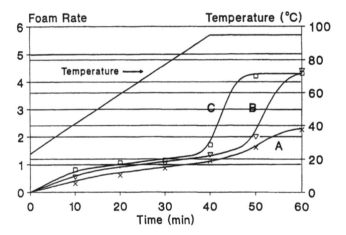

FIG. 13   Primary antifoam behavior of different hydrophilic silicas (A = FK 383DS, B = Sipernat 22S, C = Aerosil 200, all from Degussa) dispersed by 5% in silicone oil of 1000 cSt (DC 200, Dow Corning). After heating for 5 hr to 220°C and a final homogenization 10 parts of the dispersion were mixed with 90 parts granulated zeolite A (Wessalith CS, Degussa) thus forming an antifoam granulate. 2% of this granulate were mixed to a STPP-built detergent (8% linear alkylbenzene sulfonate, 4% nonionic surfactant, 16% zeolite A, 22% STPP) and immediately tested.

Microincapsulation using film-forming polymers
Impregnation of absorptive detergent constituents or specially designed carriers

The objective is the same with all three processes: immobilization of the antifoam by fixing it to a carrier and protection from the detergent constituents and the moisture in the powder. In this way, the antifoam is simultaneously "diluted" so that low feed quantities can be homogeneously dis-

SILICA		D	E	F
Type		preci-pitated	preci-pitated	fumed
BET-surface	[m²/g]	90	100	100
Mean agglome-rate size	[µm]	5	10	-
pH-value		10.3	8.0	4.0
Hydrophobicity (MeOH-wettability)	[%]	»50	‹50	‹‹50

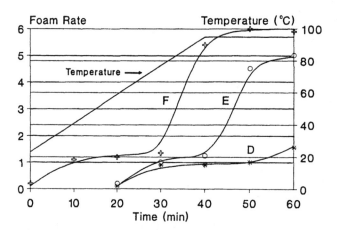

FIG. 14 Primary antifoam behavior of different hydrophobed silicas (D = Sipernat D10, E = Sipernat D17, F = Aerosil R 972, all from Degussa) dispersed in silicone oil of 1000 cSt (DC 200, Dow Corning). Measurement is carried out as described in Fig. 13.

tributed in the washing powder. Several important patents will be discussed in detail as illustrative of the processes referred to above (see Table 7).

Use of organic matrices. The first description of a silicone antifoam embedded in an organic matrix is almost 20 years old [86]. The antifoam system comprises 10–50% fumed silica and 50–90% polydimethylsiloxane (PDMS). The silica is characterized by a BET surface of at least 50 m²/g, a primary particle size of 10–20 nm and can be either hydrophilic or hydrophobic (silanized). A dispersion is produced from this and PDMS which can have a molecular weight of between 200 and 200,000. The antifoam is

TABLE 8 Compilation of Different ways of Manufacturing Silicone-Based Antifoams

Processing	Type
Embedding the antifoam in an organic matrix	
spray-drying with salts	A1
spray-cooling	A2
granulation	A3
spraying on detergent powder	A4
Microincapsulation using film-forming polymers	
precipitation out of an aqueous emulsion	B1
spray-drying	B2
granulation	B3
Impragnation of a carrier	
granulation and coating	C1
special carrier for impragnation	C2

in turn dispersed in a water-soluble or water-dispersible non-surface-active carrier. Examples of such carriers include polyethylene glycol, with a molecular weight of 4000 (PEG 4000) or a tallow alcohol with 25 mol ethylene oxide (TA25EO). Here it is supposed that due to the degree of ethoxylation the surface-active characteristics of the TA25EO are very weak. About one-third part of the system is silicone oil/silica dispersion and the remainder consists of the organic carrier. This mixture may then either be spray-cooled or be sprayed onto detergent constituents, such as tripolyphosphate, in a fluidized-bed mixer. In the latter so-called "coated agglomerates" develop. A typical agglomerate has for example the composition 5% silica, 5% silicone fluid, 40% TA25EO, and 50% tripolyphosphate. The agglomerate size is between 250 and 1500 μm.

The antifoam agglomerates produced in this manner are characterized by high storage stability in detergent powder. Good foam regulation over the entire temperature range, distinctly reduced graying of the washing and no dependence on the water hardness are said to be the advantages. Dosage of the antifoam agglomerate is adjusted in such a way that the content in the detergent powder of active antifoam is between 0.01% and 0.5%.

This antifoam system can be further developed by combining it with waxes [87]. Synergistic effects are claimed, which allow the feed concentrations of the wax and silicone to be drastically reduced. The wax component is a microcrystalline petroleum wax with a melting point of 77°C. The required dosage is up to 2%, whereas the dosage of the silicone antifoam can be reduced to between 0.04% and 0.1%. The ratio of wax to silicone antifoam can be varied within the limits 20:1 to 1:10. The silicone antifoam component is, as described above, either produced by spray cooling or by agglomeration with sodium tripolyphosphate in a fluidized bed. The wax is dissolved separately in the nonionic surfactant of the detergent (which contains 12% of a primary C_{14-15} alcohol with 7 mol ethylene oxide). This mixture is then added to the detergent slurry before spray-drying. The combination of silicone antifoam and wax is said to have a foam regulation effect both in the wash cycle and in the rinsing cycle in the case of nonionic surfactant-based detergents.

Along with pure polydimethylsiloxane polymers, modified siloxane polymers, so-called siloxane oxyalkylene alkoxysiloxane copolymers, are also used [94,96,100,101]. These compounds are derived from PDMS in which oxyalkyl groups are substituted for the methyl groups ($\underline{6}$).

$$Me_3Si\!-\!\!\left[\begin{array}{c} Me \\ | \\ -\!O\!-\!Si\!- \\ | \\ Me \end{array}\right]_p\!\!\left[\begin{array}{c} Me \\ | \\ -\!O\!-\!Si\!- \\ | \\ R \end{array}\right]_q\!\!-\!OSiMe_3$$

$$p + r = 30\text{--}120; \qquad p{:}r = 2{:}1\text{--}8{:}1 \tag{$\underline{6}$}$$

The radical R is claimed to have the following composition:

$$R = -\!X\!-\!(EO_{b/2}\!-\!PO_{b/2})OR_1$$

$X = -\!CH_2, -\!C_2H_4; b = 10\text{--}30; R_1 = -\!H, -\!CH_3;$
EO = ethylene oxide, PO = propyleneoxide

Alkoxysiloxanes have the advantage that they can improve the stability of the silicone antifoam dispersion in the organic carrier if the organic carrier is a nonionic surfactant with a portion of EO groups [100]. This stabilizing effect results in the antifoam droplets being retained in the carrier matrix and ensures effective storage stable incorporation in the detergent powder. A typical antifoam system with alkoxysiloxane as emulsifier has the following composition:

Polydimethylsiloxane (PDMS)	17.5–23.8%
Hydrophobed precipitated silica	6.0–7.9%
Alkoxysiloxane	1.7–2.0%
TA11EO	74.8–66.3%

It is manufactured by mixing the nonionic surfactants with alkoxysiloxane and then high-shear conditions. The resultant emulsion can be directly sprayed onto the washing powder (dosage 0.1–1.5% [101]) or agglomerated with other detergent constituents in the fluidized bed [94]. The advantages offered by this method of manufacturing antifoam systems are said to be high storage stability and good dispensing properties of the powder. A negative influence of PDMS on the wetting behavior of detergent powders is apparently avoided.

A system which is particularly suitable for detergents containing more than 12% surfactants is manufactured according to a multistage process [100]. Hydrophobic fumed silica (1–25%) are stirred into PDMS (75–99 %) having viscosity of 20–12,500 cSt and dispersed by means of an in-line high-shear pump (e.g., Dispax from Janke and Kunkel). When the temperature has risen to above 95°C and the viscosity to 6–10,000 cSt, high-viscosity PDMS (>25,000 cSt) in a portion of 33% to 66% is added, and dispersing is continued. The advantage of the high-viscosity PDMS is to avoid or, at least, to minimize deficiency for foam control in high active detergent compositions, presumably because the PDMS is rapidly dispersed or solubilized by the high surfactant level. This premix is added to a mixture of organic carrier and alkoxysiloxane and homogenized under high-shear conditions. The organic carrier is a twofold one consisting of a highly ethoxylated nonionic surfactant (TA25EO, TA80EO) or polyethylene glycol (PEG 6000, mp 38–90°C) and a low-ethoxylated alcohol (e.g., C_{8-24} × 2–12 EO, mp 5–36°C) at a ratio of 2–25% to 57–98%, in which 0.5–8% of the alkoxysilicone is contained. The ratio of the premix to carrier is approximately 1 : 16–25 when the antifoam dispersion is directly sprayed onto the washing powder (detergent dosage 0.25–0.65%), or 1 : 1–2, if the antifoam is to be agglomerated (detergent dosage 0.4–0.65%). The agglomerates consist to 50–85% of an inorganic carrier such as tripolyphosphate or other salts and the organic carrier/premix mixture. Pan and fluidized bed agglomeration or agglomeration according to the so-called Schugi process can be used to produce the agglomerates. This antifoam system avoids the disadvantages of other systems, such as the incompatibility of high PDMS/silica quantities in the carrier matrix based on nonionic surfactants and the too slow kinetics during transportation of the antifoam to the air/water interface in the wax/silica dispersion system.

A special matrix, called "acid soap," consists of a mixture of 40–60% lauryl acid and the residual sodium stearate or sodium palmitate having a liquid portion below 1% at 40°C [108]. In addition, this carrier has certain foam-inhibiting properties. The antifoam itself can be either a hydrocarbon-based antifoam or a PDMS/hydrophobed silica mixture. It is present at 40–45% in the matrix. The antifoam system may be manufactured by spray-cooling or by cooling the hot mixture on a band drier and subsequently grinding the flakes. Extrusion to needles, such as is known from manufacturing soap needles, is also suggested. For reasons of stability it is important that the particle size of the needles or the ground flakes is above 500 μm.

Glycerol monoesters like glycol monostearate [107] or fatty acid/fatty alcohol mixtures of compounds with 12–20 carbon atoms, such as stearyl acid and alcohol [106], are also suggested as organic matrices. Antifoam adjuncts suitable for incorporation in detergent powders are manufactured by spray cooling or in the fluidized bed with tripolyphosphate salt.

Special manufacturing processes are roller compacting [105] and the production of flakes on a belt cooler [102,110]. Roller compacting is a process in which a relatively dry powder mixture is forced under pressure (30–100 bar) through a slit with an adjustable gap width. The compacted strip produced is ground and graded into the desired particle size via screens. The porosity of the agglomerate produced according to this process is very low which contributes to the stability of the antifoam. The initial powder mixture consists of a base powder, e.g., sodium sulfate, onto which the antifoam is sprayed. A little water can also be sprayed on to increase plastication, and layered silicates or zeolite can be added to improve redispersibility.

Flakes are produced on a belt cooler, which can have a thin film of water to improve heat transfer [110]. The organic matrix comprises more than 80% of polyethyleneglycol of molecular weight of 5000–10,000 (mp 43°C), which may contain small amounts of soap (0.2–15%) [102] or TA25EO (0.25–2.00%) [110]. Both additives serve to retard release of the silicone antifoam. The antifoam is made up of 100 parts PDMS, 5–20 parts siloxane resin, and 1–5 parts of a hydrophobic fumed silica. A dispersion is prepared from the antifoam constituents which is incorporated into the molten carrier. For reasons of efficiency homogenization should be carried out in such a way that the antifoam droplets have a size of 5–30 μm before the dispersion is cooled to flakes. The ground flakes are added to the detergent at a ratio of 2.5–10% depending on the antifoam content, so that the effective antifoam level is 0.01–0.5%. The antifoam has an improved storage stability and the matrix is said to prevent migration of the antifoam in the powder.

Use of film-forming polymers. A quite different method of incorporating silicone antifoams in detergent powders is to use film-forming polymers. The term *film-forming polymers* refers to cellulose ethers, such as carboxy methyl or methyl cellulose, polyacrylate, polyvinylacetate (PVA), polyvinyl alcohol, polyvinyl pyrrolidone (PVP), and similar compounds. Here the silicone antifoam is dispersed as an oil-in-water emulsion in an aqueous solution or dispersion of polymer. The purpose of the polymer is to form a barrier layer between the antifoam dispersion and the washing powder. The film-forming polymer wraps itself protectively around the antifoam which will form a microcapsule or an agglomerate coated with the washing powder is delayed to a high degree under storage conditions which explains the stabilizing effect.

A particularly positive effect on the stability of silicone antifoam dispersions is reported when cellulose ethers are used. This is accompanied by simplification of the manufacture of the relevant antifoam emulsions. Thus it is possible to operate without using emulsifiers which is considered in many cases as being of advantage from the point of view of storage stability.

Microcapsules containing the antifoam dispersion can be produced by manufacturing a silicone oil/silica emulsion using nonionic surfactants, adding this to an aqueous solution of the sodium salt of the carboxymethylcellulose and precipitating the cellulose, simultaneously destabilizing the emulsion [98]. Precipitation is performed with multivalent cations. Aluminum salts, for example, have proved to be particularly useful. Hydrosoluble alcohols can also be used as precipitants. After precipitation washing and drying in a vacuum powders with an antifoam content of more than 50% are obtained. The resulting antifoam adjunct can be easily metered into a washing powder and has a very good storage stability.

A somewhat different variant includes spray drying [93]. The film-forming polymers (cellulose, ethers, starch ethers, polyacrylates, PVA, PVP) are allowed to swell and dissolve; the polymer concentration is 1–5%. A PDMS/silica dispersion is added (approx. 5–20%) and homogenized under high-shear conditions. The dispersion droplets should have a size of about 1–20 μm. The polymer is precipitated with 20–50% electrolytic salts, such as chlorides, sulfates, nitrates, silicates, and phosphates, to form microcapsules. This microcapsule suspension is conveyed separately from the detergent slurry to the spray tower. It is combined with the detergent slurry just before the spraying nozzle. The microcapsules are retained and are homogeneously distributed in the washing powder. High storage stability and the relative absence of weight-controlled process steps are the stated advantages. The dosage level is between 0.1% and 0.15%.

With the aid of cellulose ethers direct spray-drying of a silicone oil/hydrophobed silica-dispersion with inorganic salts and small quantities of nonionic surfactants is also possible [89]. Two percent to 8% of a PDMS/Aerosil dispersion (90–98%:2–10%), 3–8% nonionic surfactant (C_{12}—C_{20} with 4 to 25 ethyleneoxide groups), 3–8% cellulose ether, and 75–90% salt (preferably sodium sulfate) are homogenized and spray-dried. If the mixture contains more than 50% sodium sulfate, then the water required for sodium sulfate decahydrate formation is sufficient for the slurry, which has temperatures above the melting point of sodium sulfate decahydrate of 33.4°C, so that when spray drying, in principle, the cooling process for agglomerate formation is adequate and no water need be dried off. A typical composition is as follows:

PDMS	6.0%
Aerosil, silanized	1.5%
TA14EO	5.0%
TA5EO	3.0%
Na-CMC	7.0%
Na-silicate	10.0%
Na-sulfate (H_2O free)	64.5%
Water	1.5%

In addition to the good storage stability special emphasis is placed on antifoam efficiency, avoidance of fabric graying, and prevention of siloxane film formation in the machine.

A manufacturing process already known in connection with embedding of the antifoam in an organic carrier is agglomeration [88,99,111]. As a rule, an emulsion is produced from the silicone antifoam, the film-forming polymers, nonionic surfactant, or ester waxes, with which the carrier salts, such as tripolyphosphate, Na-sulfate, zeolite, and Ca-carbonate (soda ash) are agglomerated. Agglomeration can be performed in simple drum, tumbler, or plough-blade mixers (Lödige).

The use of hydrosoluble polymers in combination with waxes is considered advantageous, since this improves the homogeneity of the distribution of the PDMS/silica antifoam on the carrier [88]. A storage-stable antifoam agglomerate is produced by agglomerating 68.8 parts sodium sulfate and 10% magnesium silicate with a dispersion of 5.6% PDMS (13,000 cSt), 0.4% Aerosil, 6% TA14EO, 0.2% CMC, and 9% water in a rotary drum granulator. Seven percent of this agglomerate containing 4% silicon antifoam is added to a washing powder, which itself consists of two agglomerates, a spray-dried base powder and a sodium perborate tetrahydrate impregnated with a nonionic surfactant. The separation of a detergent into three individual agglomerates results in improved wetting and dispersing behavior of the detergent powder.

A special antifoam system is a combination comprising a high-foam anionic surfactant (e.g., alkylsulfate, alkylether sulfate, α-sulfosuccinoester) and a silicone antifoam [111]. This system is deposited on a carrier, preferably of sodium carbonate (soda ash), and cellulose ether, in a Lödige mixer and encapsulated with the film-forming polymer. What is important in this case is the use of the anionic surfactant as powder and the sequence in which the substances are mixed together. The sequence should be inorganic salts and cellulose ether, powdery surfactants, silicone antifoam and, last, the film-forming polymer. Among the film-forming polymers copolymers of vinylacetate and short-chain carbonic acids, such as crotonic acid, have proved effective. About 0.2–2.5% of the antifoam agglomerate is used; with 10% silicone antifoam in the agglomerate, this means a dosage level of 0.02 to 0.25%. The bulk density of agglomerate is around 550–600 g/L, it has a particle size distribution similar to that of a detergent, has a good granular hardness and is free flowing.

Use of absorptive carriers.   The last of the three manufacturing processes concerns the impregnation of a carrier with a silicone oil/hydrophobed silica dispersion. Here agglomeration processes use mechanical mixers such as Eirich pan granulators, Schugi mixers, fluid-bed mixers, drum and plough-blade mixers (Lödige). In the simplest case an antifoam agglomerate is produced by impregnating 60 parts of an anhydrous tripolyphosphate with 40 parts of a dispersion prepared from 70% silicone oil antifoam (PDMS: hydrophilic fumed silica = 95:5) and 30 parts water [82]. The amount of water is calculated in such a way that it is sufficient for complete hydration of the tripolyphosphate; subsequent drying is not necessary. With a 2% dosage the detergent contains 0.53% silicone oil and 0.028% silica. However, the storage stability of this antifoam is still not satisfactory.

A carrier of a quite special nature consists of an inert core material of spray-dried sucrose (25–80%) which is coated with an absorption agent (15–40%) such as starch, titanium dioxide, CMC, silica, or calcite [92]. About 5–30% of a PDMS/hydrophobic silica dispersion diluted with an organic solvent is sprayed onto a dry premix of core material and absorption agent in an Eirich pan granulator. After evaporation of the solvent the agglomerate is coated (3–30%) with paraffin (mp 35–65°C) or microwax (mp 60–63°C) dissolved in the same solvent. In this case the wax serves exclusively as a protective envelope for the antifoam agglomerate. A typical composition is 60.2% sucrose beads, 24% starch, 12% silicone antifoam, and 4.8% paraffin wax. The dosage in the washing powder is between 0.3 and 5%. The regular form of the core material is stressed as being of particular advantage since the antifoam can be more easily absorbed on this, thus making coating simpler. On the whole, this results in a storage stable antifoam system.

Gelatinized starch has been suggested as carrier for silicone antifoam [95,97,103,104]. This starch type is produced from a suspension which contains the starch particles and which is dried in a steam-heated drum to a partially hydrolized starch. It was found that the nonionic character of the starch and the relatively low pH range of 5.5–9 have a stabilizing influence on the antifoam. The small particle size of 50 to 500 μm provides an adequately large adsorption surface while the narrow particle size distribution is responsible for uniform and regular agglomerates. Particularly good storage stability can be achieved if the agglomerate is coated with a wax whose melting point is appropriate to the purpose [95], or a relatively high melting nonionic surfactant [97], such as, for instance, C_8—C_{20} with 5–20 moles ethylene oxide. An antifoam system is, for example, composed of 45–55% gelatinized starch, 35–50% of a silicone oil/silica dispersion, and 7–15% paraffin wax. It is manufactured according to well-known processes, e.g., pan granulators, Schugi mixers, or in a fluid bed. First, the starch is agglomerated with the silicone antifoam and subsequently coated with a solution of wax in an organic solvent or with molten wax. The wax can be more easily and uniformly applied as a solution in a solvent, however, this process is extremely problematical in respect of industrial-scale production since the solvent must evaporate. A variation of this antifoam system is to use two antifoams, silicone oil/silica, and hydrocarbon/silica, or hydrocarbon/alkylphosphate or hydrocarbon/diamide [103,104]. The silicone antifoam is referred to as "a low-temperature-sensitive antifoam" because it is a viscous fluid dispersion at room temperature, whereas the antifoam, due to its ongoing melting range of 50–80°C, is characterized as "high-temperature-sensitive antifoam."

Special attention is drawn to freedom from surfactants and high absorptivity with respect to the development of absorptive carriers. These carriers may be produced by spray-drying. This has been found to have a positive influence on the absorptivity. For example, a carrier produced from 15% to 65% tripolyphosphate, 15–65% sodium sulfate, 5–15% water glass, and the remainder water has a particle size of 200–800 μm with a bulk density of 500–650 g/L [90,91]. About 4–12% antifoam dispersion, comprising 86–96% silicone oil or a silicone oil/siloxane resin mixture and 4–14% silica, can be deposited on these carriers. A special carrier with a highly porous structure is based on spray-dried carbonates or sulfates (modified Burkeit) to which polymers are added [119]. This structure is characterized by a pore volume of 0.2–1 cm³/g, an average pore size of <20 μm, and an average particle size of 80–2000 μm. The absorptivity is stated to be very good and the small pore size is said to have the advantage that small antifoam droplets are released which promotes rapid transportation of the antifoam to the air-water interface. The effectiveness of the antifoam is retained during storage in the detergent powder until it is needed at the point of use. The antifoam is either a mixture of silicone and hydrophobed silica or a hydrocarbon combined with a so-called antifoam promoter such as silica, alkylphosphates, or diamides [109]. Mixtures of the silicone and hydrocarbon-based antifoams may be used. The antifoam systems are sprayed onto the carrier one after the other. It is first impregnated with the silicone antifoam and then coated with the hydrocarbon antifoam. Nonionic surfactants may be incorporated to prevent the formation of "oil spots" on fabric surfaces derived from hydrophobic antifoam oils.

V. CONCLUSION

The antifoams currently used in detergents represent the final point, for the time being, of a development which was characterized by two points of departure. On the one hand, the consumer demanded a detergent which fulfilled a wide range of requirements. Namely, for example, an improved soil removal capacity, better and milder removal of bleachable stains, applicability for the most varied fibers and types of fabric under widely differing application conditions (hand wash, all-temperature wash, soft-water conditions). On the other hand experiments were focused on attempts to better understand the physical-chemical correlations involved in the formation of foams and their stability.

Development of the antifoams used in detergents today has only been possible on the basis of the knowledge gained with respect to the antifoam effect of hydrophobic particulate materials. The concept of hydrophobic antifoam liquids (mineral oil and silicone oil) combined with hydrophobic particles has made it possible to develop numerous antifoam systems which can be adapted to any respective "foam problem." One focal point of future development will probably still be the formulation sphere. Further developments can be expected in this sector, retention of activity with long-term storage being the driving force.

REFERENCES

1. J. Falbe, Surfactants in Consumer products, Theory, Technology and Application, Springer, Heidelberg, 1987, p. 208.
2. G. Jakobi and A. Löhr, Detergents and Textile Washing, Principles and Practice, VCH, Weinheim, 1987, pp. 206
3. J. Falbe, Surfactants in Consumer products, Theory, Technologie and Applications, Springer, Heidelberg, 1987, p. 241.
4. P. L. Lagman, *Chem. Eng. News, Jan. 23*: 31 (1984).
5. H. Stache, Tensidtaschenbuch, Hanser, München, Wien, 1984, pp. 253–294.
6. J. Falbe, Surfactants in Consumer Products, Theory, Technologie and Application, Springer, Heidelberg, 1987, pp. 223–225.
7. E. Schmadel, *Fette Seifen Anstrichmittel 70*: 491 (1968).
8. J.C. Harris, *Oil and Soap 23*: 101 (1946).
9. H. L. Peper, *Coll. Sci. 13*: 199 (1958).
10. W. D. Harkins and E. K. Fischer, *J. Chem. Phys. 1*: 852 (1933).
11. E. Schmadel and C. P. Kurzendörfer, Waschmittelchemie (Henkel KGaA, ed.), Hüthig, Heidelberg, 1976, p.122.
12. ibid., p. 104.
13. W. L. St. John and W. J. Griebstein (assigned to Procter & Gamble), DE 1,056,316; January 27, 1966; filed October 27, 1956.

14. P. Krings, G. Jakobi, and J. Galinke (assigned to Henkel & Cie GmbH) DE-OS 2,243,306; March 21, 1974; filed September 2, 1972.
15. W. Leonhardt, unpublished results.
16. A. Sagredos (assigned to Unilever N.V.) DE 1,617,216; April 1, 1967; filed February 24, 1967.
17. K. Bott, H. Grossmann, and K. Kosswig (assigned to Chemische Werke Hüls AG), DE-OS 2,244,665; March 21, 1974; filed September 12, 1972.
18. O. F. Schwiegel, W. Griess, G. Ulrich, and M. Knausenberger (assigned to Unilever N.V.), DE 1,617,227; June 6, 1974; filed June 12, 1967.
19. No inventor named, (assigned to Thomas Hedley & Co. Ltd.), GB 808,945; February 11, 1959; filed May 16, 1957.
20. E. A. Schwoeppe (assigned to Procter & Gamble), DE 1,080,250; April 21, 1960; filed May 21, 1957.
21. H. Y. Lew (assigned to Chevron Research Co.), US 3,231,508; January 25, 1966; filed March 18, 1964.
22. H. Y. Lew (assigned to Chevron Research Co.), US 3,285,856; November 15, 1966; filed March 18, 1964.
23. H. J. Stimberg, J. Galinke and E. Schmadel (assigned to Henkel & Cie GmbH), DE-OS 2,043,087; March 2, 1972; filed August 31, 1970.
24. P. R. Garrett (assigned to Unilever PLC), EP 75,433; May 28, 1986; filed September 14, 1982.
25. M. N. A. Carter, P. R. Garrett, D. Giles, and A. R. Naik (assigned to Unilever PLC), EP 87,233; November 12, 1986; filed February 3, 1983.
26. G. C. Vandenbrom (assigned to Unilever N.V.), EP 126,500; November 28, 1984; filed April 11, 1984.
27. P. Schulz, J. Härer, C.-P. Kurzendörfer, F.-J. Carduck, F. W. Diekötter, U. Jahnke, and E. Schmadel (assigned to Henkel KGaA), EP 309,931; April 5, 1988; filed September 23, 1988.
28. E. Schmadel, *Fette Seifen Anstrichmittel* 70:491 (1968).
29. E. Götte and E. Schmadel (assigned to Henkel & Cie GmbH), DE 1,257,338; July 25, 1968; filed February 11, 1965.
30. E. Schmadel (assigned to Henkel & Cie GmbH), DE 1,467,620; August 6, 1970; filed December 15, 1965.
31. H.-J. Lehmann and E. Schmadel (assigned to Henkel & Cie GmbH), DE 1,617,116; October 16, 1975; filed June 25, 1966.
32. M. Berg and E. Schmadel (assigned to Henkel & Cie GmbH), DE 1,617,127; February 18, 1971; filed April 1, 1967.
33. G. Amberg and H. Saran (assigned to Henkel & Cie GmbH), DE-OS 2,333,568; January 30, 1975; filed July 2, 1973.
34. J. Glasl, H. Saran, J. Hoffmeister, and M. Berg, (assigned to Henkel & Cie GmbH), DE-OS 2,431,581; January 22, 1976; filed July 1, 1974.
35. G. Amberg (assigned to Henkel & Cie GmbH), DE-OS 2,544,034; April 7, 1977; filed March 10, 1977.
36. J. Perner, G. Frey, and H. Helfert (assigned to BASF AG), DE 2,710,355; August 10, 1978; filed March 10, 1977.

37. J. T. Thurston, *J. Am. Chem. Soc. 73*: 2981 (1951).
38. D. W. Kaiser, *J. Am. Chem. Soc. 73*: 2984 (1951).
39. J. R. Campbell, and R. E. Hatton, *J. Org. Chem. 26*: 2786 (1961).
40. E. Jungermann and H. C. Silberman, in Anionic Surfactants, Part II (W. H. Linfield, ed.), Marcel Dekker, New York, 1976, chap. 15.
41. H. Schlecht, H. Distler, and D. Stöckigzt (assigned to BASF AG), DE-OS 1,362,570; June 26, 1975; filed December 17, 1973.
42. D. W. Farren and C. W. Stuttard (assigned to Unilever N.V.), DE-OS 2,537,570; March 11, 1976; filed May 9, 1974.
43. H. Distler and P. Diessel (assigned to BASF AG), DE 2,727,382; January 4, 1979; filed June 18, 1977.
44. M. D. Key and W. G. McNee (assigned to Lever Brothers Co.), US 4,102,057; July 25, 1978; filed March 5, 1976.
45. K. Henning and J. Kandler (assigned to Hoechst AG), DE-OS 2,532,804; February 10, 1977; filed July 27, 1975.
46. W. P. Fethke, Jr. (assigned to Procter & Gamble Co.), FR 1,557,535; February 14, filed March 28, 1968.
47. H. Pöselt, A. D. Tomlinson, and H. Rabitsch (assigned to Unilever N.V.), DE-OS 3,144,470; May 10, 1983; filed November 9, 1981.
48. H. D. Hathaway and B. J. Heile (assigned to Procter & Gamble Co.), US 3,399,144; August 27, 1968; filed January 4, 1966.
49. M. A. N. Carter (assigned to Unilever N.V.), DE-OS 2,701,664; July 28, 1977; filed January 17, 1977.
50. D. Nickolls and W. B. Temple (assigned to Unilever PLC), EP 21,830; June 15, 1983; filed June 25, 1980.
51. M. Curtis, P. R. Garrett, and J. Mead (assigned to Unilever N.V.), EP 45,208; October 24, 1984; filed July 27, 1981.
52. G. Butler, M. N. A. Carter, M. Curtis, and R. M. Davies (assigned to Unilever PLC), EP 54,436; June 23, 1982; filed December 15, 1981.
53. R. L. Mayhew and F. Krupin, *Soap Chem. Spec. 38*(4): 55 (1962).
54. ibid., p. 80.
55. H. Ferch, *Chem. Ing. Tech. 48*:11 (1976).
56. Ullmanns Encyclopädie der technischen Chemie, Bd. 21, VCH, Weinheim, 1982, pp. 462ff.
57. K. Winnacker and L. Küchler, Chemische Technologie, Bd. 3, Anorganische Technologie II, 4th ed., Hanser, München Wien, 1983, pp. 75ff.
58. H. Kloepfer (assigned to Degussa AG), DE 762,723; February 28, 1942.
59. E. Wagner and H. Brünner, *Angew. Chem. 72*: 744 (1960).
60. L. J. White, J. G. Duffy, and G. L. Cabot, *Ind. Eng. Chem. 51*: 232 (1959).
61. K. Andrich (assigned to Degussa AG) DE 966,985; September 12, 1957; filed September 13, 1951.
62. A. Becker and P. Nauroth (assigned to Degussa AG), DE 1,467,019; July 27, 1970; filed March 2, 1962.
63. E. M. Allen (assigned to Columbia-Southern Chemical Corp.), FR 1,134,178; September 2, 1955; filed November 5, 1954.

64. R. Hoesch and W. A. Albrecht (assigned to Chemische Fabrik Hoesch KG), DE 1,117,552; May 30, 1962; filed April 17, 1959.

65. B. Brandt, P. Nauroth, A. Peters, and H. Reinhardt (assigned to Degussa AG), DE-OS 2,447,613; April 8, 1976; filed October 5, 1974.

66. B. Brandt, P. Nauroth, A. Peters and H. Reinhardt (assigned to Degussa AG), DE-OS 2,505,191; August 26, 1976; filed February 7, 1975.

67. H. Brünner and D. Schutte, *Chem. Ing. Tech. 89*: 437 (1965).

68. H. Reinhardt, K. Trebinger, and G. Kallrath (assigned to Degussa AG), DE-OS 2,435,860; February 12, 1976; filed July 25, 1974.

69. Roderburg (assigned to Chem. Fabrik Wesseling AG), DE-AS 1, 074, 559; February 4, 1960; filed June 25, 1959.

70. H. Winkeler, *Kautsch. Gummi Kunstst. 30*:6(1977).

71. L. O Young (assigned to W. R. Grace & Co.), DE 1,006,100; October 10, 1957; filed May 27, 1955.

72. O. Kühnert, G. Türk, and E. Eisenmenger (assigned to Degussa AG), DE 1,592,865; April 24, 1980; filed October 12, 1967.

73. A. H. Warth, The Chemistry and Technologie of Waxes, 2nd ed., Reinhold, reprint, 1960, pp. 391–393, 421 ff.

74. P. Peltre and A. Lafleur (assigned to Procter & Gamble Co.), DE-OS 2,857,155; January 3, 1980; filed June 12, 1978.

75. R. E. Atkinson and D. A. Ross (assigned to Procter & Gamble Co.), EP 8,829; March 19, 1980; filed August 28, 1979.

76. H.-U. Hempel and E. Schmadel (assigned to Henkel KGaA), DE-OS 3,115,644; November 4, 1982; filed April 18, 1981.

77. L. H. Tan Tai (assigned to Unilever PLC), EP 109,247; May 23, 1984; filed November 8, 1983.

78. J. C. Wuhrmann, W. Seiter, B. Giesen, and E. Schmadel (assigned to Henkel KGaA), DE-OS 3,400,008; July 11, 1985; filed January 2, 1984.

79. J.-P. Briand and C. C. Storer (assigned to Unilever N. V.), FR 2,559, 400; August 16, 1985; filed February 12, 1985.

80. V. B. John, G. C. Sawicki, R. Pope, and R. J. Seampton (assigned to Dow Corning Ltd.), EP 217, 501; April 8, 1987; filed July 23, 1986.

81. D. N. Willing (assigned to Dow Corning Corp.), EP 163,398; July 20, 1988; filed April 15, 1985.

82. K. W. Farminer and Ch. M. Brooke (assigned to DOW Corning Ltd.), GB 1,378,874; December 27, 1974; filed July 1, 1971.

83. G. C. Sawicki *J. Am. Oil Chem. Soc. 65*(5): 1013 (1988).

84. H. W. Fox, P. W. Taylor, and W. A. Pisman, *Ind. Eng. Chem. 39*: 1401 (1947).

85. D. Myers, Surfactant Science and Technologie, VCH, Weinheim, 1988, pp. 201, 204.

86. G. Bartolotta, N. T. De Oude, and A. A. Gunkel (assigned to Procter & Gamble Co.), GB 1,407,997; October 1, 1975; filed August 1, 1972.

87. A. C. McRitchie (assigned to Procter & Gamble Ltd.), GB 1,492,939; November 2, 1977, filed March 11, 1974.

88. A. Boeck, P. Krings, and E. Smulders (assigned to Henkel KGaA), DE-OS 2,753,680; June 7, 1979; filed December 2, 1977.

89. K. Hachmann, D. Jung, and A. Boeck (assigned to Henkel KGaA), EP 13,028; March 10, 1982 filed December 24, 1979.

90. H.-G. Smolka, H. Reuter, M. Berg, and G. Vogt (assigned to Henkel KGaA), EP 22,998; January 28, 1981, filed July 12, 1980.

91. M. Berg, G. Vogt, H.-G. Smolka and H. Reuter (assigned to Henkel KGaA), EP 36,162; September 23, 1981; filed March 9, 1981.

92. L. H. Tan Tai (assigned to Unilever PLC), EP 40,091; September 14, 1983 filed May 11, 1981.

93. H. Reuter, H. Saran, and M. Witthaus (assigned to Henkel KGaA), EP 70,491; January 30, 1985; filed July 12, 1982.

94. S. Dhanani, R. Mac Donald, J. S. Clunie, and M. C. Brooks (assigned to Procter & Gamble Co.), EP 46,342; February 13, 1985; filed July 23, 1981.

95. L. H. Tan Tai (assigned to Unilever PLC), EP 71,481; February 9, 1983; filed July 27, 1982.

96. P. A. Morgan (assigned to Procter & Gamble Co.), EP 91,802; October 19, 1983; filed April 8, 1983.

97. L. H. Tan Tai (assigned to Unilever PLC), EP 94,250; July 30, 1986; filed May 10, 1983.

98. H.-F. Fink, H.-J. Patzke, F. Spieker, and F.-J. Tölle (assigned to Th. Goldschmidt AG), EP 97,867; April 11, 1985; filed June 14, 1983.

99. H. Reuter and W. Seiter (assigned to Henkel KGaA), DE-OS 3,436,194; April 10, 1986 filed October 3, 1984.

100. M. S. Gowland, S. A. Johnson, and R. Pell (assigned to Procter & Gamble Co.), EP 142,910; May 29, 1985; filed August 21, 1984.

101. D. W. York (assigned to Procter & Gamble Co.), EP 163,352; December 4, 1985; filed May 22, 1985.

102. R. M. Baginski, B. C. Dems, L. A. Ross, and R. H. Soule, Jr. (assigned to Procter & Gamble Co.), US 4,652,392; March 24, 1987; filed June 30, 1985.

103. R. Foret and L. H. Tan Tai (assigned to Unilever PLC), EP 206,522; August 9, 1989; filed May 25, 1986.

104. W. J. Hey and J. W. H. Yorke (assigned to Unilever PLC), EP 213,953; September 20, 1989, filed September 2, 1986.

105. P. Schulz, J. Waldmann, F. J. Carduck, M. Witthaus, and E. Schmadel (assigned to Henkel KGaA), DE-OS 3,633,519; April 14, 1988 filed October 2, 1986.

106. P. M. Burrill (assigned to Dow Corning Ltd.), EP 210,721; April 2, 1987, filed May 6, 1986.

107. P. M. Burrill (assigned to Dow Corning Ltd.), EP 210,731; April 2, 1987, filed June 9, 1986.

108. P. W. Appel, F. Bartolotti, F. Delwel, A. D. Tomlinson, S. Willemse, and F. Hornung (assigned to Unilever PLC), EP 256,833; February 24, 1988, filed August 10, 1987.

109. P. R. Garrett, M. Hewitt, W. J. Iley, P. C. Knight, A. P. Pilidis, L. H. Tan Tai, T. Taylor, and J. W. H. Yorke (assigned to Unilever PLC), EP 266,863; May 11, 1988; filed August 10, 1987.

110. J. W. Revis, J. A. Sagel, and D. K. Ostendorf (assigned to Procter & Gamble Co.), GB 2,204,825; November 23, 1988; filed April 29, 1988.

111. A. Asbeck, A. Meffert, G. Rombey, and K.-H. Schmid (assigned to Henkel KGaA), EP 301,412; filed July 21, 1988.

112. P. J. Chevalier (assigned to Societe des Usines Chimiques Rhone-Poulenc), US 3,113,930; December 10, 1963. filed April 1, 1960.

113. S. Nitzsche and E. Pirson (assigned to Wacker-Chemie GmbH), US 3,235,509; February 15, 1966; filed October 3,1982.

114. R. E. Patterson, 1988 Nonwoven Conference, Tappi Press, Atlanta, 1988, pp. 39–48.

115. N. Nguyen, Evidence that Polydimethylsiloxane is a Hard Acid, M.S. Dissertation, Rensswelaer Polytechnic Institute, 1985.

116. K. W. Farminer and Ch. M. Brooke (assigned to Dow Corning Ltd.), GB 1,450,580; September 22, 1976; filed January 22, 1973.

117. K.-H. Müller, R. Tailfer and G. Türk (assigned to Degussa AG), DE-OS 3,001,573; July 23, 1981; filed January 17, 1980.

118. G. Giesselmann and K. Günther (assigned to Degussa AG), EP 62,748; May 15, 1985; filed February 16, 1982.

119. C. Atkinson, W. Jley, P. C. Knight, P. J. Russel, T. Taylor, and D. P. Hones (assigned to Unilever PLC), EP 221,776; May 13, 1987; filed October 30, 1986.

7
Antifoams for Paints

MAURICE R. PORTER   Maurice R. Porter and Associates, Sully, South
Glamorgan, United Kingdom

I.	Introduction and Historical Background	270
II.	Paint Systems	272
	A. Water-based emulsion paints for household use	272
	B. Industrial paint systems	274
III.	Antifoams for Water-Based Household Emulsion Paints	276
	A. Hydrocarbon/fatty acid/ester blends	276
	B. Hydrophobic particle/hydrophobic oil mixtures	278
	C. Water-based antifoam mixtures	281
	D. Single component antifoams	283
IV.	Antifoams for Industrial Paints	284
V.	Adverse Effects of Antifoams	289
	A. Wetting-out problems	289
	B. Loss of color	290
	C. Loss of gloss	291
VI.	Tests for Antifoam Effectiveness	292
	A. Paint manufacture	292
	B. Can filling	293
	C. Application	293
	D. Stability on storage	294
VII.	Speculations on Future Development	294
	A. Water-based emulsion paints for household use	294
	B. Industrial paint systems	295
	References	295

I. INTRODUCTION AND HISTORICAL BACKGROUND

Paint consists of solid particles and a binder dispersed in water or solvent. The particles give opacity and color. The binder binds the particles together into a coherent paint film when either solvent or water has evaporated. The main products of the industry are water-based emulsion paints for household use, solvent-based household paints, and mixed water/solvent-based paints for industrial use. The latter are used in applications such as automobile finishing and refrigerator coatings. There is a slow but steady shift from solvent-based paints to water-based paints both in Europe and the United States. Foam is an especially severe problem in both water-based and water/solvent-based paints particularly where emulsion polymers are used as binders. This account is in the main confined to the use of antifoams for water- or water/solvent-based paints.

Foam problems in paints occur in manufacture, canning, and application. During manufacture for example agitation often leads to air entrainment and loss of capacity in manufacturing plants. Thus the manufacture of paints has the following basic steps:

1. Grinding the solid particles to an appropriate size. This operation is usually carried out at high shear rate.
2. Mixing in the binder usually carried out at medium to low shear rate.
3. Mixing in the pigment and other additives, usually carried out at low shear rate.

The high shear rates required for many of these operations necessitates use of, for example, high-speed impeller mills, sand mills, ball mills, or roller mills. Although this equipment is usually designed to avoid foam formation some air entrainment is an inevitable consequence of the intense agitation involved.

Foam problems can also occur when filling paint cans. Here weight not volume is the usual measure of quantity. Weight specifications cannot be met if the paint contains excessive amounts of air.

Bubbles of air may be entrained during the application of paint which can interfere with the quality of the coating. If, however, these bubbles burst well before the paint dries, then the paint film will be unaffected. Defects in the film due to air bubbles are therefore associated with the presence of bubbles at or near the point of solidification of the coating. If bubbles burst at the point of solidification, then either small circular defects known as "pin holes" due to small bubbles or large irregular defects due to collections of polyhedral bubbles known as "craters" are found. If bubbles do not burst at all, then a reduction in the reflectivity of the coating (i.e., "gloss") may occur.

In water-based emulsion paints stabilization of foam derives mainly from the surfactants and polymeric stabilizers used in the manufacture of the paint (see Gress [1]). Thus high-foaming anionic surfactants are used in the preparation of emulsion polymers from which synthetic water-based emulsion paints are made. Industrial paints, on the other hand, often employ water-cosolvent mixtures so that the polymeric binder is rendered soluble. This combination of ingredients can yield relatively stable foam irrespective of the presence of surfactant contaminants [2].

The most generally applicable approach to foam problems is utilization of the appropriate antifoam technology. For water-based emulsion paints where foam problems derive from surfactant contaminants there exists an alternative approach based upon utilization of low-foam surfactants. However, such materials are usually unacceptably expensive.

In the early days of antifoam development for paints it was expected that one type would be developed which would be acceptable for all water-containing paints. Antifoams consisting of mixtures of hydrophobed silica and polydimethylsiloxanes (silicone antifoams) were once considered to satisfy this criterion. However, coating defects induced by the presence of polydimethylsiloxane droplets were so severe that silicone antifoams were at one time eliminated from water-based paints altogether.

The next phase in the development of antifoam technology for paints concerned discovery of the specificity of antifoams. Thus it proved extremely difficult to make an antifoam which was efficient over a wide range of paint formulations. At the same time similar problems were occurring in the textile industry where the same high-foam synthetic surfactants were used. Here foam problems were occurring particularly in the scouring of cotton and wool and the disperse dyeing of polyester fibers. The paper industry was also suffering from considerable foaming problems due to the introduction of new synthetic sizes. Both the paper industry and textile industry needs were far larger than the paint industry. Companies connected with paper and textiles developed a large number of antifoams specifically for textiles and paper. These products were then tried in paints and often worked. Thus the development of antifoams for paint, textiles, and paper developed along similar but not identical lines. Companies specializing in antifoams would offer many new products every year mainly based on blends of natural oils, fatty acids, esters, petroleum oils and waxes. Early products were difficult to both handle and disperse despite their cost effectiveness. Over the years antifoams, which in some measure overcome these handling and dispersion difficulties, have been developed. However, there remains considerable scope for further improvement and much development work is still directed at these issues.

The latest phase of antifoam development concerns the use of block copolymers of siloxanes and, for example, polyoxyethylene and polyoxypropylene compounds. These antifoams are sometimes used alone so that no mixing or blending is required. They are therefore easier to manufacture reproducibly.

All these different aspects of the development of antifoams for paints are still proceeding, but the overall driving force is cost effectiveness. The reduction of foam does not usually add any positive properties to the product, and the paint manufacturer will be looking for the most economical way to eliminate foam from his products.

II. PAINT SYSTEMS

A. Water-Based Emulsion Paints for Household Use

A typical water based emulsion paint [3] is given below. The constituents listed are mixed in the sequence shown.

	Parts by Weight
Propylene glycol	6.52
Aqueous polyacrylate solution (25%)	1.40
Antifoaming agent	0.47
Water	1.86
Titanium dioxide pigment	26.53
Water	4.66
Aqueous polyacrylate latex dispersions (46%)	50.90
Tributyl phosphate	1.71
Preservative	0.28
Antifoaming agent	0.47
Ammonia (25%)	0.20
1,2-Prophyleneglycol	2.80
Ammonium polyacrylate solution (25%) 0.93	
Water	2.20
	100.00

In special cases an additional antifoam may be added under conditions of low shear just before filling the cans. Some antifoams in particular paint systems when added at this late stage are extremely efficient at reducing foam produced by a brush or roller. The same antifoam added at an earlier stage can be quite ineffective.

Surfactant may be added to help wetting in of the solid pigment. Surfactant will also be present in the emulsion polymer (the "latex"). It is these surfactants which are mainly responsible for stabilizing foam. The thickener (polyacrylate or hydroxyethyl cellulose) will also help to stabilize any foam that is formed. Thus the more viscous the paint the more stable the foam even in the presence of antifoam (see Chapter 1). The dispersing agent will be a low molecular weight sodium polyacrylate or polyphosphate. These materials do not themselves stabilize foam, but enhance the tendency of certain surfactants to do so.

The main surfactants which cause foam are those present in the emulsion polymer. Emulsion polymers are dispersions of discrete polymer particles (0.04–3.0 micron) in water. They are prepared from monomers, water, surfactants, and initiator. Other common, but not necessarily essential ingredients, are water-soluble polymers which act as protective colloids. An excellent account of the role of surfactants in emulsion polymerization is given by Groves [4].

Surfactants used by emulsion polymer manufacturers are usually selected to facilitate the emulsion polymerization process, to achieve a desired emulsion particle size and distribution, and to enhance storage stability of the emulsion. Avoidance of foam in the finished paint will have low priority.

The nature of the paint film is determined by both the physical nature of the polymer and the size and size distribution of the particles. The choice of monomers (for example vinyl acetate, ethyl acrylate, 2-ethyl-hexyl acrylate, butyl acrylate, styrene, ethylene, butadiene) will determine the rheological and adhesive properties of the coating. The size and distribution of the particles will also determine the rheology of the paint and can affect other properties such as rate of drying, rate of absorption by the substrate, and the amount of filler which can be incorporated.

The type of surfactants normally used in emulsion polymerization are those anionic and nonionic surfactants which are typically found in detergent products. Anionic surfactants have found the most widespread use in emulsion polymerization.

Anionic surfactants yield electrical double-layer stabilization of the emulsion. Such surfactants will therefore show sensitivity to ionic strength as determined by the presence of other salts in the system. Some typical anionic surfactants used for emulsion polymerization include sodium lauryl sulfate, sodium dodecyl benzene sulfonate, sodium α-olefin sulfonate. These yield fine particle size emulsions (0.04–0.1 micron [4]) and high foam. Other anionic surfactants such as sodium 2-ethylhexyl sulfate and sodium lauryl ether (3EO) sulfate yield large particle size emulsions (0.1–3 microns[4]) and lower foam. Anionic surfactants are typically present at levels of 1–3% as a proportion of the emulsion polymer.

Nonionic surfactants stabilize emulsion particles by surrounding the particles with a hydrated headgroup layer. This layer acts as a steric barrier to particle coalescence. Nonionics are generally less efficient in stabilizing the emulsion than anionics. They are however necessary to provide stability against electrolytes and freeze-thaw cycles. Nonyl phenol ethoxylates (with 30% ethylene oxide) and lauryl alcohol ethoxylates are typical examples both of which are low foaming and yield large particle size emulsions. Typical concentrations would be 0.4–1.0% expressed as a proportion of the emulsion polymer.

In practice a combination of anionic and nonionic surfactants is used to give optimum properties. The paint manufacturer will use small particle size emulsion polymers for most paints. As we have seen, surfactants which yield small particle sizes also produce high foam.

B. Industrial Paint Systems

A typical household emulsion paint can be readily described but a typical industrial paint cannot because industrial paints vary in composition to a far wider degree than household paints. The essential features, however, remain the same, namely a pigment, a binder, and solvent or mixed water-solvent medium which is usually volatile. The binder can be an emulsion polymer as described above. More usually, however, it is a solvent soluble polymer. The paint film can be formed from a thermoplastic polymer, an air-drying polymer or a heat cured thermosetting polymer. Industrial coatings known as varnishes are clear transparent films which contain no pigment or very low levels of pigment.

Thermosetting water-based paint systems tend to give the most severe foam problems during application particularly if they are cured by heat or catalyzed to cross-link very quickly. Examples of such systems are based on resin binders such as

Water-reducible alkyds*
Epoxy resins
Acrylic colloidal emulsions
Polyesters

Each of these resin binders can be cross-linked with an amino-formaldehyde compound using a reactive proton on the resin and a methylol ($-CH_2OH$) group on the amino-formaldehyde compound. The system can be catalyzed with an organic acid (*para*-toluene sulphonic acid) so that the reaction may be accelerated and a coating can be dried in seconds. If foam occurs during application, it will have very little time to collapse so the essential function

---

*Alkyds are cross-linked polyesters produced from natural oils reacted with dibasic acids and polyols.

of any antifoam is to prevent foam from forming rather than ensuring any foam formed collapses quickly.

A typical composition of a thermosetting water-based paint is given by Nielsen [5]. Thus a water-reducible alkyd resin is mixed at low speed with other ingredients to give a composition.

	Parts by weight
Medium oil length alkyd resin	100.5
Water	201.0
Ammonium hydroxide	8.0
Ethylene glycol butyl ether	11.0
Manganese catalyst	2.1
Titanium dioxide	287.7
Antifoam	10.0

The important feature of this formulation is a mixed solvent, ethylene-glycol butyl ether/water, which is necessary to keep the alkyd resin in solution. Depending upon the application method and the alkyd resin a wide variety of solvent compositions are used in practice.

Water-based industrial paints using cosolvents differ from water-based household paints in that the polymeric binder is dissolved in the former but not in the latter. With such industrial paints Kushnir et al. [2] have shown that the resin itself is the agent responsible for foaming. Thus at a certain ratio of water to ethylene glycol butyl ether cosolvent a maximum in foam stability is found. This correlates with both a maximum in viscosity and the point of incipient phase separation of the polymeric resin.

Foam control in industrial paints is in general more difficult than with household paints. This arises from a variety of causes. One of which concerns rapid drying films. Other difficulties derive from more severe criteria concerning the quality of the paint film where a smooth continuous film is often essential. Defects are therefore more noticeable. Moreover, the performance required of the paint with respect to for example corrosion resistance is more critical. Therefore paint film defects due to either foam or even antifoam are much less acceptable than with household paints.

Industrial paints are generally applied using more vigorous methods than are household paints. Thus industrial paints are applied by for example spraying, flood-coating, electrocoating, dipping, curtain coating, and machine reverse roller coating. Paints to be sprayed must be carefully formulated to avoid foam. Machine roller coating can give problems due to foam slowly being generated in the machine and then being suddenly transferred to the reverse roller. The result is a coating which is subject to disastrous deterioration in quality.

In addition to all these factors other difficulties for the antifoam formulator arise because of the greater variety of industrial paint types which are often tailored to specific applications with a wide variety of substrates (marine coating or car finishing, for example).

III. ANTIFOAMS FOR WATER-BASED HOUSEHOLD EMULSION PAINTS

Development of antifoams for aqueous emulsion paints has followed essentially empirical lines. As we have seen, there has been a need for formulation of specific antifoams for specific paints. Often these formulations are complex and contain many components. However a number of features are common to most. Thus the antifoam ingredient must be insoluble and well dispersed in the medium for which foam control is desired. In the case of aqueous media the active antifoam ingredient must also be hydrophobic (see Chapter 1). Typically hydrocarbons, siloxanes and fluorocarbons form the basis of these hydrophobic ingredients. Often, however, these ingredients are difficult to disperse and they are therefore sometimes predispersed to form emulsions in water. Such emulsions are readily mixed with the foaming medium to produce a fine dispersion of the active hydrophobic antifoam. Surfactants are used to assist preparation and stabilization of these emulsions. Dilution of the antifoam in this manner, however, increases handling costs and can produce storage problems. Dispersal difficulties are therefore often overcome by adding a suitable surfactant to the antifoam ingredient so that it may be more readily directly dispersed in the foaming medium. Surfactants used to prepare emulsions or assist direct dispersal of antifoam can, however, contribute to the foaming problem which the antifoam is supposed to combat.

A. Hydrocarbon/Fatty Acid/Ester Blends

Blends consisting of hydrocarbons with fatty acids and various other ingredients formed the basis of the early antifoams used in the paint industry. These materials were formulated more or less without regard to any scientific principles. They were often prepared as oil-in-water emulsions. The earliest antifoams were usually pastes or soft solids which were difficult to disperse directly for a given application often requiring a predispersal step before addition to a system. Later products tended to be essentially liquid. The rheological properties of these antifoams were largely dominated by the physical state of the hydrocarbon. Thus liquids contain liquid hydrocarbons and the pastes contain hydrocarbon waxes.

A typical example of an early paste was described by McGinn [6] over 40 years ago. This was based upon an empirical finding that a mixture of a mineral wax and a partial glyceride of a fatty acid was a good antifoam in paper manufacture but that addition of a ricinoleic acid gave improved performance. The composition for paints derived by McGinn [6] from this consisted of

	Parts by weight
Scale wax	20
Candelilla wax	2
Glyceryl monostearate	15
Ricinoleic acid	5

This was mixed until uniform and then the following added at 50°C and stirred

	Parts by weight
23.5% sodium hydroxide	2.3
Water	55.5

Finally 0.2 part by weight of formaldehyde was added as preservative.

The waxes are readily recognized as the main hydrophobic components of this antifoam blend. The ricinoleate soap, formed in situ, is a dispersal aid, and the glyceryl monostearate probably functions as a hydrophobic component of the antifoam. This early example illustrates the complex nature of antifoam blends used in emulsion paints. A similar patent by McGinn[7] teaches the use of glyceryl monostearate and methyl oleate.

Ten years later Snook [8] modified the same basic composition to enhance fluidity. This was done by first mixing the following components at 60°C.

	Parts by weight
Scale wax	5
Paraffin oil	5
Glyceryl monostearate	5
Methyl ricinoleate	10
Monostearate	10

To this blend was added

	Parts by weight
Water	64
40% aq. formaldehyde	0.4

Later developments produced clear liquid products from which water was excluded. An example, due to Schott and Ward[9], consists of

Paraffin oil	61.5
PPGEO	9.75
Peanut oil/diethylene glycol reaction product	9.75
Polyethylene glycol (200 mol wt.) coconut fatty acid	11.6
Tall oil	5.85
Triethanolamine	1.55

Here PPGEO is a butyl-blocked polypropylene glycol ester of oleic acid.
This concoction is typical of antifoams formulated in the sixties. Many similar concoctions are used today because of advantages of cost effectiveness and storage stability. Modern products often contain polysiloxane polyglycol copolymers which can boost performance as described by Gruenert et al. [10].

With such complex concoctions it is doubtful whether even present producers have a clear understanding of the function of each component and even whether it is really necessary. These concoctions are however very easy to manufacture, capable of an enormous number of variations and are extremely difficult to analyze.

B. Hydrophobic Particle/Hydrophobic Oil Mixtures

That finely divided hydrophobic particles may significantly enhance the antifoam performance of hydrophobic oils represents an important discovery. Silica had been incorporated into silicone antifoams as far back as 1953 [11] but the importance had not been appreciated. Boylan in the early sixties [12] found that hydrophobic silica increased the antifoaming efficiency of both polydimethylsiloxane and organic hydrophobic liquids in a dramatic manner. A typical example is given by Liebling and Canaris [13] where mixtures of polydimethylsiloxane (1–6 parts by weight) and silica (8.9 parts by weight) were mixed to a dry powder and heated to 196°C to form hydrophobed silica. An effective antifoam was prepared by adding this to mineral oil (89.5 parts by weight).

An essential feature of this type of antifoam is that the silica surface be rendered hydrophobic. There are many ways of achieving this. The method described by Liebling and Canaris [13] needs prolonged heating to high temperatures. Other methods using different reactive species can involve less severe conditions. Examples of such species include

Polysiloxane fluid with Si—OH end groups (see Sullivan [14])
Organo silicon halides[15]
Cyclic polysiloxanes[16]
Trimethylsilylamines[17]
Hexamethyldisilazane[18]

Other finely divided solids will also function in this context. Examples include

Polyvinyl chloride [19]
Polypropylene [20]
Etylenebis(stearamide) [21]
Amide treated with chlorsilanes [22]
Talc [23]
Long-chain aliphatic ketones and secondary alcohols [24]
Hydroxy stearyl alcohol fatty acid esters [25]

Small particle size and a strongly hydrophobic surface are essential requirements for these materials if they are to be effective ingredients in admixture with hydrophobic oils for aqueous systems. The most effective and versatile particles have so far proved to be silica hydrophobed with polydimethylsiloxane. However, other types of particles have often found application for paints. Thus for example the ester of hydroxystearyl alcohol with saturated fatty acid (of chain length C_{18}—C_{22}) finely dispersed in mineral oil [25] forms a particularly effective antifoam for emulsion paints. It seems possible that many of the early antifoam compositions based on hydrocarbon/fatty acid/ester blends (Sec. III.A.) are in fact variations of the simple concept of hydrophobic particle/hydrophobic oil mixtures modified by the addition of dispersing agents and other ingredients of sometimes dubious usefulness. A detailed description of the properties of mixtures of hydrophobic particles and oils as antifoams together with hypotheses concerning their mode of action is given in Chapter 1.

Mixtures of different finely divided hydrophobic solids are also described in the patent literature. An example is talc and ethylenebis(stearamide) mixed with a hydrocarbon oil. Thus Curtis and Woodward[23] teach a formulation consisting of

	Parts by weight
Liquid hydrocarbon	74.8
Vinyl acetate/fumerate ester of tallow alcohol	7
Talc	15
Etylenebis(stearamide)	2
Silicone oil	0.2

All the components are heated together at 160°C, cooled and homogenized. This patent claims that the bis(stearamide) (which is insoluble) helps to slow down the settling of the talc and enhance the activity of the antifoam. Particle settling is, however, still a major problem with this type of defoamer. Vinylacetate/fumarate copolymer acts as both dispersing agent for the particles and as thickening agent for the liquid hydrocarbon (which inhibits settling).

Use of hydrophobic silica with other finely divided solids also apparently produces efficient products. Sinka et al. [26] compare a combination of hydrophobic silica/ethylenebis(stearamide) in mineral oil to that of hydrophobic silica alone in mineral oil in latex paints when mixed on a paint shaker. Results are presented below.

Product	% entrained air
Blank (no antifoam)	17.00
Hydrophobic silica	1.37
Hydrophobic silica/	0.55
ethylenebis(stearamide)	

Although antifoams based upon hydrophobic silica/ethylenebis(stearamide) were found to be more efficient it was found that the hydrophobic silica would begin to settle out and the viscosity would change with time. Improved stability was obtained by addition of polyoxypropylene glycols with butyl end groups and organophosphate esters. However, the most stable products were generally inferior in antifoam efficiency to less stable products.

Apart from enhanced antifoam efficiency another advantage associated with addition of hydrophobic silica to ethylenebis(stearamide) concerns the rheological behavior of the antifoam. Thus antifoams with high concentrations of ethylenebis(stearamide) in mineral oil are very efficient but are generally of very high viscosity and are therefore difficult to handle and pump. It has been claimed by Michalski and Youngs [27] that hydrophobic silica can act as a viscosity reducing agent in such systems.

It is clear then that a major problem with hydrocarbon oil antifoams containing finely divided solids is the stability of the composition. During storage solids can settle out. This process is sometimes accentuated by decrease in viscosity. Ihde [28] describes the shelf stability problems of silica-based antifoams and the addition of oil soluble organic resins such as coumarone-indene, petroleum hydrocarbon, terpene resin, or hydrocarbon-formaldehyde resins to stabilise and inhibit settling of the silica.

There have been many patents filed on the use of the thickening and gelling agents. However, when complex multiphase systems are manufactured then the method of mixing, rates of heating, rates of cooling and the type of equipment used all become relevant in determining the efficacy of the antifoam. The following example by Lichtman and Rosengar [29] illustrates the complexity of some products. Their preparation commences with a mixture of

	Parts by weight
Paraffin oil	6
Castor oil	1
Ethylenebis(stearamide)	4

which is heated at 140°C. To this mixture is then added vinyl acetate/fumarate (3.5 parts by weight) as gelling agent. This concoction is maintained at 140°C for 15 min and then added rapidly with agitation to paraffin oil (85.3 parts) at 15°C. The resulting mixture is cooled to 26°C and subsequently homogenized at a certain pressure. The temperature is then raised to 50°C for 4–12 hr with no agitation followed by cooling to 25°C with agitation. Silicone oil (0.2 part) is then added. It is claimed [28] that this composition yields a good antifoam performance in latex paints.

Despite difficulties concerning stability antifoam systems based upon mixtures of hydrophobic particles and hydrocarbon oils are extremely efficient and relatively cheap. Many combinations of the basic ingredients are possible. Patents are still being filed with novel combinations (see, for example, Kaychok and Boylan [30], Flannigan[31], and Wuhrmann et al. [32]).

C. Water-Based Antifoam Mixtures

Water-based antifoam systems are usually oil-in-water emulsions where the main advantage over use of the neat antifoam concerns ready dispersion in the medium to be defoamed. A difficulty with this approach concerns preparation of a stable emulsion which when diluted becomes unstable in order to drive the active antifoam ingredient (the oil) to the air-water surface and not remain in the bulk of the system. Mixtures of hydrophobic particles and hydrophobic oils do in fact show a tendency to form water-in-oil emulsions (see Chapter 1). Water-based antifoam systems claimed to be suitable for latex paints are also prepared in the form of such inverted emulsions[33]. Here water-in-oil emulsions presumably disperse less readily in water than the neat antifoam and incur higher storage and transport costs per unit weight of active ingredient. Stability to sedimentation on storage may, however, be greater than for the neat antifoam.

An example of an early oil-in-water antifoam is [34]

	Parts by weight
C_{16} fatty alcohol	35
Hexadecane	5
White paraffin oil	3
$C_{17}H_{35}COO(EO)_{10}H$	2
Water	65

Using suitable emulsifying techniques the mixture gives an oil droplet size distribution which is optimal for both shelf stability and antifoam efficiency. This system has however proved more successful in papermaking applications than in paints but represents an important development showing the need for the correct droplet size distribution.

Later oil-in-water emulsions incorporated hydrophobed silica. Details of a method for preparation of water-based antifoams from silicone oil and hydrophobed silica are, for example, given by Raleigh [34].

Water-based products have generally simply involved emulsification of oil-based antifoams which are known to perform efficiently. Although the approach has been effective the resulting formulations have now become so complicated that it is often difficult to elucidate the function of each component. For example, some water-based formulations now contain

Active antifoam ingredients
Mineral oil
Hydrophobic silica
EO/PO copolymers
Silicone fluid

Dispersing agents
Polyethylene glycol monoleates and dioleates
Naphthalene sulfonates
Glyceryl stearates
Nonyl phenol ehthoxylates

Stabilizers (to give the emulsion storage stability)
Xanthan gum
Polyacrylic acid salts
Vinyl acetate/maleate copolymers

Preservative
Formaldehyde

An example of such a complex formulation is given by Schmidt and Gammon [33]. These inventors claim a water-in-oil antifoam blend suitable for latex paints. This is prepared by first mixing

	Parts by weight
Paraffin oil	2
Ethylenebis(stearamide)	2.5
Polymethacrylate	3
Silicone surfactant	1.5

and heating to 150°C. The hot mixture is then added to paraffin oil (21 parts by weight) at 25°C and the whole cooled to 50°C. This is then added to

	Parts by weight
Paraffin/naphthenic oil	5
Paraffinic oil	6
Hydrophobic silica in oil	8
Castor oil + 15EO	1
Water	49.0
Formaldehyde	0.1

The resulting mixture is homogenized and silicone oil (0.1 part) added. Careful specification of the order of addition of components, rates of heating and cooling at various stages, and shear rates during mixing all suggest that these factors affect the performance of the resulting antifoam. Rapid cooling, for example, can have a significant effect on the particle size of any component which precipitates out of solution. This will in turn affect the antifoam efficiency and storage stability (settling rates) if, for example, that component is the hydrophobic particle component of the antifoam.

Sedimentation on storage is still a potential problem with these multiphase antifoam concoctions. In an attempt to eliminate this problem, Gammon [36] prepared oil-in-water microemulsion antifoam concoctions. These are, of course, optically clear liquids which do not exhibit phase separation and therefore present no potential storage problems. They consist of a water insoluble hydrophobic organic antifoam material (ranging from triglycerides to hydrocarbons) mixed with an ethoxylated nonionic surfactant, a "coupling" agent (such as an alkylene glycol), water, a base, and an antigelling agent. Gammon [36] claims that such concoctions are effective antifoams for paints. However, few microemulsion products have appeared on the market to date.

D. Single Component Antifoams

Potential storage problems are, as we have seen, always possible with multiphase antifoam systems. Use of a single-phase system avoids many of these problems.

The microemulsions described by Gammon [36] are clearly representative of such an approach. Other examples involve the use of single component antifoams. In the main this means polymeric materials containing a significant proportion of polyoxyethylene or polyoxypropylene or both as block copolymers. When dispersed in solution at temperatures above their cloud points, drops of cloud phase separate out which function as antifoams (see Chapter 1).

Simple polyoxyethylene/polyoxypropylene block copolymers and their derivatives with a hydrocarbon chain have found application for many years in industrial fermentation, industrial detergents, and machine dishwashing (see Chapter 9). They have, however, not proved very successful as paint antifoams. This may well concern the effect of the significant concentrations of anionic surfactant, commonly present in emulsion paints, on the cloud points of these materials. Thus it is well known that the cloud points of ethoxylated materials increase with the addition of anionic surfactant [37]. If the cloud point is increased above the application temperature, then any foam control will be lost.

These problems can be largely overcome with copolymers of polyoxy-alkylenes and polysiloxanes. Such polyalkoxypolysiloxane copolymers will function as antifoams for emulsion paints but can give paint film defects similar to those found with polydimethylsiloxanes (see Sec. V). Steinberger and Werner [38] claim to have produced polyalkyoxypolysiloxane copolymers which do not give such defects. However, these materials require the presence of hydrophobed silica in order to enhance antifoam performance in much the same manner as is found with hydrocarbons or polydimethylsiloxanes. They therefore cannot be regarded as single component antifoams.

Fink et al. [39] describe another polyalkoxypolysiloxane which can be used alone. These materials, however, only provide deaeration and do not effectively destroy foam in emulsion paints.

Polyalkoxypolysiloxanes have not so far replaced the older mineral oil based antifoams such as for example the mineral oil/wax/emulsifier/hydrophobic silica combinations for emulsion paints. They are both more specific in application and more expensive. These materials are, however, proving more successful in industrial paints (see Sec. IV).

IV. ANTIFOAMS FOR INDUSTRIAL PAINTS

Here we remember that the industrial paints for which foam problems are most severe are those consisting of mixtures of water and a cosolvent. These water-based industrial paints are distinguished from household emulsion paints in that the binder is usually dissolved in the water-cosolvent mixture. Variations in solvent composition will then affect the viscosity and surface properties of the solution and therefore the foam properties and resistance to antifoam. Kushnir et al. [2] have, for example, examined the foam behavior of a water-based coating prepared with an acid functional acrylic polymer (typical of those used in flow coat/dip coat formulations) cross-linked with melamine. Most of this work was done with varnishes (i.e., paint without pigment) using a typical formulation of 18%–20% resin, 7%–12% cosolvent (ethylene glycol butyl ether) dimethyl-ethanol amine mixture and 65%–75% water. Interestingly it was found that the cosolvent level had a marked effect on foam behavior. Thus foam stability was very low at high cosolvent levels (very soluble systems) then increased as the cosolvent level was reduced, passed through a maximum and then decreased again with further reduction of cosolvent. This foaming maximum correlated with a viscosity maximum and an abrupt change in dynamic surface behavior as revealed by the phase angles of dynamic dilational moduli. These authors attribute this behavior to the onset of phase separation of the resin binder as the cosolvent level is reduced and where large aggregates are forming in solution. This gives rise to enhanced viscosity, slow foam film drainage rates and an enhanced resistance to antifoam as revealed by tests with seven different antifoams. Dramatic increases in the phase angles of dynamic dilational moduli will be associated with the formation of such large aggregates [40]. Kushnir et al. [2] state that defoaming a coating system near this foaming maximum is impractical.

Other authors have emphasised the importance of the solubility of the antifoam in the paint. Thus Schnall [41] carried out tests using unspecified antifoams on an air-drying alkyd, a baking polyester-melamine, and a baking acrylic-melamine where each formulation had five different ratios of water to organic solvent. In most cases it was shown that antifoam efficiency varied with the solvent/water ratio so that the less soluble the antifoam the higher the efficiency. Schnall [41] notes that if the antifoam dissolves it may even enhance foam stability. Nielsen [42] also finds that changes in the water/solvent ratio in a water-reducible alkyd could improve the antifoam efficiency. The same author [5] compared eight different antifoams in a water-reducible alkyd with different cosolvents and found that in many cases the relative efficiency of the antifoams was dependent upon the cosolvent employed. It is not clear whether this is due to changes in antifoam solubility or is due to changes in viscosity as found by Kushnir et al. [2].

The most successful antifoams for industrial paints are silicone based. The most common and widely used silicones are the polydimethylsiloxane fluids which generally, but not always, contain hydrophobic silica. In the past these materials have given severe paint defects. There have however been significant improvements in silicone-based products to avoid these defects. These improvements have involved three different approaches.

The first utilizes a better knowledge of the relationship between the chemical structure of the silicone and the observed paint defects. Much of this information is confidential to the antifoam manufacturers. Dimethylpolysiloxanes, of molecular structure shown in Fig. 1, give the most efficient defoaming but can also give the worst paint defects. They are available in a wide range of molecular weights with n varying from 1 to several thousand. Antifoaming efficiency becomes greater as n increases but so do paint defects. By choosing the optimum value of n for particular paint formulations, the paint defects can be reduced while retaining antifoam efficiency. The optimum value of n will vary depending upon the paint system in gen-

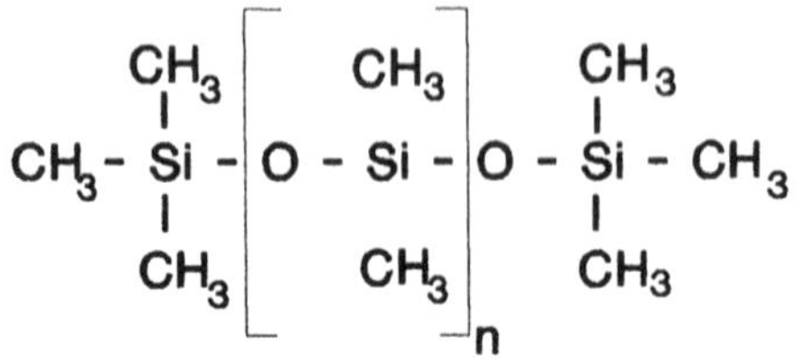

$$\text{CH}_3\text{-}\underset{\underset{\text{CH}_3}{|}}{\overset{\overset{\text{CH}_3}{|}}{\text{Si}}}\text{-}\left[\text{O}-\underset{\underset{\text{CH}_3}{|}}{\overset{\overset{\text{CH}_3}{|}}{\text{Si}}}\text{-}\right]_n\text{O}-\underset{\underset{\text{CH}_3}{|}}{\overset{\overset{\text{CH}_3}{|}}{\text{Si}}}\text{-}\text{CH}_3$$

FIG. 1   Polydimethylsiloxanes.

eral and the solvent system in particular. Although it is difficult to generalize, the lower the molecular weight, the fewer the problems. However there is a lower limit of molecular weight where the silicone does not act as an antifoam, this is at about 40 cS. Careful formulation with other non-silicone antifoams (e.g., fatty acid esters) can give excellent results.

A second approach utilizes dimethylpolysiloxane emulsions with controlled particle size. Most paint defects attributable to silicones in water-based systems are due to aggregation of the silicone droplets to form spots which are the source of nonwetting. Particles greater than 50 microns in size will give problems. If the silicone emulsion has a very small particle size which is stable and does not aggregate in use, then silicones are less prone to give problems. However, if the particles are too small then antifoaming efficiency falls away. Thus there must be a consistent controlled particle size generally between 2 and 50 microns. This is achieved by using a combination of the correct choice of surfactants and technique in emulsification. Photomicrographs are shown in Figs. 2 and 3 showing a small particle (2–30 micron) silicone emulsion compared to a coarse particle size (2–100 micron). Not only are the particles smaller in Fig. 2 but there is a smaller spread of particle size [43].

The third approach to improvement of silicone antifoams for industrial paints involves modification of the siloxane polymer to obtain better compatibility with the resin system. Thus silicone copolymers with polyesters, amino-resins, urethanes, vinyls, and polyglycols show better compatibility with resin systems which have a common chemical grouping. For example, a siloxane copolymerized with a polyester is likely to be more compatible with a polyester than the parent siloxane.

The most common type of modified siloxane are those block copolymers formed from polydimethylsiloxane and polyethylene oxide or polypropylene oxide/ethylene oxide. As we have seen, these materials find occasional use in water-based household paints. They are also used in industrial paint systems. Thus, for example, Berger et al. [44] describe their use as deaerating agents for low cosolvent and cosolvent-free hardenable resin paints.

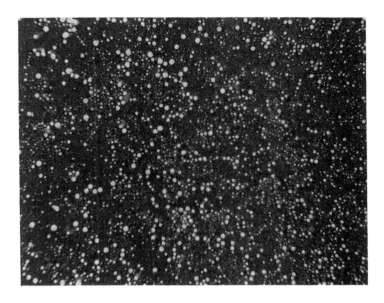

FIG. 2 Photomicrograph of small-particle silicone emulsions (aqueous phase dyed with methylene blue, magnification 90×).

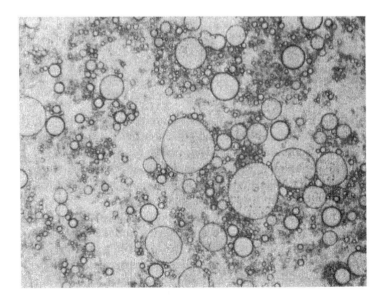

FIG. 3 Photomicrograph of large-particle silicone emulsions (magnification 90×).

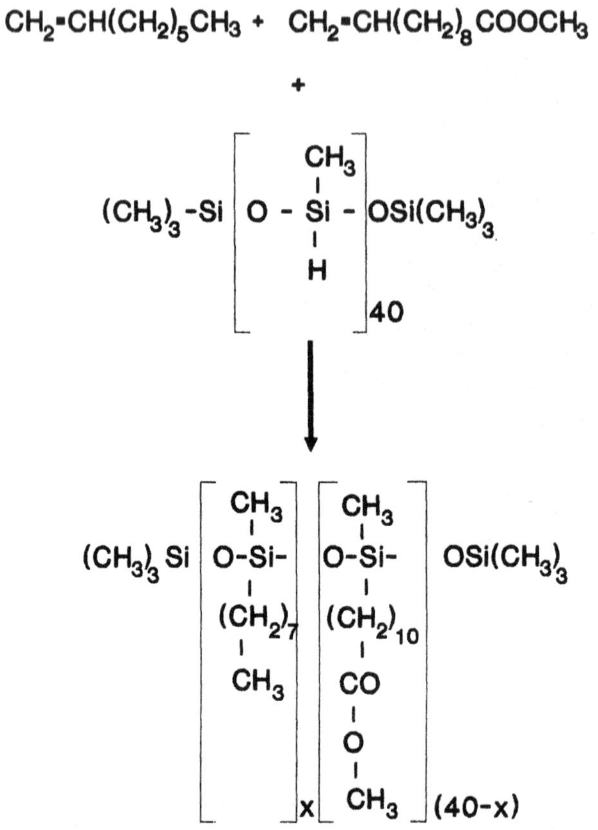

$$CH_2\text{=}CH(CH_2)_5CH_3 \ + \ CH_2\text{=}CH(CH_2)_8\,COOCH_3$$

$$+$$

$$(CH_3)_3\text{-Si}\left[\begin{array}{c} CH_3 \\ | \\ O - Si - \\ | \\ H \end{array}\,OSi(CH_3)_3\right]_{40}$$

$$\downarrow$$

$$(CH_3)_3\,Si\left[\begin{array}{c} CH_3 \\ | \\ O\text{-}Si\text{-} \\ | \\ (CH_2)_7 \\ | \\ CH_3 \end{array}\right]_x\left[\begin{array}{c} CH_3 \\ | \\ O\text{-}Si\text{-} \\ | \\ (CH_2)_{10} \\ | \\ CO \\ | \\ O \\ | \\ CH_3 \end{array}\right]_{(40-x)}\,OSi(CH_3)_3$$

FIG. 4 Modified polysiloxanes.

Modified siloxanes may, in principle, be tailored to produce surface and solubility characteristics which are compatible with any cosolvent/water-based paint system. Thus antifoam behavior may be optimized and paint film defects, due to preferential wetting of the substrate by the siloxane, minimized. For industrial cosolvent/water-based coatings hydrophobed silica is not usually mixed with these modified siloxanes.

An example of the use of this approach for industrial paints is given by Hempel et al. [45], who describe a typical method of introducing long-chain alkyl groups, ester groups, and polyglycols into a polysiloxane chain. The procedure is illustrated in Fig. 4. Here polydimethylhydrogensiloxane is reacted with a mixture of olefin and unsaturated ester using a platinum catalyst. The ester group may then be transesterified with a monoalkylated polyethylene glycol. A large number of variations are possible so that the solubility and surface characteristics can be modified by varying either the olefin, ester or polyglycol or their relative proportions. These materials are claimed by Hempel et al. [45] to be effective antifoams for an air-drying water-reducible alkyd, a polyacrylate, and a baking enamel.

V. ADVERSE EFFECTS OF ANTIFOAMS

A. Wetting-Out Problems

Antifoams added during manufacture of paint should ideally reduce foam problems during manufacture, reduce aeration during can filling and reduce foam problems during application. However, defects in the final dry paint film can be caused by antifoams giving wetting-out problems in the wet coat during application and subsequent drying. These defects can apparently look very similar to defects in the paint film caused by air bubbles. Thus, defects in the paint film can be caused either by the antifoam not working efficiently so that bubbles are present or by the antifoam giving wetting-out problems. There is considerable confusion in the literature concerning these defects. Thus they are variously described as "cissing," "crawling," "fish-eyes," "craters," and "pinholes." These terms are not clearly defined. They are generally ascribed to wetting-out problems, but this is not always true, as small pinholes and larger "craters" can be due to air bubbles.

Water-based paints are replacing organic solvent—based coatings in both household decorative paints and industrial coatings. Such water-based coatings not only give significantly more foam but also less readily wet substrates. The author has found that this latter problem increases in severity in going from emulsion paints to semigloss emulsion paints to gloss emulsion paints and then to water-based industrial coatings.

A basic condition required for proper substrate wetting is that the surface tension of the liquid be equal to or lower than the surface tension of the substrate so that the contact angle measured through the paint is as close to zero as possible. The concept of "critical surface tension" of a substrate was developed by Zisman [46], where a liquid with a surface tension higher than the critical surface of the substrate will not completely wet the substrate (to give a zero contact angle). Water, with a high surface tension, will therefore not wet substrates as readily as organic solvent solutions which have low surface tensions. Water-thinned coatings are therefore more likely to give poor wetting properties on low-energy surfaces. Nevertheless emulsion paints contain surfactants used in polymerization, which will lower the surface tension to enable the wetting of most surfaces such as wood, metals, and most paint surfaces.

When emulsion paints were first produced and the problem of foaming was encountered, silicones were tried as antifoams in such systems. These early silicone antifoams gave excellent antifoam behavior but caused severe problems in the application of the paint film itself [47–51]. Thus paints containing silicone can give uneven areas which are often known as fish-eyes or cissing. These are usually substrate spots where the paint has not wetted the surface so that holes in the paint film form after drying. The author has found that other antifoams can also give fish-eyes.

Microscopic examination of fish-eyes usually reveals a particle in the paint at the center of the eye. The particle is often, but not always, an antifoam in the paint film which wets out on the substrate and gives an area of the substrate which aqueous solutions of higher surface tension will not wet. The antifoam is insoluble in the wet paint [2,47] and will wet out on the substrate if the surface tension of the antifoam is lower than that of the critical surface tension of the substrate. Silicone oils and mineral oils can meet this criteria with many substrates. The particle size of a silicone droplet is important as large droplets tend to give fish-eyes, while very small silicone droplets will not [48]. The fish-eyes produced by antifoam contamination on a substrate are surprisingly large compared to the size of the contamination.

A similar phenomenon has been examined by Wilkinson et al. [52] and Aronson et al. [53] who studied an oil drop on a solid substrate partially submerged in a thin film of water of varying thickness. Decreasing the thickness of the film of water initially causes no change in the oil drop diameter while the oil/water/substrate contact angle is increasing from its equilibrium to its advancing angle; once this is reached, the oil drop spreads rapidly until, at a critical water-film thickness, it disrupts or separates from the water phase, to form an oil drop surrounded by bare substrate. The process is depicted in Fig. 5. It is a possibly analogous to that by which a small oil drop can give a defect in a paint film. The system studied by Wilkinson et al. [52] and Aronson et al. [53] involves a homogeneous substrate, and the situation could become more complex if the substrate itself varies in surface properties.

B. Loss of Color

The pigments included in emulsion paints to give opacity are usually titanium dioxide (rutile) and cheaper materials such as talc. Other pigments add color. Such pigments are usually insoluble in water and are contaminated with surfactant to facilitate dispersal throughout the paint. When a pigment is added to an opacifying pigment, there can be a redistribution of surfactant leading to flocculation. The paint can then apparently lose color. It has been found that antifoams with their associated surfactants can take part in this process [48,49,51,54–56].

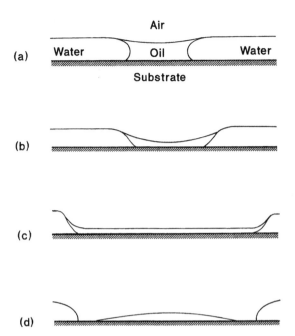

FIG. 5 Schematic illustration of dewetting of aqueous film induced by bridging droplet [52]. (a) Formation of configuration. (b) Water height lowered, contact angle increases to its advancing angle. (c) Oil forced to spread at its advancing contact angle displacing water. (d) Dewetting of substrate. (From Ref. 52. Reproduced by permission of Academic Press Inc.)

An antifoam can rob the pigment stain particles of surfactant, and the pigment or stain can agglomerate, resulting in a change of color usually going to a paler shade. This process is slow, taking up to 48 h to complete. The effect of antifoams on the loss of color of a paint formulation should therefore not be tested for at least 48 h after mixing the paint. Delileo [54] gives a good account of the practical problems encountered in color stability and the adverse effects of antifoams.

C. Loss of Gloss

Antifoams can often cause a loss in the gloss of emulsion coatings particularly if there is a high concentration of hydrocarbon oil in the antifoam. On matt emulsion paints this is not important but on the newer semigloss and now high-gloss water-based paints this can be a serious defect and therefore must be evaluated in any test procedure.

VI. TESTS FOR ANTIFOAM EFFECTIVENESS

Foam problems in surface coatings occur in manufacture, can filling, and application. Testing of antifoams should occur under conditions relevant for each of these processes. Tests for the effectiveness of antifoams under application conditions should therefore take into account the possibility of the loss of antifoam effectiveness on storage (see Kelly et al. [57]. Thus paints can often be stored for periods of greater than a year before use. During this time antifoam may adversely interact with other paint ingredients and may even be slowly solubilized [58].

Evaluation of antifoam in paints can be misleading if the paint already contains an antifoam. This is invariably the case if the sample is obtained from a normal large-scale manufactured batch. It is therefore usual to make paint samples in the laboratory.

A. Paint Manufacture

Two separate stages in paint manufacture can give rise to foam problems. These are during pigment dispersion at high shear and during addition of binder to the pigment dispersion at low shear. Laboratory tests for antifoam efficacy involve air entrainment followed by measurement of the volume of foam generated. Agitation of the paint with a high-speed stirrer is a common method of air entrainment [48,58]. Other methods include air injection devices [59–61] and agitation in the standard paint shaker used for mixing color tints into household paints. Use of a perforated disk which moves up and down in a cylinder [62] is, however, believed by the author to simulate manufacturing conditions more closely than air injection, stirring, or shaking. It is shown in Fig. 6. Here it should be stressed that different methods of air entrainment can give rise to different rankings of the efficiencies of a series of antifoams.

Quite often 20 or more antifoams will be tested at various concentrations on a single paint. The apparatus used for testing should therefore be simple, easy to clean, easy to use, and yield reproducible foam measurements. Ease of cleaning is particularly important because trace amounts of certain antifoams (silicones, for example) can have a significant affect on foam behavior.

In these test methods the foam is readily observed and easily measured visually by volume in a suitable graduated vessel. Elaborate techniques employing for example photoelectric devices are seldom justified for the evaluation of antifoams for addition at the manufacturing stage. The rate of decay of the foam is sometimes measured by taking observations at times up to 1 hr [58]. In comparing the effects of different antifoams on foam decay, it is, however, desirable to compare complete decay curves rather than integrated foam-time curves as described by Kushchnir et al. [2]. Integration

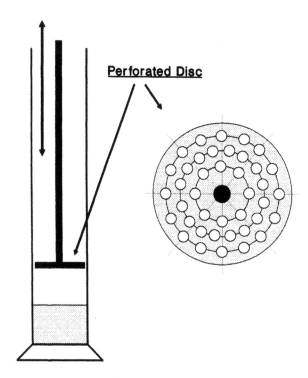

FIG. 6 Foam generation using perforated disk in cylinder. (From Ref. 62).

can be misleading because foams with radically different decay profiles may give identical values of the decay integral.

B. Can Filling

Weight, not volume, is the dominating factor in filling cans of paint. It is impossible to meet weight specifications if the paint contains dispersed air. Specific gravity measurement is therefore usually used for evaluating anti-foams designed to be effective in preventing problems in can filling. It is a quick and easy measurement [57] which is readily done on those samples aerated in the various ways described above (Sec. VI.A.). It is particularly useful when the amount of foam is small and not easy to measure visually by volume.

C. Application

The most important tests concern examination of the effect of the antifoam on the properties of the final dry paint film. If the antifoam gives no defects on the paint film, then it is selected on the basis of the tests described above. On the other hand if the paint film shows defects which can be attributed to the antifoam then no matter how efficient the antifoam (as measured by those tests) it will not be used in the paint. Instead a less efficient antifoam which gives no defects will be selected.

For household emulsion paints tests on the film forming properties of the paint simply involve coating the paint by brush and/or roller on a substrate, drying the paint, and visually examining the dry paint film for defects. Various workers use different substrates and methods of evaluation [1,5,48]. The defects observed are due either to particles of antifoam disrupting the wetting of the substrate by the coating or to the presence of air bubbles arising from antifoam inefficiency. An excellent description of these defects is given by Koerner et al. [48]. Application testing methods for industrial paints are essentially similar to those for household emulsion paints. However, industrial paints are often applied from baths which are under recirculation. Therefore it is desirable to run a simulated recirculation test to check antifoam efficiency over prolonged time intervals as described by Neilsen [42].

D. Stability on Storage

All foam control agents in paint systems lose some efficiency in storage. Newsome [58] attributes this phenomenon to interaction of surfactants in the paint with particles of antifoam. Storage tests are often accelerated by using elevated temperatures of 40–45°C. However, Gress [1] suggests storing paints for four–six weeks at room temperature. Newsome [58], on the other hand, uses gentle agitation of the paint for periods of about 60 min to simulate aging. The usual tests of foam volume after air entrainment, specific gravity and evaluation of surface defects are done on samples of paint before and after the storage test.

VII. SPECULATIONS ON FUTURE DEVELOPMENT

A. Water-Based Emulsion Paints for Household Use

Emulsion paints will constitute the largest proportion of household paints in the future. Cost will be the main criterion in deciding the approach to foam control in such paints. Foam control will therefore not be obtained using low-foam surfactants unless they become as cheap and efficient as those which are used at present. Here we note that acetylenic derivatives, fluorinated surfactants, and siloxane copolymers are now being produced which give excellent emulsifying properties combined with low foam. These low-foam surfactants could be utilized either directly in the paint and/or in the manufacture of emulsion polymers and other resin components. Undoubtedly research will continue in this area to find surfactants which give the required surfactant properties as well as keeping foam to a minimum. However, this type of surfactant is significantly more expensive than those at present used in emulsion polymers. It seems unlikely then that such surfactants will replace conventional high-foaming surfactants to any significant extent for the foreseeable future.

Foam control will in the main continue to rely upon the antifoam added by the paint manufacturer. Low-cost hydrocarbon/ester/soap products will undoubtedly still be used. Ways may even be found of utilizing hydrophobed silica/polydimethylsiloxane mixtures without producing coating defects. It is possible that polyalkoxypolysiloxane siloxane copolymers will become increasingly utilized in this area, but they will be more often used as admixtures with other components rather than used alone.

Hydrophobic silica will still be used extensively as an antifoam promoter for both silicones and hydrocarbons. Improved methods of hydrophobing will, however, be developed to yield better stability against problems of sedimentation and chemical interaction.

B. Industrial Paint Systems

The main practical advantage of a low-foam paint system is that application can be carried out more easily and rapidly. This will mean that cost consideration may sometimes be compromised to secure low-foam behavior. Similar types of antifoam products to those mentioned above will predominate with the main emphasis on silicone-modified polymers as these have so far shown most promise. Again they will be utilized in blended systems rather than used on their own because of lower cost and more versatile performance.

REFERENCES

1. W. Gress, *Polym. Paint Colour J. 174*: 452 (1984).
2. P. Kuschnir, R. R. Eley, and F. Louis Floyd, *J. Coat. Tech. 59*: 75–87 (1987).
3. R. Heyden, A. Asbeck, M. Eckelt, M. Petzold, and G. Uphues (assigned to Henkel Kommanditgesellschaft auf Aktien), US 4,094,812: June 13, 1978; filed December 20, 1976.
4. R. Groves, Industrial Applications of Surfactants (D. R. Karsa, ed. Royal Society of Chemistry, 1987, p. 73.
5. A. C. Nielsen, *J. Water Borne Coat. May*: 9 (1980).
6. E. P. McGinn (assigned to Nopco Chemical Company) US 2,563,856: August 14, 1951; filed October 12, 1945.
7. E. P. McGinn, (assigned to Nopco Chemical Company) US 2,563,857: August 14, 1951; filed October 12, 1945.

8. C. E. Snook (assigned to Nopco Chemical Company) US 2,715,614; August 16, 1955, filed March 31, 1955.

9. J. H. Shott and J. E. Ward (assigned to Nopco Chemical Company) US 2,849,405: August 26, 1958; filed December 28, 1953.

10. M. Gruenert, H. Hempel and H. Tesmann (assigned to Henkel AG) EP 0036 597: February 15, 1984; filed September 30, 1981.

11. C. C. Currie (assigned to Dow Chemical Co.) US 2,632,736: March 24, 1950; filed August 22, 1946.

12. F. J. Boylan (assigned to Hercules Powder Company) US 3,076,768: February 5, 1963; filed April 5, 1960.

13. R. Liebling and N. M. Canaris (assigned to Nopco Chemical Company), US 3,207,698: September 21, 1965; filed February 13, 1963.

14. R. E. Sullivan (assigned to Dow Corning) US 3,383,327: May 14, 1968.

15. Assigned to Witco Corporation GB 1 490 393; filed January 9, 1975.

16. F. J. Boylan, C. Porter, and O. D. Bruno (assigned to Drew Chemical Corp.) US 4,377,493: March 22, 1983; filed June 26, 1981.

17. Assigned to Shinetsu Chemical Company, GB 1 468 896; March 30, 1977; filed June 7, 1974.

18. W. J. Raleigh (assigned to General Electric Company) US 4,005,044: January 25, 1977; filed June 20, 1975.

19. I. A. Lichtman and F. E. Woodward (assigned to Diamond Shamrock) US 3,697,440: October 10, 1973; filed June 27, 1969.

20. F. J. Boylan (assigned to Hercules Incorp.) US 3,705,859: December 12, 1972; filed September 23, 1970.

21. H. J. S. Shane, J. E. Schill, J. W. Lilley (assigned to Hart Chemical Company Ltd., Canada) US 3,723,342: March 27, 1973; filed December 21, 1971.

22. D. W. Suwala (assigned to Diamond Shamrock Corp.) GB 1 480 717: July 20, 1977; filed November 3, 1975.

23. J. H. Curtis and F. E. Woodward (assigned to Diamond Shamrock) US 3,673,105: June 27, 1972; filed April 11, 1969.

24. R. Heyden, A. Asbeck, and E. Echelt (assigned to Henkel KGA) US 4,087,398: June 2, 1978; filed November 29, 1976.

25. R. Heyden and M. Eckett (assigned to Henkel and Cie GmbH), US 3,919,911: November 11, 1975; filed February 28, 1973.

26. J. V. Sinka and I. A. Lichtman (assigned to Diamond Shamrock Corp.), US 4,021,365: May 3, 1977; filed August 18, 1975.

27. R. J. Michalski and R. W. Youngs (assigned to Nalco Chemical Co.), US 3,923,683; December 2, 1975; filed January 1, 1974.

28. F. J. Ihde (assigned to Diamond Shamrock Corporation) US 4,123,383: October 31, 1978; filed September 19, 1977.

29. I. A. Lichtmann and A. M. Rosengart (assigned to Diamond Shamrock Corp.), US 3,677,963: July 18, 1972; filed September 30, 1970.

30. R. W. Kaychok and F. J. Boylan (assigned to Drew Chemical Corp.), US 4,626,377: December 2, 1986; filed September 10, 1984.

31. W. T. Flannigan (assigned to Imperial Chemical Industries PLC), EP 76 558: April 13, 1983; filed June 7, 1982.

32. J. C. Wuhrmann, H. Mueller, K. D. Brands, A. Asbeck, and J. Heidrich (assigned to Henkel Kga) EP 147 726: December 12, 1984; filed January 2, 1984.
33. W. T. Schmidt and C. T. Gammon (assigned to Diamond Shamrock Corporation) US 4,225,456: September 30, 1980; filed November 6, 1978.
34. Assigned to Badische Anilin- and Soda-Fabrik Aktiengeselschaft GB 1,402,597: August 13, 1975; filed November 17, 1971.
35. W. J. Raleigh (assigned to General Electric Co.), US 4,005,004: January 25, 1977; filed July 20, 1975.
36. C. T. Gammon (assigned to Diamond Shamrock Corp.), US 4,208,301: June 17, 1980; filed July 7, 1978.
37. C. Manohar and V. K. Kelkar, *J. Colloid Interface Sci. 137*: 604 (1990).
38. H. Steinberger and C. Werner (assigned to Bayer Aktiengesellscaft), US 4,562,223: December 31, 1985; filed August 4, 1983.
39. H. F. Fink, O. Klocker, G. Koerner, G. Rossmy, and C. Weitemeir (assigned to Th. Goldschmidt AG) US 4,520,173: June 28, 1985; filed April 7, 1983.
40. J. Lucassen, *Faraday Disc. Chem. Soc. 59*: 76 (1975).
41. M. J. Schnall, Proc. 10th Water Borne and High Solids Coating Symp., Louisiana, 1983, p. 46.
42. A. C. Nielsen, *J. Water Borne Coat. Feb*: 17 (1980).
43. Basildon Chemicals, private communication.
44. R. Berger, H. F. Fink, and O. Klocker (assigned to Th. Goldschmidt AG), EP 0 257 356: March 2, 1988; filed August 3, 1987.
45. H. U. Hempel, G. Grunert, H. Tesmann, and H. Muller (assigned to Henkel KGA) US 4,504,410: March 12, 1985; filed June 16, 1982.
46. W. A. Zisman, *J. Paint Technol.* 44(6):41 (1972).
47. M. A. Ott, *Mod. Paint Coat. Aug*: 31 (1977).
48. G. Koerner, F. Fink, R. Berger, and W. Heilen, World Surfactant Congress, IV, 211 (1984).
49. R. D. Nicholson, Austral. OCCA Proc. *News Aug*: 9 (1983).
50. R. T. Maher and E. M. Antonucci, *Mod. Paint Coat. Dec*: 43 (1979).
51. L. Kelley *Mod. Paint Coat. June*: 33 (1975).
52. M. C. Wilkinson, A. C. Zettlemoyer, M. P. Aronson, and J. W. Vanderhoo, *J. Colloid Interface Sci. 68*(3): 508 (1979).
53. M. P. Aronson, A. C. Zettlemoyer, and M. C. Wilkinson, *J. Phys. Chem.* 77(3): 318 (1973).
54. L. M. Dilelio, *Resin Rev. 20*: 29 (1977).
55. E. M. Antonucci, Water Borne and Higher Solids Coatings Symp. Feb. 12–14 New Orleans, LA, 1979 p. 234.
56. E. M. Antonucci, P. Cosetino, R. W. Stangs, and T. F. O'Farrell, *Water Borne Coat. Nov*: 12 (1977).
57. E. L. Kelley, R. W. Harrison, and B. B. Harris, *Polym. Paint Colour J. Aug*: 734 (1978).
58. G. Newsome, *Paint and Resin*, March/April (1982).
59. Deutsches Institut fur Normung DIN 51381.
60. American Society for Testing and Materials ASTMD -892-72.
61. Deutsches Institut fur Normung DIN 51566.62. Deutsches Institut fur Normung DIN 53902 Teil I.

8

Surfactant Antifoams

TREVOR G. BLEASE, J. G. EVANS, AND L. HUGHES ICI
Surfactants, Wilton, Middlesbrough, United Kingdom

PHILIPPE LOLL ICI Surfactants, Everberg, Belgium

I.	Introduction	300
II.	Types of Polyoxyalkylene Surfactant Antifoam	300
	A. Polyoxyethylene/polyoxypropylene block copolymers	300
	B. Ethylene diamine–based polyoxyethylene/polyoxypropylene block copolymers	301
	C. Polyol-based polyoxyalkylenes	301
	D. Fatty alcohol–based polyoxyalkylenes	302
	E. Fatty alcohol polyoxyalkylene alkyl ethers	302
III.	Ecotoxicological Aspects of Copolymer Surfactant Antifoams	303
	A. Biodegradation of copolymer surfactant antifoams	303
	B. Aquatic toxicity of antifoams	305
IV.	The Use of Copolymer Antifoams in Machine Dishwashing	305
	A. Machine types and operating conditions	306
	B. Selection of surfactant antifoams	307
	C. The defoaming performance of polyoxyethylene/polyoxypropylene copolymers under dynamic conditions	309
V.	Use of Copolymer Antifoams in Sugar Beet Industry	311
	A. Introduction	311
	B. Description of a sugar plant	314
	C. Causes of foam in the sugar beet industry	317
	D. Composition of the antifoam	318
	E. Toxicological implications	321
	References	321

I. INTRODUCTION

Other chapters in this volume have described some important industrial applications of antifoams. This contribution focuses on two particular applications, namely dishwashing and sugar beet processing, for which polyoxyethylene/polyoxypropylene–derived antifoams have played an important role in satisfying the foam control demands of the industry. For dishwashing a distinction will be made between domestic usage and industrial/institutional application due to their differing performance requirements. Ecotoxicological issues surrounding the polyoxyalkylene copolymer products have been an important catalyst for new product development in this area over recent years and these matters will be covered in a separate section.

It is important to note at the outset that the antifoaming power of the surface-active agents covered in this chapter is temperature dependant. Above the cloud point of the surfactant in the application medium (the temperature at which hydrogen bonding is insufficient to maintain the surfactant in solution) the antifoaming power is considerably enhanced [1].

II. TYPES OF POLYOXYALKYLENE SURFACTANT ANTIFOAM

In the application areas here described only a few classes of polyoxyalkylene surfactant are commercially significant. For this chapter a relatively short description, paying particular regard to the antifoaming aspects of the structures, will suffice. A more general treatment of the compound classes and their properties can be obtained from other texts [2,3].

A. Polyoxyethylene/Polyoxypropylene Block Copolymers

These products, simply consisting of polyoxyethylene and polyoxypropylene blocks, were first commercialized in the early fifties by Wyandotte Chemicals Corporation [4]. The generic structure is typically

$$HO(EO)_a(PO)_b(EO)_aH,$$

where

EO $=$ —CH_2CH_2O— ethylene oxide group
PO $=$ —$CH_2CH(CH_3)O$— propylene oxide group

Their utility is evidenced by their widespread use for industrial antifoaming to this day. Both "normal" (above) and "reverse" block structures, that is,

$$HO(PO)_a(EO)_b(PO)_aH$$

FIG. 1 Structure of ethylene diamine–based polyoxyethylene polyoxypropylene block copolymer.

are employed. The terminal secondary hydroxyl group of the reverse block structure offers some chemical stability benefits compared to the primary hydroxyl functionality of the normal block products. Optimal antifoam efficiencies, for both structural types, are obtained at polyoxypropylene: polyoxyethylene molar ratios between 4:1 and 9:1 and molecular weights greater than about 2000 amu [2]. A product for which $a = 2$ and $b = 32$, with a cloud point in 10% aqueous solution of about 17°C, would be considered typical of the normal block antifoam products available from several manufacturers. By varying the molar ratio a continuum of cloud points up to greater than 100°C can be obtained.

B. Ethylene Diamine–Based Polyoxyethylene/ Polyoxypropylene Block Copolymers

These products have been available in normal (hydrophile end-block) and "reverse" (hydrophobe end-block) structures, as illustrated in Fig. 1, since the 1950s. For useful antifoaming performance, molecular weights are somewhat higher than for the polyoxyethylene/polyoxypropylene block copolymers, typically greater than 3500 amu [2]. As for the products described in Sec. A, the cloud point can be continuously varied from 10°C to greater than 100°C.

C. Polyol-Based Polyoxyalkylenes

Many polyols are employed as starting bases for a wide variety of more specialised polyoxyalkylene antifoams. Among the polyols most commonly employed presently are

Glycerol
Sorbitol
Pentaerythritol
Trimethylolpropane

Such products can yield relatively better antifoam performance than conventional polyoxyethylene/polyoxypropylene block copolymers. Moreover, polyol-based products can be used as antifoams for applications in which multifunctionality is desired. Multifunctionality can be achieved by various partial postreactions of the polyolpolyoxyalkylene.

D. Fatty Alcohol–Based Polyoxyalkylenes

This class of polyoxyalkylene surfactant antifoams, of generic structure

$RO(EO)x(AO)yH$

where

R   = C_{8-18} linear (or substantially linear) alkyl chain
EO = ethylene oxide group
AO = propylene oxide or butylene oxide group,

is assuming increasing importance nowadays. The awareness of the enhanced biodegradability possibilities of these surfactant antifoams compared to other compound classes (see Sec. III) has led to replacement of the traditional polyoxyalkylene antifoam products by fatty alcohol–based polyoxyalkylenes in many important applications, for example dishwashing (see Sec. IV).

In order to obtain the optimum biodegradability performance these molecules must be kept small, that is x and y generally in single figures [5]. Within this relatively limited scope, seemingly small but highly significant structural variations must be played, involving, for example,

The choice of fatty alcohol
The numbers of polyoxyalkylene moles added
The internal arrangement of the polyoxyalkylene moles (block, normal or reverse, or random)
The choice of propyleneoxide or butylene oxide as the hydrophobic polyoxyalkylene

in order to best meet the specific antifoaming performance requirements of the application.

E. Fatty Alcohol Polyoxyalkylene Alkyl Ethers

Evolving from the fatty alcohol–based polyoxyalkylene antifoams in the 1980s have been the etherified, so called end-capped, derivatives [6,7]. These molecules are of generic structure:

$RO(EO)x(AO)yR'$

where R, EO, and AO are already defined and R' is generally a small ($\leq C_4$) linear alkyl chain.

The acquisition of highly biodegradable hydrophobicity, via the addition of the linear alkyl end cap, can allow reduction or even complete replacement of the hydrophobic polyoxyalkylene component (i.e., y tends to zero) for the equivalent antifoaming performance. Additionally the substitution of an ether linkage for the chemically comparatively labile hydroxyl functionality is advantageous for industrial applications involving prolonged exposure to stringent chemical conditions, most notably the extreme pH environment of segments of the industrial and institutional detergency market.

III. ECOTOXICOLOGICAL ASPECTS OF COPOLYMER SURFACTANT ANTIFOAMS

A. Biodegradation of Copolymer Surfactant Antifoams

1. Polyoxyethylene/Polyoxypropylene Block Copolymers

Amongst the earliest surfactant antifoams were the ethylene oxide/propylene oxide copolymers end blocked with either the polyoxyethylene hydrophile or the polyoxypropylene hydrophobe. When used as antifoams these copolymers have molecular weights typically greater than 2000 amu. They are poorly biodegradable because the polyoxypropylene moiety confers a high degree of bioresistance [8].

The ease of biological degradation of these antifoam copolymers is determined by their structure. For ethylene oxide/propylene oxide copolymers, both the hydrophylic (polyoxyethylene) and hydrophobic (polyoxypropylene) entities influence the degree and extent of degradation. Oxypropylene groups in these copolymers have a negative effect on rate and degree of biodegradation. Poor degradability of oxypropylene chains is usually associated with restriction of the hydrolytic sequence due to steric hinderance by pendant methyl groups. Other factors such as the lower solubility of the polyoxypropylene moiety compared to polyoxyethylene may further limit the bioavailability of these structures to microorganisms.

The hydrophylic polyoxyethylene chain on the otherhand degrades more readily but at a rate dependent upon molecular weight. Thus ethylene glycol and its oligomers have been shown to exhibit biological degradation in laboratory models [9]. Several polyoxyethylene degrading bacteria capable of growing on ethylene glycol with chain lengths in excess of 1000 amu have been isolated from the environment [9]. While polyoxyethylene chains in excess of 1000 amu are considered to be less readily biodegradable, eventual aerobic degradation may occur [10]. Anaerobic degradation of polyoxyethylene by microorganisms has also been reported [11]. Polyoxyethylene chain degradation in ethylene oxide/propylene oxide copolymers is considered to occur by initial cleavage at the α carbon of the hydrophobe. This is followed by sequential hydrolytic, and oxidative breakdown of the end glycol groups by the β-oxidation pathway [12]. Biodegradation of unsubstituted polyethylene glycols is reviewed in detail by Swisher [13].

Legislation in particular within the European Community has limited the application of ethylene oxide/propylene oxide antifoams for certain detergent applications. Failure to meet the legal requirement of a minimum 80% primary biodegradation for this class of compounds has led to the search for other antifoam structures which do. Typically, the degree of primary biodegradation which usually represents a limited breakdown of the substance but not total demineralization to water and carbon dioxide in standard screening and confirmatory tests is of the order of 10%–30% for ethylene oxide/propylene oxide block copolymers [14]. Primary biodegradation is a measure of the removal of the molecule from the aqueous phase in contrast to ultimate biodegradation—the complete mineralization to inorganic products such as carbon dioxide and water.

The importance of structure in determining the biodegradation of surfactant antifoams is well understood. Much attention has centred on the hydrophobic character of the propylene oxide group, the total carbon chain length, chain length distribution, branching and substitution. The hydrophilic ethylene oxide portion, although less influential, may become a determining factor particularly in relation to chain length.

2. Fatty Alcohol Polyoxyethylene/Polyoxypropylene Derivatives

The development of fatty alcohol alkoxylates for antifoam applications has centered on the C_8 to C_{15} fatty alcohol hydrophobes. In most cases, the length and type of fatty alcohol has only a minimal effect on biodegradability unless a high degree of branching is present [14]. Propylene oxide groups have a dominating effect on the degree of biodegradability compared to the influence of the corresponding ethylene oxide groups [14]. Limiting values for propylene oxide of between 7 and 10 units within alcohol alkoxylates for compliance with European Community primary biodegradation requirements have been reported [5].

Other studies of primary and ultimate biodegradability using semicontinuous activated sludge (SCAS) bioreactors revealed that the insertion of propylene oxide groups into the polyoxyethylene chain of fatty alcohol ethoxylates decreases the rate and extent of biodegradation. The magnitude of the effects were proportional to the propylene oxide block size and the extent of branching in the alcohol [15]. Alcohol alkoxylate antifoams based on a C_{13}—C_{15} oxoalcohol with varying polyoxypropylene and polyoxyethylene groups assessed in standard tests [16] show moderate ultimate degradation

TABLE 1 Polyoxypropylene and Polyoxyethylene Block Copolymers Acute Fish Toxicity

		Molecular weight	Rainbow trout 96 hr LC50 (mg/L)
Polyoxyethylene(5)	Polyoxypropylene(32)	2090	203
Polyoxyethylene(10)	Polyoxypropylene(32)	2400	>1000
Polyoxyethylene(25)	Polyoxypropylene(32)	2900	315
Polyoxyethylene(152)	Polyoxypropylene(32)	8350	>1000
Polyoxyethylene(16)	Polyoxypropylene(48)	3450	>1000

Note: Figures in parentheses refer to number of moles of oxypropylene or oxyethylene.
Source: Ref. 17.

after 28 days [17]. This compares with block copolymer antifoams which have a low degree of ultimate biodegradability in the same tests.

3. Fatty Alcohol Polyoxyethylene Alkyl Ethers

Defoaming agents based on fatty alcohol polyoxyethylene alkyl ethers are reported [18] to give excellent primary (98%) and ultimate biodegradation (75%) in standard screening tests [16].

B. Aquatic Toxicity of Antifoams

It is well known that polyoxyethylene/polyoxypropylene block copolymers are considered to be low-risk compounds for impact on aquatic life. Typically, the degree of acute aquatic toxicity, determined as the 96-hr LC_{50} value (the chemical concentration lethal to 50% of fish exposed over 96 hr), is greater than 100 mg/L [14]. Studies on a range of polyoxypropylene/ polyoxyethylene block copolymers for acute toxicity to rainbow trout have confirmed the low toxicity of this antifoam class (see Table 1). By contrast fatty alcohol polyoxyethylene polyoxypropylene compounds introduced as alternatives to the ethylene oxide/propylene oxide block copolymers for bio-degradation reasons display LC_{50} values of 0.7–2.9 mg/L toward a number of freshwater fish species [17]. The toxicity to freshwater invertebrates, par-ticularly Daphnia magna, is found to be comparable to the toxicity range of 0.25 to 4.4 mg/L expected for typical alcohol ethoxylates [17]. Thus for these fatty alcohol derivatives improved biodegradability with respect to block copolymers is associated with increased aquatic toxicity.

IV. THE USE OF COPOLYMER ANTIFOAMS IN MACHINE DISHWASHING

The aim of automatic dishwashing is the complete cleaning and drying of crockery and utensils, leaving them ready for immediate use. This is achieved by manual removal of large food particles prior to loading in to the machine, followed by freshwater prewashing, washing, rinsing, and drying cycles. Both batch and continuous dishwashing machines are used. One common feature of automatic dishwashing, using either type of machine, is the relatively higher energy levels compared to manual operation. Foam is perceived as a benefit in the latter operation for aesthetic reasons and as an indication of effectiveness. By contrast the efficiency of circulation systems in automatic machines is significantly reduced when foam is present. Thus the surfactant requirements for the low-foam operation necessary in machine dishwashing are very different to those of hand dishwashing.

Foam generation during automatic dishwashing is influenced by several factors, including the machine type, the washing program selected, and the type and level of soils present.

Degradation of proteinaceous soils such as dairy products under the conditions found in the washing cycle (high temperature, high pH) leads to the formation of surfactants and to the generation of unacceptable levels of foam. Increased use of surfactants as emulsifiers in the preparation of convenience foods also contributes to the problem.

A. Machine Types and Operating Conditions

1. Batch Machines

There are many types of domestic and small-scale industrial batch dishwashing machines. The ware to be washed is placed in a tray or rack which is loaded into the dishwasher. It remains static during the cleaning operation which is achieved using rotary spray arms. A typical machine is illustrated in Fig. 2. Typical operating conditions are given in Table 2.

2. Continuous Machines

This type of machine is illustrated in Fig. 3. It differs from the batch equipment previously described in several ways:

1. Ware is loaded on to a moving rack which carries it through the successive cleaning and drying cycles.
2. Throughput times are lower (less than 1 min).
3. Capacities are much higher, e.g., over 7000 plates per hour.
4. Pump rates and pressures are significantly higher than batch operated systems (2.5–3.0 bar).

As with batch industrial machines the final rinse water is recycled back into the prewash and wash tanks to conserve energy. This introduces residual rinse aid into the washing cycle. The nett result is a greater demand on the defoaming performance of the surfactants used.

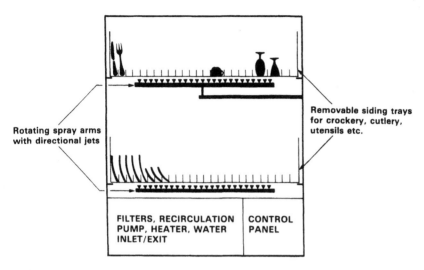

FIG. 2 Schematic of batch dishwasher.

B. Selection of Surfactant Antifoams

Surfactant foam control in the automatic dishwashing process is achieved by the incorporation of surfactant antifoams into the rinse aid and/or the main wash detergent. Levels of surfactant in main wash detergents are typically ≤ 5%, whereas in rinse aids they are considerably higher, typically 16–25% by weight.

Careful selection of the surfactant antifoam(s) is necessary to achieve a balance between the defoaming characteristics of surfactant with low aqueous

TABLE 2 Typical Operating Conditions for Domestic and Industrial Autodishwashers

	Domestic	Industrial
Cycle times	60–80 min	Typically 1–3 min (up to 30 on some machines)
Temperatures (°C)		
Prewashing	Ambient	40
Washing	55–65	55–65
Rinsing	60–70	80–85
Drying	Up to 90°C	Optional
Recycle of water?	No	Yes
Spray pump pressures	Up to 1 bar	1–2 bar

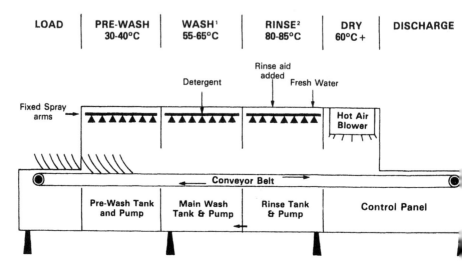

FIG. 3   Schematic of continuous dishwasher.

cloud points (i.e., less than 30°C) and the better wetting performance obtained with products of higher aqueous cloud points. This is particularly important for the operation of continuous machines, wherein a surfactant or surfactant combination must wet effectively at 80–85°C in the rinsing section and also perform satisfactorily as a defoamer in the washing section at around 60°C—two opposing requirements.

Examples of typical formulations are given in Tables 3 and 4 [19–23]. These tables reveal the type of components likely to be found in automatic dishwashing formulations. The permutations of these and other components of dishwashing formulations are almost endless and therefore cannot be dealt with here.

Historically, low-foaming performance has been achieved by the use of polyoxyethylene/polyoxypropylene copolymers. Key features of these molecules are their relatively high molecular weights (up to 2500 amu) and high polypropylene oxide content, which can be up to 90% by weight of the molecule. This combination allows them to perform efficiently as antifoams and wetting aids.

However with the introduction of European Community legislation [24] which was due to take effect at the end of 1989, radical changes to molecular compositions have had to take place for this market, since most of the older products are not sufficiently biodegradable. As a consequence, at the time of writing much product and formulation development work is still being

TABLE 3 Formulations for Domestic Dishwashing

Powder	%w/w	Rinse aid	%w/w
Nonionic surfactant antifoam	3.0	Nonionic surfactant antifoam	25.0
		Ancillary components (hydrotropes, water softeners, etc.)	30.0
Inorganic salts, including sodium metasilicate, sodium tripolyphosphate, sodium carbonate, sodium sulfate,	95.0	Demineralized water preservative	Balance
Bleach (chlorine-releasing agent)	2.0		

This type of formulation is to be found in many patents (see, for example [19–23]).

carried out aimed at providing the necessary low-foaming (and wetting) performance while still complying with the biodegradability requirements.

C. The Defoaming Performance of Polyoxyethylene/ Polyoxypropylene Copolymers under Dynamic Conditions

Several laboratory methods for the assessment of the foaming tendency of surfactant systems are available. However, to best simulate the dynamic conditions found in dishwashing machines, the Association Francaise de Normalisation method (AFNOR) [25] (Chute et Recyclage) has been used.

TABLE 4 Formulation for Industrial Dishwashing

Powder	%w/w	Rinse aid	%w/w
Nonionic surfactant antifoam	2–5	Nonionic surfactant antifoam	30.0
Inorganic salts (as domestic)	50–55	Ancillary components (hydrotropes, water softeners, etc.)	30.0
Sodium or potassium hydroxide	15–20	Demineralized water/ preservative	Balance
Chlorinating agent/ fillers/water	Balance		

An example of an industrial powder is given in Ref. 23 and an industrial rinse aid in Ref. 22

TABLE 5 Defoaming Performance of Polyoxyethylene/Polyoxypropylene Copolymers

Product	Cloud point (°C)	Biode-gradable [24]	Foam height (mm) 30°C	50°C	70°C
No defoamer present	—	—	>600	>600	>600
Polyoxyethylene/polyoxy-propylene Block copoly-mer 20% ethylene oxide	22[a]	No	479	805	68
Ethylene diamine-based polyoxyethylene/polyox-ypropylene copolymer	36[a]	No	328	38	43
Fatty alcohol-based po-lyoxyethylene/polyoxy-propylene copolymer	25[a]	No	77	42	46
"	40[a]	No	>600	208	66
"	<5[a](36[b])	Yes	50	58	60
"	25[a](48[b])	Yes	366	80	85
"	34[a](56[b])	Yes	>600	425	90
" with terminal alkyl group	28[a]	Yes	510	58	65
" with terminal alkyl group	41[b]	Yes	116	64	55

[a]Cloud point measured as 1% w/w aqueous solution.
[b]Cloud point measured as 10% w/w in 25% w/w butyl diglycol.
Source: Ref. 17.

The apparatus consists basically of a water-jacketed glass tube and recirculating pump. Surfactant antifoam solution is pumped to the top of the column and allowed to drop freely to the bottom.

The method has been used to screen a number of candidate antifoams and examples of results are presented in Table 5. Evaluation is carried out in an alkaline solution containing milk powder as the soil to simulate the cleaning operation. These results demonstrate

1. The necessity and effectiveness of an antifoam in the dishwash operation.
2. The measure of success achieved in developing sufficiently biodegradable fatty alcohol–based molecules which meet the low-foaming requirements.
3. The need to select products with the correct cloud point so that a balance between defoaming and wetting performance can be obtained. In extremis, a product with a low aqueous cloud point (<10°C) will work well as a defoamer but will give very poor wetting performance, and vice versa when the product's cloud point exceeds 30–35°C.

The effect of operating temperature on antifoaming efficiency can be determined by measuring the speed of rotation of a dishwashing machine spray arm under clean conditions (pure water) and in the presence of proteinaceous soils. The efficiency can be calculated as follows:

$$\text{Efficiency} = \frac{\text{speed of rotation (soil conditions)}}{\text{speed of rotation (water)}} \times 100\%$$

Typical results for a series of fatty alcohol alkoxylates, in which the cloud point is altered by changing the ethylene oxide/propylene oxide ratio can be seen in Fig. 4 [17, 26].

V. USE OF COPOLYMER ANTIFOAMS IN SUGAR BEET INDUSTRY

A. Introduction

In contrast to cane sugar manufacture, beet sugar factories need large quantities of antifoams. The natural foaming tendency of aqueous sugar solutions, combined with the high pressures and/or high flowrates, otherwise leads to unacceptable foaming and overflowing. Engineering modifications can decrease the foaming at some stages of the process, but in general the use of antifoaming agents is necessary. The selection of appropriate antifoams is a complex matter since a product added at one point in the process can have effects downstream. This complexity has led to a few specialized companies, with considerable expertise and experience, supplying antifoam products to the sugar beet industry.

Antifoams are required at different stages of the process, but the most important are (1) the beet transport and washing (2) the diffusion (sugar extraction) (3) postdiffusion (see Fig. 5). The antifoams employed for (1) are completely different from those for (2) and (3).

Consumption of antifoams varies from site to site and year to year and is also greatly dependent on the type of antifoam employed. It has significantly decreased over the last decade due to extensive use of more sophisticated pumping devices and more efficient products (vide infra). Despite certain papers mentioning relatively high consumptions [27, 28, 29, 30] a rough average total process consumption with efficient modern antifoams can be estimated at 100 g/tonne of sugar beets. This assumption leads to a Western European market of some 11,000 tonnes of antifoams.

Many of these antifoams are proprietary mixtures, and therefore the exact composition is not known. However they are usually complex blends of several components having different functions [31]. Some of them exhibit synergistic antifoam effects. Other components have in fact no antifoam

Antifoaming Efficiency (50°C)

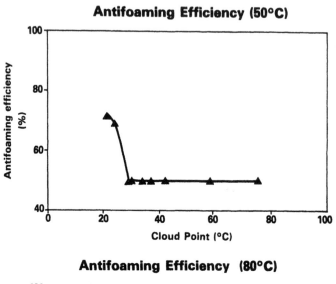

Antifoaming Efficiency (80°C)

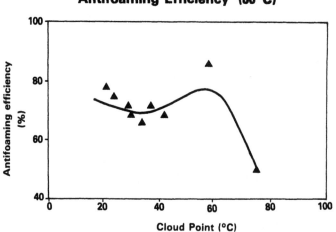

Cloud point measured as 1% w/v solution in
demineralised water

FIG. 4 Fatty alcohol alkoxylates defoaming efficiency variation with cloud point
and temperature (cloud points altered by changing ethylene oxide/propylene oxide
ratio). (From Refs. 17 and 26.)

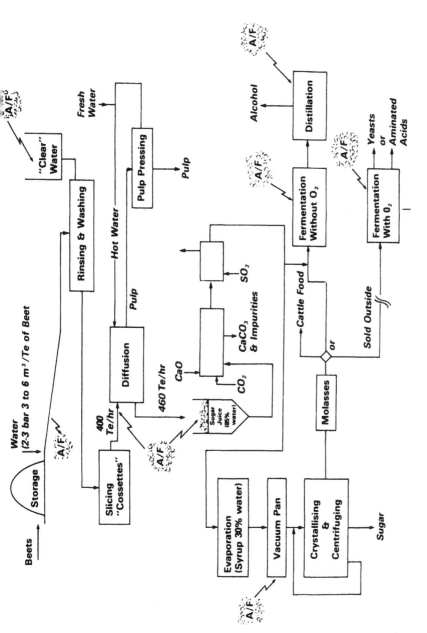

FIG. 5 Schematic of sugar beet processing, highlighting antifoam (A/F) application areas.

TABLE 6   Percentages of Ground Stones Adhering to the Beets Delivered to the
Sugar Plants

	% soil + stones (of total weight)
Dry campaign	8–18
Normal campaign	15–25
Rainy campaign	30–50

function. They serve, for example, as diluents, pourpoint depressants or
spreading agents. However major antifoam components in most sugar beet
antifoam mixtures are polyoxyethylene/polyoxypropylene copolymers.

B.   Description of a Sugar Plant [32–34]

A typical flow diagram of a sugar plant is shown in Fig. 5. Stages in the
process where antifoam addition is usually necessary are indicated in the
figure.

1.   Washing of the Beets

Sugar beets are crops harvested in most of the United States, Canada, and
Europe between September and November. On delivery to the sugar beet
factory, significant amounts of soil and stones can adhere to the beets, as
illustrated in Table 6.

a.   Transportation.   At the sugar plant, transport of the beets from the
storage area to the washing area is increasingly performed by a mechanical
process (conveyor belt). However, hydraulic transportation is still dominant.

In this latter case, water guns project recycled water under pressure onto
the beets to transport them in the hopper. Water flowrates of 3 to 6 m^3/t of
beets are necessary to transport about 250 tonnes/hr of beets in 60-cm-wide
hoppers. Typically a sugar plant handles 10,000 tonnes/day of beets on
which 30,000 to 60,000 m^3/day of water are sprayed at a pressure of 2–3
bar. When hydraulic transportation is used, there can be a lot of foam in
the hoppers after the water guns. Thus an antifoam is added to the water
supply of these guns.

The replacement of hoppers by a conveyor belt will obviously affect the
need for antifoams. However, in such a case the washing section must have
a bigger capacity as the hydraulic transportation also contributes to the washing.

b.   Washing and Rinsing.   Several pieces of equipment permit the removal
of the earth, stones and vegetation to obtain washed beets. Basically clean
water is introduced into the last rinse and flows to the proceeding stage so

as to be in countercurrent with the circulation of the beets. As there exist several potential foaming problems during this washing/rinsing process, an antifoam is added into the clean water tank. This is a preventive action as this water is used upstream. However, it is sometimes insufficient and some antifoam must be added to avoid foaming at some specific parts of the circuit.

Once the beets are washed, they are cut into slices (2–3 mm thick, 10–15 cm long) called cossettes.

2. Sugar Extraction: The Diffusion

Sugar is extracted by a continuous countercurrent liquid-solid extraction performed in a diffuser with hot water at 75°C maximum. Essentially the cossettes enter at one end of a very large, long cylinder, called the diffuser, while the hot water enters at the other end. The sugar juice which emanates from the diffuser is basically a 15% sugar solution in water, although some impurities are present at this stage. Many different types of diffuser, nowadays universally known by the names or initials of the original manufacturers, are used in the industry (e.g., DDS, Olier, BMA, Buckau, De Smedt, and RT) differing in their countercurrent geometry. To illustrate the application of antifoam agents in the diffusion stage one type of diffuser, the RT diffuser, will be described in some detail.

a. RT Diffuser. A representative flowchart of a RT diffuser is shown in Fig. 6. A lot of diffusers are now fitted with a heat exchanger which will use the heat from the hot sugar juice to prewarm the cossettes.

As the diffuser rotates, large quantities of juice are falling in the tank. At this stage, due to the relatively high temperatures involved (70–75°C) and to the mechanism of action of the copolymers (see Sec. I), it is not too difficult to obtain a low level of foam.

It is, however, more difficult to obtain the same result with the cold juice falling from the heat exchanger because its temperature is lower. Thus higher quantities of antifoam are required. It is often preferred to use the same antifoam for both temperatures, and the choice of the right copolymer(s) is of importance.

The antifoams are usually added to the hot juice and cold juice tanks. Some products give better results when directly added onto the cold cossettes. Small amounts may also be introduced in the hot water coming from the pulp processing.

In total, an RT diffuser requires about 20–40g of an efficient antifoam mixture/tonne of beets.

b. BMA and Buckau Wolf Tower Diffusers. In these diffusers an antifoam is mainly needed in the cossette mixer. Cossettes are mixed with the hot juice transferred from the diffuser, which in this case is a vertical tower, by a pump. The antifoam will also help the draining of the juice through the

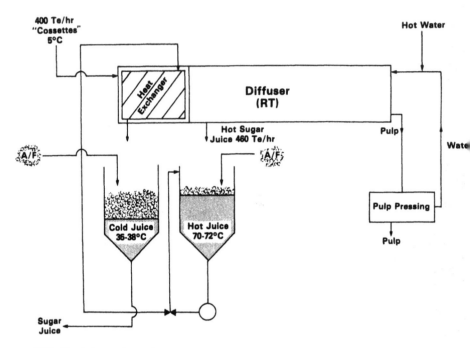

FIG. 6   Schematic of RT sugar beet diffuser.

perforated disks and thus reduce the power consumption. A BMA diffuser would require 10–20g of antifoam/tonne of beets.

c. Desmet Diffuser.   This diffuser consists of a 33-m-long conveyor screens which moves a 0.6- to 1-m thick layer of cossettes below a series of juice sprayers. Basically each sprayer is fed with the juice collected after the preceding one. The foaming is particularly a problem at the exit of the diffuser where the juice is most concentrated. Thus the antifoam is often added in the last part of the diffusion and it is common to need 60 g/t of beets.

3. Purification and Crystallization

The sugar juice is screened and then a preliming occurs. The lime used is often in the form of an aqueous solution known as milk of lime. The next step is the addition of the balance of liming agent and continuous bubbling of carbon dioxide. This will precipitate the lime as small insoluble calcium carbonate crystals on which most of the impurities will adhere. This operation is called carbonatation and is usually divided into a first and second carbonatation. There can remain enough antifoam within the sugar juice to be active at this stage. However, it is common to add antifoams into the lime milk tank (up to 5 g/t of beets), sometimes into the preliming vessel (up to 10 g/t) and rarely into the first carbonatation.

The juice is then filtered and sent to evaporators to obtain a thick juice containing about 65% dissolved solids. Evaporation to a supersaturated sugar solution is performed in vacuum pans whereupon crystallization is induced by seeding. Small amounts of antifoams are used in the vacuum pans. These products will also reduce the viscosity and the surface tension and improve the crystal growth. Several centrifuge stages are employed to extract the maximum sugar. The remaining liquid, called molasses, can be used to produce alcohol by fermentation or sold to animal feed, yeast, citric acid, or amino acid manufacturers.

C. Causes of Foam in the Sugar Beet Industry

1. Transport and Washing Stage

At the transport/washing stage damage to the skin of the sugar beet root often occurs. This leads to release of surface-active compounds like saponins which are primarily located just below the skin of the root. Foam problems which occur at this stage may be in part attributed to such compounds. They are also exacerbated by the presence of surface active vegetable components derived from the soil which adhere in significant quantities to the beets (see Table 6). In rainy periods, soil contamination is greater, the concentration of vegetable impurities is higher and foaming more severe. This leads to higher antifoam demand.

2. Diffusion and Postdiffusion

a. Components of Beet Liquors. There exists a tremendous number of organic compounds in the sugar juice [29]. In addition to sucrose, by far the greatest one, the following components have been identified:

Nitrogen-free substances.

Other carbohydrates in small amounts, mostly glucose, fructose, rafinose, and kestoses
Traces of arabinose and gelectose
Various mono-, di-, and tricarboxylic acids, both unsubstituted and hydroxy acids
Small quantities of pectic material
Saponins
Traces of numerous other organic compounds (vegetable fats made up of lecithin, oleic, palmitic, and erucic acids)

Nitrogen compounds. As illustrated in Table 7, protein fragments are the largest fraction and betaine is the largest individual component.

It has often been mentioned that saponine, a glycoside of oleanolic acid, is a prime cause of foaming. Some tests [35] showed that the addition of

TABLE 7 Nitrogenous Materials in Beets

	% on beets	% of total
Amino acids and others	0.042	21.0
Amides	0.015	7.5
Ammonia	0.005	2.5
Nitrates	0.002	1.0
Proteins and peptides	0.115	27.5
Betane and choline	0.020	10.0
Purines	0.001	0.5
Total	0.200	100.0

Source: Ref. 33 and 34.

saponine and/or oleanolic acid to a normal production sugar did increase the foaming tendency. It was however, concluded that the main effect was due to the presence of peptides (amphoteric surface-active agents). Proteins of the sugar juice are strongly surface active and give rise to foam [34].

b. *Concentrations.* The saponin contents of the beet peeling and the crown were found to be 570 and 120 mg/kg, respectively [36]. In a diffusion juice, it could reach up to 3673 mg/kg of dry substance [37], thus about 580 mg/L of sugar juice. Some papers [37] even mention higher levels, up to 9193 mg/kg of dry substance.

c. *Surface Tension.* It was found [38] that the surface tension of saponin solutions dropped very quickly down to 42 mN/m when the concentration reached 25 mg/L. Another study [39] mentions that the surface tension of diluted molasses could decrease down to 39 mN/m. The surface tension of liquors from sugar beets harvested 1 hr before the test can vary between 60 and 53 mN/m [40] from the beginning to the end of the campaign. In addition, it was also shown that the surface tension can significantly decrease when the storage time of the beets does increase. Both factors may explain why the foaming difficulties at end of campaigns are usually more stringent than at the beginning.

D. Composition of the Antifoam

As indicated above, sugar beet antifoams are proprietary mixtures, usually complex blends, and the exact composition is not generally known.

1. Transport and Washing of Beets

Water temperatures during the beet harvest season are low (usually 5–15°C). Copolymers will only function effectively as antifoam ingredients at tem-

TABLE 8 Typical Composition of a Beet-Washing Antifoam

	Before 1975	Nowadays
Mineral oil	50%	50%
Animal and/or vegetable acid oils	50%	15–25%
Copolymer	—	15–25%
Alcohol	—	5–10%

Source: Ref. 17.

peratures above their respective cloud points. At temperatures below their cloud points they are too water soluble and may even exacerbate a foam problem. Therefore copolymers used for transport/washing of beets should have low cloud-point temperatures. Esterification of copolymers with cheap fatty acid will reduce the cloud points if the copolymer itself is unacceptable in this respect.

In the past, the antifoam agents used for the beet washing did not contain copolymers components. Since 1975 such polyalkyleneglycol copolymers have become important ingredients, representing some 15%–25% of oil-based formulations. The improvement in antifoam efficiency has allowed consumption for this stage to decrease down to 20–60 g/tonne of beet, halving the previous dosage level.

The few patents which deal with sugar antifoams usually do not refer to beet washing. However a typical composition is shown in Table 8.

The alcohol employed is usually isopropanol or isobutanol permitting a homogeneous composition with a low pourpoint to be obtained. The antifoam added into the supply water of the water guns or in the beet washers must have an immediate action. The high pressures/flowrates at these points will permit the recipes given in Table 8 to disperse. However, the antifoam added in the clean water tank must have a preventative action. It must disperse in water in order to be carried and be active at the critical points downstream. Thus the formulator may add up to 5% of a nonionic surfactant to the above-mentioned recipe.

2. Sugar Extraction: The Diffusion

The diffuser is also a critical part of the sugar extraction process as far as foam is concerned due to the natural foaming tendency of the sugar juice and due to the high volume and flowrates involved in some diffusers. In the patent literature several types of surfactants are mentioned as being used in the sugar diffuser antifoams.

In the United States there are only a limited number of surfactants allowed by the Food and Drug Administration [41] to be used in sugar antifoam

TABLE 9   Typical Composition of a Diffusion Antifoam

	Before 1975	Nowadays
Acid fish oil	70%	—
Acid vegetable oil	20%	—
Fatty acid	—	30%
Mineral oil	10%	—
Copolymer	—	70%

Source: Ref. 17.

recipes. These are restricted to

A n-butoxy polyoxyethylene/polyoxypropylene copolymer
Polyoxyethylene (400) or (600) dioleate
Polyoxyethylene (600) monoricinoleate
Polypropylene glycol (MW 1200 to 2500)
Propylene glycol mono- and diesters of fats and fatty acids
Polysorbate 80 (polyoxyethylene (20) sorbitan monooleate).

As a consequence one patent [42] refers to compositions such as 40%–60% sugar cane oil + 5%–10% emulsifier + 30%–50% mineral oil, a preferred emulsifier being polyethylene glycol 400 monooleate. Another patent [30] covers blends such as 2%–5% ethylenebis(stearamide) + 5%–10% polyethyleneglycol (MW 400 to 600) and up to 15% fatty alcohol ethoxylate (C_{11-15} + 7 EO) in mineral oil.

In other countries, blends containing up to 80% of alcohol alkoxylates, like $C_{12/18}$ + 5 EO + 13 PO, have been suggested [43]. Another patent [25] covers the following composition 65% of a polypropyleneglycol (MW 2000) + 15% of polyethyleneglycol dioleate + 10% of a polypropyleneglycol monooleate + 8% isopropanol + 2% polyethoxylated sorbitan monooleate.

Esters of fatty acids with polyoxyethylene/polyoxypropylene copolymers are also mentioned [44], and the use of esters of nonylphenol ethoxylates with fatty acids is claimed.

Esters of fatty acids with polyoxyethylene/polyoxypropylene copolymers initiated on ethylene diamine would also be effective [45]. Reaction products of a carbohydrate, a fatty acid, and an alkylene oxide [46] have also been claimed. More recently [47], the use of an alkyl end-capped alcohol alkoxylate has been patented. A typical composition of a diffusion antifoam is given in Table 9. The "old" recipe (used up to 1975) had to be heated to be handled, but this is no longer the case with the formulation used today.

3. Other Stages

a. Carbonatation. Antifoams are sometimes used during the preparation of the milk or lime, during the preliming, and rarely during the carbonatation. In most cases it is common to use the same antifoam as the one used at the diffusion stage.

b. Vacuum Pans. Small amounts, less than 5 g of a product per tonne of massecuite (the concentrated sugar syrup) are added before applying vacuum. This product will avoid the formation of large air bubbles which would burst at the surface and generate some losses. Some of them are claimed to reduce the viscosity and surface tension of the massecuite and will generate improved crystal growth. The products [48,49] which are used are coconut oil, methyglucoside coconut oil ester, fatty alcohol alkoxylates, etc.

c. Fermentation. Some plants carry out a fermentation of the molasses on the sugar juice with yeasts to produce alcohol. A medium plant will produce up to 100,000 L of alcohol/day and a large one up to 200,000 L/day. Copolymers blended with vegetable fatty acids are also largely used as antifoams. The amounts of antifoams used during the fermentation will be about 100 g/100 L of produced alcohol if molasses are used, and about 150 g/100 L with sugar juice.

During the distillation about 20 g/100 L of produced alcohols can be necessary.

E. Toxicological Implications

In the past some producers may have used spent engine oils as a component of their sugar antifoams [29]. This practice is now proscribed by legislation [34,50]. Due to the intensive purification of sugar, toxicity risks are minimal. In fact, a first purification is achieved during the carbonatation. This is then completed by a crystallization process which is remarkably efficient in reducing nonsugar contamination by a factor of some 200 to 300 [35]. Some studies [27] have shown that the remaining antifoam concentration in the sugar is very low (less than 0.4 ppm).

Extensive use of high molecular weight polyoxyethylene/polyoxypropylene adducts at the diffusion and post diffusion stages is also a significant factor which contributes to minimise the toxicity risk. Polyoxyethylene/polyoxypropylene copolymers do in fact exhibit very low toxicity, which explains their wide use in the pharmaceutical industry. The copolymers used in the sugar industry usually exhibit LD50 > 2000 mg/kg [51,52].

REFERENCES

1. M. N. Fineman, G. L. Brown, and R. J. Myers, *J. Phys. Chem. 56*: 963 (1952).

2. M. J. Schick (ed.) Nonionic Surfactants, Marcel Dekker, New York, 1967.
3. N. Clinton, P. Matlock, and S. D. Gagnon, in Encyclopaedia of Polymer Science and Technology, 2nd ed. (J. I. Kroschwitz, ed.), Wiley-Interscience, New York, 1986, pp. 225–307.
4. L. G. Lundsted (assigned to Wyandotte Chemicals Corp), US 2674619; filed April 6, 1954.
·5. K. G. Henkel, Fatty alcohols: Raw materials, Methods, Uses, Düsseldorf, 1981, pp. 193–225.
6. K. H. Schmid, H. Pruhs, T. Attenschöpfer, and R. Piorr (assigned to Henkel KGAA), EP197434; filed October 15, 1986.
7. L. Person and B. Mourrut (assigned to ICI France SA), EP276050; filed July 27, 1988.
8. A. Hettche and E. Klahr, Tenside Dets. 19: 127 (1982).
9. D. P. Cox, Ann. Rev. Microbiol. 23: 173 (1978).
10. L. D. L. Jenkins, J. Appl. Bact. 47: 75 (1979).
11. R. Dwyer, Appl. Env. Microbiol. 46: 185 (1983).
12. E. Scholberl, J. Kunkel, and K. Espeter, Tenside Dets. 18: 64 (1981).
13. R. Swisher, Surfactant Biodegradation, Marcel Dekker, New York, 1987.
14. K. Boch, L. Huber, and P. Schoberl, Tenside Dets: 25, 86 (1988).
15. C. Naylor, F. Castalds, and B. Hayes, J. Am. Oil Chem. Soc. 65: 160 (1988).
16. OECD Guidelines for Testing of Chemicals, OECD, Paris, 1981.
17. ICI Chemicals and Polymers Ltd, Internal data.
18. P. Gode and W. Guhl, Seifen Ole, Fette, Wasche 87: 421 (1985).
19. F. V. Ahmed, M. A. Camara, J. A. Kaeser, C. E. Buck, and J. F. Cush (assigned to Colgate Palmolive Co), GB2206601; filed January 11, 1988.
20. F. V. Ahmed, M. A. Camera, J. A. Kaeser, C. E. Buck, and J. G. Cush (assigned to Colgate-Palmolive), EP330,060; filed August 30, 1989.
21. C. R. Barrat, J. R. Walker, and J. Wevers (assigned to Procter and Gamble), EP75986; filed May 13, 1987.
22. D. Hemm and N. Schindler (assigned to Henkel KGAA), EP186088; filed November 15, 1989.
23. T. M. Kaneko (assigned to BASF Wyandotte), US4272394; filed June 9, 1981.
24. EEC council directives 73/404/EEC and 82/242/EEC.
25. Association Francais de Normalisation (AFNOR) method NFT 73–412.
26. T. G. Blease, Speciality Chems. 9: 448 (1989).
27. R. J. Michalski (assigned to Nalco Chemical Co) SA6706246, filed June 24, 1978.
28. O. S. Malmros and J. Tjebbes, Sucrerie Belg. 99: 21 (1980).
29. E. Greulich, Zuckerindustrie 104: 1031 (1979).
30. R. J. Wachala and R. E. Svetic (assigned to Nalco Chemical Co) US3990905; filed November 9, 1976.
31. I. Lichtman and T. Gammon, in Kirk-Othmer Encyclopaedia of Chemical Technology, 3rd ed. Vol. 7, 1979, pp. 430–448.
32. La Raffinerie Tirlemontoise in Manuel de Sucrerie, 4th ed. 1984.
33. P. C. Hanzas and R. W. J. Kohn, Am. Soc. Sugar Bett Technol. 11: 519 (1961).

34. R. A. McGinnis, in Kirk-Othmer Encyclopaedia of Chemical Technology, 3rd ed., Vol. 21, 1983, pp. 904–920.
35. R. A. McGinnis, in Beet Sugar Technology, 2nd ed., Beet Sugar Development Foundation, 1971.
36. J. F. T. Oldfield, and J. V. Dutton, *Int. Sugar J. 70*: 7–9, 40–43 (1968).
37. T. Shiga, H. Okuyama, and N. Nosaka, *Proc. Res. Soc. Jpn Sugar Ref. Technol. 14*: 12 (1964).
38. J. R. J. Johnson, *Am. Soc. Sugar Beet Technol. 11*: 201 (1960).
39. P. Joos and R. Ruyssen, in Chemistry, Physics and Application of Surface Active Substances, Vol. 2, Gordon and Breach, New York, 1967, p. 1143.
40. A. Vanhook and W. F. Biggins, *Int. Sugar J. 54*: 7–10 (1952).
41. M. L. H. Herquet, *Ind. Agr. Alim. 67*: 117 (1950).
42. 21 Code of Federal Regulations 173 340, Office of the Federal Register of National Archives and Records Administration, U.S. Government Printing Office, Washington, 1988.
43. S. E. Kent (assigned to Hodag Chemical Co), US2762780; filed February 10, 1984.
44. R. Heyden and A. Asbeck (assigned to Henkel KGAA) DE2164907; filed May 30, 1984.
45. H. Hedtrich, H. Kreiss, and E. Schneider (assigned to Nopco Muenzing GMBH), DE2250975; February 24, 1977.
46. B. Mourrut and J-J. Martin (assigned to Atochem), GB2135985; filed September 12, 1984.
47. P. Jacques (assigned to Raffinerie Tirlemontoise), GB1216987; filed December 12, 1970.
48. W. Dietschne, K. Lorenz, C. Vamvakaris, and A. Hettche (assigned to BASF AG) EP180081; filed May 7, 1986.
49. P. D. Berger, *Sugar Technol. Rev. 3*: 241 (1976).
50. Y. Oyama, Y. Matsuo, and H. Nishi (assigned to Riken Vitamin Oil Co Ltd), US4427454; filed January 24, 1984.
51. Arrete du 6 Fevrier 1989 fixant la liste des auxiliaires technologigues pouvant etre utilises en sucrerie, *J. Officiel Republ.* Francaise.
52. P. H. Elworthy and J. F. Greon, in Nonionic Surfactants (M. J. Schick, ed.), Marcel Dekker, New York, 1967, p. 923.
53. Toxicity and Irritation Dated on Pluronic Polyols, Wyandotte Chemicals Corp., 1961.

Index

Alkoxysiloxane copolymers (*see also* Polyalkoxypolysiloxane copolymers, Polysiloxane polyglycol copolymers, Silicone glycol copolymers, Silicone polyethers, and Siloxane oxyalkylene copolymers), 256

Alkyl polyacrylates, 143

Aluminum distearate, 205

Amide wax particles, in oil—based antifoams, 165

Antiflatulents, 181

Antifoam synergy, hydrophobic particles and oils, 66–67

Automotive fuels, 127

Bridging coefficient, 92

Bridging oil lens, in foam film, 89

Brownstockwasher antifoams, 165
  oil-based, 168
  water-based, 168

Cinematographic studies, high speed, 36, 39

Cloud point, 18, 212, 283, 300, 301, 308, 310, 319

Contact angles:
  air—water, 31, 33, 241
  oil—water, 98, 102

Cream, 164

Critical film (rupture) thickness, 10, 24, 26, 27, 138

Critical surface tension, 289

Critical wetting tension (*see* Critical surface tension)

Crude oil, 21, 120, 139
  production of, 121

Dearation, of paints, 284

Detergent builder, 222, 229, 230

Diamides, 233

Diffusion, of gas:
  interbubble, 139

Dilational modulus (*see also* Dynamic dilational modulus, Surface Dilational elasticity, Surface dilational viscosity, and Surface elasticity), 142

Dimethicone, 180
  specification of, 182, 184
  toxicological studies of, 181

Disjoining pressure, 9

Duplex film, 13, 15, 19, 27, 83

Dyeing machines (*see also* Jet dyeing machines), 198
Dyeing processes, antifoam attributes for, 203
Dyes and foaming, 194–198
Dynamic contact angles, 36, 38, 65
Dynamic dilational modulus (*see also* Dilational modulus, Surface dilational elasticity, Surface dilational viscocity, and Surface elasticity), 21, 284
Dynamic surface tension, 12, 13, 20, 21, 120
Dyspepsia, 178

Electrostatic double layers, 16
Emulsions, 30, 31, 102
  antifoam for paints:
    oil-in-water, 281
    water-in-oil, 282
Equation of state for a foam, 134
Entry coefficient, 13, 14, 16, 17, 19, 86, 121
  initial, 14
  semi-initial, 14
Ethylene bis (stearamide) based antifoams (*see also* Ethylene bis [stearic amide]):
  and hydrocarbon oil, 279
  and mineral oil, 280
Ethylene bis (stearic amide), 217, 218

Fatty acids, 33, 63, 168, 169, 204, 229
  alkaline earth salts of, 217
  effect of melting on antifoam behavior, 65
  ethoxylates, 204
  and hydrocarbon:
    antifoam for paints, 276
Fatty alcohols, 16, 20, 168, 169, 204
  ethoxylates, 204
  polyoxyalkylene alkyl ethers, 302
    biodegradability of, 302, 304
  polyoxyalkylene derivatives, 302
Flotation froths, 30

Fluorosilicones, 143
Foam film drainage, 24, 54
  and resistance to antifoam, 285
Foam film rupture by particles, rate determining step, 57
Foam lifetime measurement, 131
Foaminess index, 130

Gas–oil separator, 123
  three phase, 124
  physical methods of foam control in, 123
Gel particle antifoams, 169
Gibbs elasticity, 6, 7, 8, 12
Gibbs (adsorption) equation, 86
Glass microspheres, hydrophobic, 59
Granular temperature, 48

Handling of antifoams, 171
Hydrocarbon wax (*see also* Wax), effect of melting on antifoam behavior, 63, 65
Hydrocarbons, spreading on water, 84
Hydrophobed or hydrophobic silica (*see also* Silica), 66, 180, 278
  antifoam with hydrocarbon, 66, 165, 242
    incorporation in detergent powder, 245
  antifoam with mineral oil, 165, 204

Industrial paints:
  foam problems in, 274
  silicone antifoams for, 285
Intestinal gas, 177

Jet dyeing machines, 199
Jet dyeing processes, antifoams for, 211

Kraft brownstock, 153, 162
Kraft pulp mill, 152
Kugelschaum, 3, 128

Laplace equation, 34
Lube or lubrication oils, 11, 24, 27, 126

Lubricants, 120

Machine dishwashing, copolymer antifoams in, 305
Machine dishwashers, 306
Magnesium stearate, 218
Marangoni:
  effect, 22
  flow, 12, 96
  spreading, 22, 78
Marginal regeneration, 54, 55
Methanol wettability test, 240
Microcrystalline waxes, 242
Microemulsions, oil-in-water antifoams for paints, 283
Milk, 164
Monoamides, 232

Neumann's triangle, 89
Nonaqueous foams, 21, 99, 119, 127

Oil lenses, 14
Oscillating jet technique, 12, 13, 20, 21

Paints:
  emulsion,causes of foam in, 272
  testing of antifoams for, 292
  thermosetting water based, foam problems in, 274
Paint film defects, 284
  caused by air bubbles, 270, 289
  caused by antifoam oils, 289, 290
  and silicone antifoam, 285, 286
Paper, 151
Papermaking furnish, 155
Paraffin wax (see also Wax), 63, 217
Particles:
  bridging mechanism, 33
  effect of edges, 40
  geometry, 39, 41
  hydrophobic, 30, 31
  rough, 49, 96
Petroleum jelly, 242
Pickering emulsion, 30, 98
Phosphoric acid esters, 204, 205, 235

Poiseuille flow, 24, 54
Polyalkoxypolysiloxane copolymers (see also Alkoxysiloxane copolymers, Polysiloxane polyglycol copolymers, Silicone glycol copolymers, Silicone polyethers, and Siloxane oxyalkylene copolymers), 284
Polydimethylsiloxanes, 82, 143, 164, 208, 246
  and intestinal gas, 177
  manufacture of drugs and, 177
  mixtures with hydrophobed silica, 67, 204, 246, 271
  molecular weight and antifoam efficiency, 286
  spreading behavior on water, 82–84
  spreading rates on water, 27
Polyederschaum, 3
Polyethylene glycols, 168
Polyhedral foam, 3, 128
Polyol-based polyoxyalkylenes, 301
Polyoxyethylene, degrading bacteria, 303
Polyoxyethylene–polyoxypropylene block copolymers, 204, 283, 300, 308
  antifoams from, 300
  aquatic toxicity of, 305
  biodegradation of, 303
  defoaming performance in dishwashing machines, 309
  ester with fatty acid, 205
  ethylene diamine based, 301
  structure of, 300
  in sugar diffuser antifoams, 321
Polysiloxanes (see also Polydimethylsiloxanes):
  modified, 177, 179, 286
  molecular weight and viscosity, 179
Polysiloxane polyglycol copolymers (see also Alkoxysiloxane copolymers, Polyalkoxypolysiloxane copolymers, Silicone glycol copolymers, Silicone polyethers,

and Siloxane oxyalkylene co-polymers), 278
Polytetrafluoroethylene (PTFE), 33
Prevention of foam, by physical means, 161
Pulp, 151

Reynolds' equation, 54
Rigid liquid foam film, 24, 54, 55
Rupture times, particles in films and, 56

Silica (see also Hydrophobic silica):
  fumed, 180
  manufacture, 237, 238
  precipitated, 180
Silica hydrophobing:
  by alkylchlorosilanes, 239, 240
  by fused—on waxes, 240
  by polydimethylsiloxanes, 240
Silicone antifoam:
  compounds, 67, 208
  preparation of, 248
Silicone antifoam emulsions:
  controlled particle size for paints, 286
  particle size of for textiles, 210
  for textiles, 209
  thickener system for, 210
Silicone antifoams in detergent pow-der, 253
  absorptive carriers for, 261
  deactivation processes of, 253
  film forming polymers for, 259
  organic matrices for, 254
Silicone glycol copolymers (see also Alkoxysiloxane copolymers, Polyalkoxypolysiloxane co-polymers, Polysiloxane poly-glycol copolymers, Silicone polyethers, and Siloxane oxy-alkylene copolymers), 143, 204, 205, 211
  structure of, 211
Silicone glycol with silica antifoams, nonionic emulsifiers for, 213

Silicone glycol and silicone antifoam compounds, mixtures for jet dyeing, 215
Silicone polyethers (see also Alkoxy-siloxane copolymers, Polyalk-oxypolysiloxane copolymers, Polysiloxane polyglycol co-polymers, Silicone glycol co-polymers, and Siloxane oxyalk-ylene copolymers), 211
Siloxane oxyalkylene copolymers (see also Alkoxysiloxane copoly-mers, Polyalkoxypolysiloxane copolymers, Polysiloxane po-lyglycol copolymers, Silicone glycol copolymers, and Silicone polyethers), 256
Siloxane resins, 247
Simethicone, 180
  emulsions, 185
  oral suspension, 185
  specification, 183
  tablets, 184
  toxicological studies, 181
Soaps, 229
  calcium, 33, 229
  Krafft temperatures, 230
  sodium, 33
Solubility of antifoam in paint, 285
Solubility products, and antifoam ef-fectiveness, 32
Solubilization, 94
  of antifoam oil, 11, 13
Soya bean oil antifoam, 28
Spotting of textiles during dyeing, an-tifoam effects, 209–215
Spreading agents, 79
Spreading coefficient, 14, 16, 86, 121
  initial, 14, 88, 89
  semi-initial, 14
Spreading pressure, 23
Spreading rate on water, 27
Staining problems of textiles by sili-cone antifoams during dyeing, 209, 212, 214
Storage of antifoam:
  settling of particles and, 280

gelling thickening agents for, 280
stability and, 283
Sugar antifoam recipes permitted by
USFDA, 320
Sugar beet industry:
causes of foam in, 317
copolymer antifoams in, 311
Sugar beet washing antifoam, 319
Sugar diffuser antifoams, 320
Surface dilational elasticity (*see also*
Dilational modulus, Dynamic
dilational modulus, Surface
elasticity, and Surface dilational
viscosity), 140
Surface dilational viscosity (*see also*
Dilational modulus, Dynamic
dilational modulus, Surface
elasticity and Surface dilational
elasticity), 140
Surface elasticity (*see also* Dilational
modulus, Dynamic dilational
modulus, Surface dilational
elasticity, and Surface dilational
viscosity) 19, 21, 22
Surface shear viscosity and crude oil
foam stability, 142

Surface tension gradient, 4–6, 8, 11,
19, 20, 22, 23, 29, 30
Synergistic oil–particle antifoams (*see*
Antifoam synergy)

Textile dyeing auxilliaries, 194
formulating antifoams into, 206
Trialkylmelamines, 234
Triglycerides, 63

van der Waals forces, 8
Viscosity:
and foam lifetime of crude oil, 139
and foaming of paints, 284
of solution and antifoam requirement, 55

Washing machines:
front loaders, 225
top loaders, 223
Water–based antifoam systems for
paints, 281
Water extended antifoams, 165
Wax, microcrystalline petroleum, 255
Wood pulp (*see* Pulp)
Work of emergence of particle:
into air–water surface, 44
into oil–water surface, 106

Milton Keynes UK
Ingram Content Group UK Ltd.
UKHW020018071024
449327UK00032B/2838